Don't Skip This

This book wants to be a history of the land.

Because this history wants to be useful, it starts at the beginning. It starts with the Big Bang. It may turn out that the Big Bang isn't the beginning, but it's as close as we've gotten.

The book starts at the beginning because, if you start anywhere else, a confusing divide arises between the earlier part that you don't talk about (i.e., that you take for granted, as given), and the later part that you do talk about. The part you don't talk about becomes a *stage* for the part you do talk about. The part you do talk about unfolds as . . . *action*.

Because the action seems to unfold in the part you talk about, the part you don't talk about doesn't seem to play any role at all.

In fact, though it may be invisible because it's so taken for granted, the part you don't talk about plays *most* of the roles.

Since so many of the actors are invisible, the actions of the actors we *can* see are hard to make sense of. Perhaps calling this divide "confusing" understates it. "Mystifying" is more like it.

It's easy to locate this mystifying divide. The action part we call *historical*. History is where the *action* is. The rest of it—the stage—is bracketed as . . . *pre*historic. Geology begins with the Cambrian: everything that came before is . . . *Pre*cambrian. History starts with written records: everything that came before is . . . *pre*historic. Science starts in the 17th century: everything that came before is . . . *pre*scientific.

But if the stuff that came before—what we think about as the stage—really does fill most of the roles, then we really can't understand *any* part of the story without acknowledging it. But to acknowledge it, we have to erase the divide between "before" and "after." This obligates us to understand that there is no . . . *pre*science, *pre*history, *Pre*cambrian.

Stories really only make sense when everything's included.

This is true of even the slightest stories. This book, for instance: it has *its* origins in the Big Bang. Of course the book itself didn't materialize then. Before the book could materialize atoms had to evolve, and stars and planets and life-forms, among them humans, and humans had to evolve writing and printing and the book business as we know it today.

Without the atoms though . . . *no book*, and without the Big Bang . . . *no atoms*.

In saying these things I'm not talking about just any stars, I'm talking about the Sun. I'm not talking about just any planets, I'm talking about the Earth. I'm not talking about just any writing, I'm talking about English. *Real* things happen in *real* places. Real things can only happen *locally*.

There are books that try to evade this reality. They're perfused with an Olympian air. So universal is their perspective, so detached, so free, so untainted by their inevitably local origins, it's as though they were written, not even on Mt. Olympus but . . . *nowhere at all*. It's an illusion. Even those books supposed to have been dictated by angels had to have been written down by people, and people are always *some*where, nailed to the Earth by gravity if nothing else. Writing is inescapably marked by its local origins. It was produced *here*, in *this* reading room, in *this* library (like Marx's *Capital*); or *there*, in *that* summer house just over the hill (like Twain's *Huck Finn*); or in *this* chilly room of *that* drafty palace (like Du Fu's "Spending the Night at the Palace Chancellery in Spring").

Being written in *this* chilly room of *that* drafty palace, "Spending the Night at the Palace Chancellery in Spring" had also to have been written in Chinese and 1247 years ago. (It's enough to say where to say when: gravity nails us then, not now and then.) "Spending the Night at the Palace Chancellery in Spring" had to have been brushed with ink on paper. The poem includes references to jade bridle pendants and the emperor and the Ninth Heaven and the Moon. "Spending the Night at the Palace Chancellery in Spring" could have been produced in no other time and at no other place. "Spending the Night at the Palace Chancellery in Spring" was a product of its time and place as surely as was the millet eaten by its author, as surely as was the wind sweeping through the room.

The poem came from the land it came from.

The land wrote it, we could say, in a figure of speech, but it is not a figure of speech to say that the land made it. There is nothing made that doesn't come from the land. Poetry is not privileged nor is music nor is your speech.

We know the land by its form, its swells and dips and sudden leaps; by

Five Billion Years of Global Change

Five Billion Years of Global Change

A HISTORY OF THE LAND

Denis Wood

THE GUILFORD PRESS
New York   London

© 2004 The Guilford Press
A Division of Guilford Publications, Inc.
72 Spring Street, New York, NY 10012
www.guilford.com

Printed in the United States of America

This book is printed on acid-free paper.

Last digit is print number: 9 8 7 6 5 4 3 2 1

Library of Congress Cataloging-in-Publication Data
Wood, Denis.
  Five billion years of global change : a history of the land / Denis
Wood.
       p. cm.
Includes bibliographical references and index.
  ISBN 1-57230-958-X (pbk.: alk. paper)
  1. Earth–History. 2. Land use–History. 3. Human settlements–
History. I. Title.
  QB631.W66 2004
  508–dc22
                                                    2003022301

for
Christine Baukus and Irv Coats

its skies, its unending rain or snow, its cloudlessness or its sunsets; by its trees or its lack of trees or its fields of tobacco or its rows of olives or its paddies of rice; by its villages and towns, by its cities; by the language on its signboards and the sounds in the mouths of its people; by its music, by its poetry.

It all comes from the land.

This book, for example, which pretends to be a history of this land, arises from it like the mist down along the branch on a cool morning in the fall. It could only have been written by me, and I could only have written it in Raleigh, in North Carolina, at the turn of the millennium.

I don't mean to confuse you by my use of "could." I could have written other books (I have, I will), and others elsewhere could have written a book much like this (they have, they will), and of course this book could have been quite different. Indeed, it has been: there's no page, no paragraph, that hasn't been written and revised and trashed and written again (and revised and trashed and written again), and so had I died, say, before getting this far, others could well have published a version less frequently revised.

Yet everything I'm saying about this book would have been true of that book. It too could only have been written by me, and I could only have written it in Raleigh, in North Carolina, at the turn of the millennium.

It would have been a different book, but no less a product of its time and place.

This isn't about probabilism or possibilism or determinism or fate. This is not a book of philosophy. It's a book of history recounting from the beginning of time a sequence of events that has brought us to this: my hand and your face in these pages. This is the book I wrote, and except that I came a quarter of a century ago to North Carolina State University to be the social scientist on a faculty of landscape architects, I never would have done so.

Just as, say, Antonio Gramsci wouldn't have written *Letters from Prison* if he hadn't been in prison.

It's that simple, really: you can't write a letter from a prison you're not in.

This is not to say that if you're in prison you *have* to write letters, though most do, especially polymaths like Gramsci. It's to say that his *Letters from Prison* flow from his being in prison as his being in prison flowed from his resistance to the rise of fascism in Italy, as the rise of fascism in Italy flowed from Italy's situation in Europe in the aftermath of World War I, as World War I flowed from. . . .

⌈None of these events is inexplicable. They flow from their circumstances as water flows downhill. They take the simple pathways things do.⌋

Not that there's only one path.

Everything arises from the land (which is to excuse nothing): peace, war, political philosophy, the industrial proletariat, peasants, the food they eat, the plants that provide it, the soil the plants grow in, the rock from which the soil develops, the crust crumpled into the rock that the rains off the Atlantic beat and lash, the widening basin the Atlantic fills, the drifting continents, the spinning planet, the flying Milky Way. . . .

It all comes from the land (which is not to excuse it) *and from what the land comes from.* . . .

Yes, it's hard for me (in any case) to think of Gramsci (or of anyone) without thinking of the Milky Way (of the universe), and you too I hope by the time you've finished this book—no! well *before* you've finished it, when you're only partway through—for that's its premise, that we are all and everything *organic sprouts* (and this too is to excuse nothing), which is to say, *products of history*: prisons, books of poetry, virgin forests, canals, growing cities, dying towns, mountain ranges, coral reefs, fields of amber waving grain, space shuttles, rivers like strong brown gods, green grass and igneous sunsets, land grant universities, landscape architecture departments, history courses, their texts—

I'd been a freshly minted PhD in geography teaching high school in Worcester, Massachusetts, when I was asked to apply for a job—there's plenty of opportunity for freedom in this story (it's *not* about determinism)—with the Department of Landscape Architecture in the then School of Design at North Carolina State University in Raleigh. It interested me that landscape architects were taught design, but not about the land. How could they know what they were doing, I wondered, with no insight into the context within which they were doing it? I took the job—the azaleas in April! the dogwoods!—to find myself teaching a history of landscape architecture that obligated me to inaugurate a course in the history of the land that my students over the next quarter of a century whipped from a more or less traditional historical geography course with its emphasis on field systems and settlement patterns into the straight-from-the-Big Bang version you've got in your hands with *its* emphasis on . . . *first principles.*

They just kept asking questions: *what great students!*

The course gave up being about appearance and turned into one about practice: What are . . . *native* plants? What exactly did we mean by . . . *nature*? What could it mean to design . . . *with* nature? If humans evolved like other organisms, how can what we do be . . . *artificial*? If we aren't *in* . . . *harmony* with nature, when did we fall out of it?

These and the other questions like them weren't part of the curriculum, but they were begged by the history I was teaching, for I tried to make do without "nature," "landscape," "balance," "native," moving straight from the as-concrete-as-I-could-be (clouds, streets, trees, songs) to the as-abstract-I-could-get (gravity, friction, metabolism, interaction) with as few intermediate-level concepts as I could get away with. This was a program, of course, not an accomplishment; but by stretching for this freedom from received truth I was able to beg the question, "What exactly *is* the land anyway?"

Which is the point of teaching—isn't it?—to beg the question. . . .

Geez, they were great students!

Dick Wilkinson had the vision to hire me. Art Sullivan had the strength to keep me on. They were both people I could talk to about the land, and I'm sure there's more of both of them in this book than I could possibly credit. The late horticulturist J. C. Raulston sat in on the course and gave me in return much for which I cannot thank him. He also made it possible for me to work through some of these ideas at a Davidson Horticultural Symposium.[1] Will Hooker gave me a similar opportunity at the first North Carolina State University Spell of the Land Conference.[2] Wilbur Zelinsky offered me a chance to elaborate some of my argument in an article for *The Oxford Encyclopedia of Global Change*.[3] The courses I taught at Duke University in their Comparative Area Studies Program in 1994 and 1995 helped me understand what I was getting at when I insisted that "globalization" was old hat. At one time or another I've hashed out most of these ideas with my indispensable friend, Arthur Krim, not least the Neolithic, the music, and the history of the landscape. He, Wilbur Zelinsky, Tom Koch, Stacy Amaral, and Ingrid Wood reviewed the early part of the book when I had crazy concerns about what I was doing. Diana Wu said useful things about a later section.

Peter Wissoker asked for the book and Joanie Johnson assured me I could do it. That was important, Joanie! I couldn't have written the book had I not received a Christine Baukus and Irv Coats Starving Writers Award. Cindy Levine and Eric Anderson were always there to solve my research problems. Irv kept leaving relevant books on my kitchen table. Jonathan Nyberg gave me one when I really needed it. Arthur and my mother, Nancy, sent me on-target clippings from the bewildering range of complementary things they read. Panama, Mike Evans, Dread, Grayson, Carmine, Chris, KP, PK, Freedom, and Shy—each in his own way—helped me keep it real. Mary Woltz gave me specific encouragement precisely when it was most needed.

Chandler was always warmly anxious about my progress. Randall read and commented on the early part of the book and was ever eager to hear me stumble through whatever difficulties I was having. And Ingrid heard *way* too many times the rehearsal of my doubts about the point of it all.

Contents

1

Missing the Global in the Local and the Local in the Global

Were I to run a probe, let's say from outer space, down through the stratosphere (through the ozone layer), through the tropopause and the troposphere, through whatever clouds of water vapor (is it raining?), down through the branches of a *flamboyán* (the raindrops breaking up on the leaves, bouncing and dripping and slipping off), through the zinc-plated roof beneath it, into a room, through a person, perhaps through his head and then his hand, through his *cuatro* (he's hunched over it, play-ing it against the sound of the rain on the roof), down through the floor boards, through the space beneath the house, through the shallow, red, friable, crumbling clay beneath it, through the rest of the lithosphere, through the asthenosphere, through the outer liquid core of nickel–iron into the solid mass that may lie there at the center of the Earth, at no point—*at no point*—would I have encountered a system at rest, a ground zero, a firm place on which to stand from which to observe the motion of everything else: it's *all* in motion.

The man's name happens to be Rubén. He happens to be a peasant farmer, a *jíbaro*. The *cuatro* he happens to be playing is a sort of guitar with 10 strings in five double courses. His house happens to stand in the central highlands of Puerto Rico. The *flamboyán* through which the rain drips is also called the royal poinciana. The royal poinciana (*Delonix regia*) is said to be a native of Madagascar.

Once we head down this road there's nowhere to stop.

Take Rubén, for instance: Rubén is a member of the ninth or 10th generation of his family to live and die in the highlands of Puerto Rican. Prior to that most of his people hailed from the coast outside San Juan. Who all his people were before that gets harder and harder to say, but certainly many of them would claim to be descended from ancestors born in Spain. They would claim to be Spanish.

The Spanish in the part of Spain Rubén's people came from—the coast of the south—had ancestors among, working backward, the Berbers, Moors, Visigoths, Alani, Vandals, Romans, Carthaginians, Greeks, Phoenicians, and earlier North African populations. Without going back any further, Rubén's Spanish ancestors came from Europe, Africa, and Asia.

If today the *cuatro* has 10 strings in five double courses, early in the last century it had nine strings in four double courses plus a fifth single string. The name *cuatro* is Spanish for "four." It arose in the 19th century, from the fact that then the *cuatro* had just the *four* double courses. This was in distinction to the guitar, which had *six* single strings. The *cuatro* had evolved *from* the guitar. The guitar had evolved in 16th-century Spain from the *vihuela*. The *vihuela* had evolved over the preceding five or six centuries from the Arab *'ud*. Again we're all over the planet.

The clay beneath Rubén's house evolved through the deep chemical weathering of a granitic regolith. This regolith had evolved from the weathering of magma intruded from pockets deep below the Earth's surface. Puerto Rico is an elevated chunk of a tectonically active ridge of a plate that broke free of the Pacific Plate 80 million years ago. (The Pacific Plate's the one Los Angeles sits on.) For 80 million years this splinter, which today we call the "Caribbean Plate," has been working its way eastward . . . a couple of centimeters a year.

To say "weathering" is to refer to the exposure of rock to the atmosphere in the presence of water. The rain dripping from the *flamboyán* was evaporated from the Atlantic Ocean. It was squeezed out of the atmosphere, as rain, when the air was forced up over the mountains by the prevailing trade winds. The trade winds are the surface face of a giant doughnut of gyring air girdling the planet.

It's all in motion: Rubén—via the Puerto Rican coast and the southern coast of Spain—from the three continents of the Old World; the *cuatro*, via the guitar by way of the *vihuela*, from the *'ud* of Iraq; the *flamboyán* from Madagascar; the soil from magma from deep within the Earth's interior; the rain from the Atlantic; the island from the Pacific; the air from . . . everywhere.

From everywhere but *here* at the moment. In *motion* but at the moment *impaled* on my probe like an olive and a piece of cheese on a toothpick.

For This Local Scene to Exist, the Whole World Had to Be Just So

⌐It may all be in motion but it's whole too, everywhere integrated. The motions of the crust, of the atmosphere, the weathering of the rock, the migration of humans and music create in every place a seamless whole.⌐

For instance, the *cuatro*'s not an adventitious . . . *add-on*. The *cuatro grew* in Puerto Rico. It grew in Puerto Rico along with the *jíbaro;* along with Rubén's *jíbaro* house (raised from the ground on its *zocos*, roofed in zinc); along with the size of his fields and the pattern they make on the slopes; along with the crops he grows in them; along with the coffee in the shade along the road and the *flamboyán* in the yard. None of this would have grown where it did had there not been this roughly dissected topography of high relief; and had the island not been settled in the trades by the time that sugarcane—in *its* slow, globe-girdling migration west from the islands of the Pacific—been poised to leap the Atlantic.

If the trades had failed or shifted, or sugarcane been elsewhere in its migration, or the Caribbean Plate not broken off, the history of Spain been other than it was, or the *'ud* not been carried west from Iraq—had any of these or a thousand other things not unfolded as they did—this man with his *cuatro* wouldn't be here, his fields wouldn't be here either, there would be no house, no red scar in the earth, no—

⌐What I'm saying is that for this intensely *local* scene to exist, this islanded moment of a man hunched over his *cuatro* playing against the rain on his metaled roof, the whole world had to be . . . *just so.*⌐

There's nothing special about Rubén. Rubén could be Mosammat Anowara Begun. Anowara happens to be standing at the edge of a banana grove adjacent to the cluster of houses that is the village of Chamurkhan. Chamurkhan is in Bangladesh. All the houses here have metal roofs like Rubén's. Anowara happens to be a widow. She happens to be wearing a sari and speaking on her cell phone, a Nokia. It happens to be the only phone in the village. Anowara happens to sell . . . phone calls. For *this* intensely local scene to exist, this villaged moment of a woman standing at the edge of a banana grove with a cell phone in her hand, her head covering pulled back so she can hear, the whole world had to be . . . *just so.*

Anowara could be Anton Letsoin. Anton *used to live* on the island of Ternate. Ternate is one of the old Spice Islands in the Laut Maluku, the Molucca Sea, a little dot really, off the coast of Halmahera. This all happens to be in Indonesia where Anton is a retired army sergeant. It happens to be night now and Anton's standing talking with others in a deserted rattan factory in the harbor of Bitung. Bitung's in the northwest corner of Sulawesi across the Laut Maluku from Ternate. It's hot and humid and Anton's shirt sticks to him. Anton's here because his house on Ternate was burned down by Muslims in an act of war that is tearing the Spice Islands—tearing the Moluccas—up. Anton is a Christian. For *this* intensely local scene to unfold, *this* islanded moment of a man standing beneath a bare bulb in an old factory commiserating with other refugees like himself, the whole world had to be . . . *just so.*

Anton could be Kathy Gilje. Kathy happens to be perched on a stool at the counter between her kitchen and the rest of the loft. She happens to have a glass in her hand. There's wine in it. The loft happens to be on the sixth floor of a building in Soho in Manhattan not far from where the World Trade Center towers used to stand. A vase on the counter—glass, or perhaps crystal—cradles a bunch of long-stemmed tulips some of which bow gracefully toward the counter. Kathy's an artist, an oil painter. One of her paintings leans against the wall to her right. She's talking about art, about the art scene in New York, to a person on the stool beside her. For *this* intensely local scene to exist, this apartmented moment of a woman beside a vase of tulips leaning toward a companion at a counter discussing art, the whole world had to be . . . *just so.*

Just so. . . .

The Local Is the Global at Work

It's seamless, whole, integral . . . one thing, but at the same time every part is in its *own* motion. Everything grows from, is rooted in, the local. *Where else?* But what else *is* the local but the intersection in the here-and-now of *global* processes, masses of air in motion, tectonic plates, the gradual migration of seed, the ceaseless conjugation of bacterial DNA, the ebb and flow of capital, the exchange of songs? Because the local is everywhere, and is everywhere different, and because in the local every global process intersects with, chats with, meets up with, every other, to think about the local is always to think about the global—and vice versa—because in the local *everything* touches, everything pushes, everything leans on, everything *tickles* everything else. The local is where the global hits the rubber. The local is the global at work.

It's easy, in the apparent insularity of our individual lives, to imagine this isn't so, to imagine that what you or I do is too little, too minor, too isolated to affect anything else, to effect any kind of larger change, to make any kind of difference: it's just local. But the fact is change is almost *always* the sum of a lot of very little things. Single big things, like asteroids plowing into the planet, are relatively rare.

Except for collisions with asteroids, the really big changes, the *huge* ones—like continents forming supercontinents and supercontinents breaking apart and moving around the planet (significant global changes like that)—they're the consequences of little nothings, twinges really, scarcely to be noticed. How fast did I say the Caribbean Plate has been moving? A couple of centimeters a year? Yet thanks to those centimeters what had once been in the doldrums of the Pacific is now in the trades of the Caribbean, and few global facts are of greater significance for Rubén.

Rubén's kind of farming moved to the island more rapidly than the island moves, but still it moved little by little. From its origins in the Middle East, agriculture seems to have spread west at something better than a kilometer a year before stalling out along the Atlantic for 6,000 years. Agriculture spread at roughly the same rate from its origins in the Americas, but with no Atlantic to cross it reached the island sooner. Rubén's practice represents a fusion of these two streams of domesticated grains and pigs, tubers and fowls.

The '*ud* moved even faster, if still not at the speed of light. It's hard to imagine that when Tariq ibn Ziyad led his Berber army across the Strait of Gibraltar into Andalusia in 711 he gave much thought to what musical instruments he was carrying. What could have been of . . . less moment? Yet the guitar that was to evolve from the '*ud* has traveled farther faster, and penetrated more deeply into world culture, than Islam. Radio, television, and the movies may have helped spread the guitar (and Elvis and the Beatles), but it's still one kid after another picking it up.

Every second 2.7 people are born, every 10 seconds 27, 162 a minute, 9,720 an hour, 233,280 a day, over 80 million a year. *That's* global change. Driving it? Among the most private, most intimate, most intensely local acts of all . . .

Little by little. But because in the local everything is hooked up, because everything is moving around *through* the local, *nothing* is isolated, every act is consequential, and all change is global.

Act Global, Think Local

Global change is Puerto Rico centimetering along; it's agriculture across thousands of years seeding itself around the globe; it's guitars in mere hundreds of years twanging their way into every corner of the planet; it's men inseminating women and thereby changing the balance of humankind to the rest of life on Earth. This is not new: global change has been how change has taken place ever since the Earth spun itself out of chaos. That was 5 billion years ago: 5 billion years of global change.

Which is why "Act local, think global" has it backward. You *can't* act locally. It's not an option. Because everything's hooked up, you can *only* act globally. It's not something you need to think about doing. Choice doesn't enter the picture. Breathe in, breathe out: the world breathes with you.

Analogously you can't *think* globally. You can think *about* the globe all you want, but what's going to *happen* depends on what *all* the actors are up to: the other seven billion people, the essentially uncountable plants, the infinitely more numerous bacteria, the as yet incalculable interactions we think about as climate, and every now and then . . . an asteroid. The global is an integration of these and everything else. *Trying* to affect it is sublimely presumptuous.

On the other hand, you *can* think locally. This is to say, *you* can think about *your* situation, about what *you* need, about what *you* want. Let me suggest a straightforward if sort of infuriating example: let's say you want clean air. Frankly, it's a bizarre thing to want: *it's too abstract.* But let's assume it's a way of saying succinctly that you're tired of sucking in air that smells and tastes so bad you instinctively breathe shallowly to avoid taking in too much of it. Maybe it's also a way of articulating the anxieties aroused by the disappearance of the woods you grew up with, the frustrations generated by longer and longer waits at lights. Perhaps it's no more than an issue raised in the last electoral campaign that's stuck in your head. You're at some public event, there's a table or a booth, you sign a petition, you pick up a brochure, a bumper sticker. Now above your tailpipe it says, "Clearer Cleaner Safer Greener."

Okay: you can't clean up the world's air—*forget that*—but you *can* work on getting the air around you cleaner. You can give up your car, for example. Your car is a *major* contributor to the smell and taste of the air around you. You sit there warming the car up on a cold morning, and it stinks. Or when you're idling at a long light in the afternoon: when the sun hits that pool of exhaust gases, the low-level ozone just knifes into your skull. What I'm saying is: if you don't like the smell of the air around you, the car's *gotta* go. You could do other things too. You could make sure any land you personally control works as hard as it can to keep the air around you clean: you could let your lawn revert to meadow, for example, or woods, or whatever was there before it was planted in grass. There are a *lot* of things you can do.

This is not to say that the coal-burning steam-turbine plant upwind doesn't have to go too. It does. But if your efforts have anything to do with its going, they will have been driven by the same kind of thinking that led you to get rid of your car: *thinking about your local situation.* You *know* what you have to do to clean up *your* air. So why don't you? One reason is that *it's so hard to believe that what one person does counts.* And one reason that's hard to accept is because we fail to understand—we fail to *accept*—that *this* is how global change takes place: one twinge, 1 centimeter, one little nothing after another (we've relegated all that to . . . *prehistory*).

As important as clean air may be to you, the fact is, you're more likely to reflect on the pain—and the *cost*—of car *ownership* than you are on dirty air: when you write out the check to pay off the BP card, for example, and the collision insurance, and the car loan payment, or when you're searching your pockets for coins for the parking meter, or circling the block looking for a place to park. You probably aren't going to get rid of your car to clean up the air but you *might* to clean up . . . *your life.*

However you reached the decision it wouldn't be *acting locally.* It would be *considering your local situation*, but *acting globally*, because whatever you do is automatically connected to everything else, is inescapably a change to the globe (which is how the globe changes, one little nothing at a time). You *will* have an effect on global atmosphere, but you'll have it because you thought about . . . *what a pain it is to own a car.*

Let me suggest another outrageous example: slowing the growth of world population. *Gimme a break!* Who's thinking about world population pressure—*that* unembraceable abstraction—when he or she has got his or her arms around you, the flesh is quivering, and—

Nuh uh: thinking about world population pressure's not going to stop anybody from doing anything. But thinking about the shame of having a baby outside marriage *might* (it might not), or about dying in childbirth (or the days lost at work!), or about how much work it's going to be to *feed* a kid, or about what it would feel like *not* to be able to feed it, or about having to sacrifice other things to clothe a kid, to send a kid to school, thoughts of having to work harder, longer, of not being able to—

These kinds of thoughts, thoughts about these kinds of personally important things, can at least *nudge* someone toward fumbling for a condom. Fear of world population growth can't: *it's too abstract,* too inconceivable, too (frankly) meaningless. You *will* have an effect on the world's population, but because you thought about how much better care you can render to fewer kids. Or how little you want the responsibility of children. Or how much you enjoy the singles scene. Whatever.

"Infuriating" I called my example of the car, "outrageous" the one about population. *Why?* Because in them I sort of imply there might be something *easy* about making these changes. Sure, you *could* just give up your car, just walk away from it; but you couldn't do it easily—even if the law let you—because your car is rooted in so much you do. There's a lot of noise right now about fast food, about how much of it we eat (and how this has affected the American family), about its effect on our health, on our labor practices (teenage labor), on how we farm. *But try to imagine fast food without the car.*

Try to imagine American high school without the car. Malls. Shopping. Cineplexes. Our homes! Dating. Visiting grandma. After-school soccer. Getting to work. Television (those advertisements!). Our ideas about what constitutes success, freedom, status. Our very imaginations. Abandoning your car changes your relationship to every one of these domains, which is to say, it changes your relationship to everything (to all the systems integrating the globe).

It's *not* easy to do.

The same thing is true about the number of kids one has. Whatever this is it is not a foolish, arbitrary number: it's grown out of the same comprehensive soil the rest of our lives has grown out of, just as Rubén's *cuatro,* the patterns his fields make, and the way the *flamboyán* shades his house grew from the soil of his life.

Changing these things isn't easy precisely *because* everything's connected, tied together with strings of mutual and multiple reciprocity. The shopping mall is no more adventitious in our lives than the *cuatro* in Rubén's, the cineplex is precisely as native as apple pie (maybe more so). But for the same reason, it's why changes in these behaviors have such wild consequences: *everything* changes with them.

It's this people seem to have a hard time getting their heads around: walking away from the car is not just walking away from the car, it's walking away from the *world* the car facilitates and that facilitates the car.

It's not easy to see this interdependence of the car and the world for a lot of reasons. Some of the ways the car and the world are connected are . . . simply hidden. They're *concealed.* Others are merely hard to see. They're less hidden than . . . *imperceptible.* But underlying our inability to see these connections is our predisposition to imagine they don't exist: the car is not *connected* to the world because it *is* the world. It has become part of the *stage* on which we act out our lives. The car is a given, like the Moon and the Sun. As with the rest of the stage, its origins are . . . *lost in prehistory.*

The real block is this way we have come to conceptualize our relationship to the rest of existence: it's set, we come on stage, and only then does the play begin. It is a way we have been caught up in the amber of a story we tell and retell without ceasing. It's *this story* this book wants to look at, and . . . *rewrite.*

The First Veil: Concealed Connections

To say "concealed" is to say ... *purposefully hidden*; but when I say "purposefully hidden" I'm not talking about corporations illegally venting pollutants under the cover of darkness, or dumping PCBs along the highway at night, or burying drums of chemical wastes in their backyards. Certainly such things are purposefully hidden. But not much is hidden this way.

I'm not talking about the extended chains of buying and selling either, those that blur our connections to the cow whose udders were stripped for the milk on our cereal; or that obscure our relationships with the *braceros* whose backs were bent for the lettuce on our sandwich; or the rest of those many, *many* hands that intervene—those of the graders, for instance, and the truckers, the wholesalers, the long-distance haulers, the stockboys, the checkout clerks, and the baggers—as hard as it may be to keep them in mind while forking food to mouth.

It's hard enough to keep them in mind even without taking account of the efforts grocery stores make to divert our attention, grocery stores, the food sections of the daily papers, the cookbooks, the glossy magazines at the checkout counters: "Don't think about where the food came from," these say—that is, don't think about what global processes resulted in the intersection of us and the food in this aisle of the supermarket—"Think about how it's going to look on your table. Think about how it's going to taste in your mouth!" And *this* is what I'm talking about when I talk about *concealed* connections, the way our attention is *diverted* from the hand to the mouth, from labor to pleasure.

Just as grocery stores fail to advertise the poorly paid stoop labor bottled up in their jars of pickles, so department stores fail to decorate their walls with posters of the *maquiladoras* (where the alluring mounds of soft and brightly colored shirts and sweaters are assembled) or the cardboard shacks (in which so many live who work in the *maquiladoras*) of, say, Juárez, Nogales, or Matamoros. No, it's pictures of *us in the clothes* the stores splash around, *our* attractiveness wearing the sweat of others (the sweat has evaporated ... in transit). As again it's *us* (always) behind the wheel in the ads for cars and gasoline, *idling* in front of the opera, *prowling* the outback, *ferrying* the kids to soccer, doing anything but *polluting* the atmosphere (smiling faces, no shots of tailpipes).

Since so few read the labels on the fruit they buy, few *experience* the mutual dependence of the United States and Mexico, Chile, and New Zealand. But then how many actually experience their relationship to their employers? I mean, how many paychecks separate out the hours worked for the boss from the hours worked for the laborer? Who could make sense of such break-outs anyway, swathed as everything is in clouds of mystification: about *merit*, about the benefits of *competition*, about the heroic ascent of *individuals*.

When I'm talking about concealed, this is what I'm talking about: the willful sleights, the elisions, the . . . *failures* of memory. I'm talking about the way we hide the forests we clear-cut behind tasteful buffers of green. I'm talking about the way we fence off our mines. I'm talking about the way we daub with camouflage paint the gas and oil pipelines too big to bury. I'm talking about the way we site hog farms and power plants . . . as far as possible from those who consume their output.

It's all done by magic!

The Second Veil: Connections That Are Hard to See

Further obscuring these impediments to understanding our place in the world is the fact that . . . *not everything is visible in the first place*. The wastes pouring from a tailpipe, for instance, are not easy to see. The air isn't labeled. It doesn't carry tags, "Dirtied in St. Louis," "Fouled by the TVA," "Made unbreathable by Smithfield Meats."

Air is hard to get our heads around. Phenomena we now recognize as different faces of the atmosphere—wind, for instance, and changing temperatures, seesawing pressure, and rain—were once accepted as wholly distinct things, among others by me when I was younger. The idea that there's a giant ocean of air took a while to sink in. That it presses on me with 64 pounds per square inch is a fact I still know only in my head, not in my bones. That it's all *one* ocean of air—and endlessly circulating— remains difficult to handle. That something can waft to me through the air from far away, like perfume from a body I can't see, or germs from a sneeze I don't hear, there's something . . . *untoward* about it, about how it works, about what it means. That smoke from my chimney or your tailpipe could materially modify this (all but invisible) ocean: *"C'mon, who are you kidding?"*

Yet air is but the most pressing (literally) of a host of . . . elements, compounds, mixtures, cycles, systems, *spheres* caught up in the knitting of our daily lives. Water is only marginally easier to see and understand than air. In its gaseous state, water's *part* of the air. As a solid, water's rock. As a liquid, it courses *through our bodies*. Like air water too is everywhere; and like air it also lacks the labels that might warn us, "Heavy concentrations of mercury," "High coliform bacteria count," "Laced with PCBs." Amplified (invisibly) in aquatic food chains, these substances can be precisely as devastating as their causes can be mysterious.

Still less easy is it to see how air and water *interact*. Carbon dioxide may be one of the components of air but it's also readily dissolved in water. This means the atmosphere and oceans are constantly passing it back and forth. So whatever else is going on with carbon dioxide, its presence in the atmosphere is powerfully affected by the oceans. The rate at which it's exchanged depends in complicated ways on temperature and pressure and on its concentration, but it also depends on things like photosynthesis. Along these pathways carbon circulates, not only between air and water but among all life-forms, down into the earth even, initially in the form of the carbonate skeletons and the shells marine organisms sediment in their growth and decay; but subsequently, as the rocks these sediments settle into, as limestone, as dolomite. These in turn, when thrashed about, metamorphose into marble. Here is where *most* of the Earth's carbon is, on the ocean floors and in sedimentary rocks. Only when these rocks are exposed and weathered can their carbon reenter the atmosphere and oceans.

In this cycling of carbon—as in that of water, phosphorus, nitrogen, oxygen, sulfur, manganese, potassium, and other elements and compounds—the atmosphere, hydrosphere, lithosphere, and biosphere *all* participate.

We participate. All these things move *through us* (*are* us, *become* us). *We* are part of *their* motion. So figure dissolves into ground. Ground resolves as figure. What was stage penetrates, *turns into*, actors. Actors sink to stage.

The Third Veil: Predisposing Us Not to See These Things . . .

It is not easy to see all this. Much of it happen too fast. Or it takes place too slowly. Things are too large to see (they're continents). Or too small (they're atoms). Carbon dioxide combines with water in the presence of sunlight in the leaves of plants to make oxygen and glucose in a series of reactions that take place in minute *parts* of seconds. It takes millions of years to lift marble into air. Ground that was bare is suddenly full of seedlings. We could sit before a wall of dolomite and watch it every minute of our lives and see no change at all. Stage and actor interpenetrate, but not in ways that are immediately obvious to the actors we call human.

And because we can't see it change—which is to say, *watch it act*—because it's too fast or too slow, or too big or too small, we dismiss it all as stage. *We* act. *We* move around. The stage just sits there. *We* change. It doesn't. So that when it does, when it suddenly stands up and acts, we're startled, confused (*"Where'd it come from?"*).

And since the only action that *really* matters is our own—I mean *yours*, *mine*—in effect the stage is the world *we're* born into, the world *we* grow up in. The world of our childhood becomes conflated with "the way things have always been." Ultimately it's conflated with *prehistory*: "once upon a time," "that was then," "back in the day."

It turns into a kind of paradise.

From which we imagine we've fallen.

Thanks to the snake.

Or *some* intruder in paradise, *some* outsider. Never us. We can't *see* how *we* could muck up paradise because we can't *see* how the world works, and we can't *see* how the world works because we can't *see* how it got here.

We need the whole story, not just the little fragment we anoint as history. We need to realize there is no stage, that it's *all* action, and that every actor speaks. We need to hear what they've been saying. It's time to listen.

Because . . . we are the globe.

2

The Idea of Prehistory Makes It Hard to Think about Global Change

Prehistory is the name of a deceit. Prehistory wants both to announce a beginning, the beginning of history (which comes after it), and to claim a place in a stream of history that has its spring elsewhere (sometime earlier, in a *pre*prehistory). Prehistory wants to say: this is history *and* it is before history.

Certainly prehistory feels like history. Like history it constitutes itself as a narrative of events, a chronicle, a story. It records and analyzes the past.

What's the difference? What's the difference between a *history* that records and analyzes the past and a *pre*history that records and analyzes the past?

Humans. Humans make the difference. They obligate this nonsense.

At the same time that humans require that history concern itself with humans, and indeed with only *some* humans (the "historical" peoples, those with a "written" history), their curiosity drives them to unravel . . . *the rest of the story*, what in a screenwriting age we have learned to call the back-story, the story . . . *before the story.*

Screenwriting has also given us prequels. Prequels teach that the story before the story is just as much a story as the story itself, insist to us, rub our faces in, a certain contingency of inaugural events. Flashbacks had long since suggested this, suggested that though the *telling* of the story began *here*, the story *itself*—as unveiled in the flashback—began else-where. Flashbacks implied that the story might have begun . . . *much ear-lier*, that the flashback could easily have had its own flashback. Flashbacks forced us to accept that we just . . . *picked the story up* where we picked it up, forced us to acknowledge that there was nothing but a . . . certain con-venience that made us do so.

Of course these conventions were not born in screenwriting. They had long since—for centuries—been staples of storytelling, indeed of his-tory. Speech must begin in time. Where to pick up a thread that appears to evade it?

Where to begin?

Perhaps once, when a dominant story some humans told postulated a nearly coterminous beginning for everything, it made sense to begin a his-tory with humans, or nearly with humans. In a Bible at hand, for instance, over 800-pages long, and which very much begins *in the beginning*, humans are rushed on stage two-thirds of the way down the first page. That is, on day 6 of the history of everything.

To many this story is no longer creditable history. Yet its essential characteristic, the inauguration of *history* with the appearance of human-kind, remains securely ensconced in the distinction between history and *pre*history (in the distinction between the humanities and science). Of course, the ratios have been reversed. If prehistory were 800-pages long, *history* would begin two-thirds of the way down the . . . *final* page.

In construing history the way we do we ignore almost all of it.

And since—whenever our appearance—it is all *our* history, this means we are ignoring almost all of *our* history. Which means we don't have a clue where we're coming from.

Which means we don't know where we are.

If we don't know where we are, we're not in a great position to con-template where we want to be.

Where we are is a spatial metaphor for *when we are*. It has the advan-tage of drawing attention to the shift over the last couple of thousand years that has occurred in how we think about *where* we are. Where once we imagined ourselves at the center of a plane surface, a square or disk (wherever on the planet we might have been), surrounded by an aqueous or airy void over which the Sun and the Moon and the stars rose and set, we came to see ourselves somewhere on the surface of a sphere. Where once we imagined that sphere to be in the center of a system of Moon and planets and stars that revolved around it (and so around us), we came to see this sphere (and so us) revolving around the Sun. Where not all that long ago we imagined the Sun to be the center of the field of stars, we have come to understand the Sun as a minor star in an undifferentiated arm of an ordinary galaxy, one among hundreds of millions in the universe.

A parallel displacement in time has yet to take place. Looking around us in time there is foregrounded a relatively recent history of presidents and kings, moguls and emperors. There are banners, and flags waving, and gun smoke, shields and arrows, temples and pyramids and burial mounds, further off, more dimly, Neanderthals in animal skins, and behind them dinosaurs in an encircling mist, the temporal equivalent of that aqueous or airy void that surrounded the flat world we used to think we lived on.

History is bright (sort of) and the rest is . . . *prehistoric*.

Why Do We Want a Full History?

Why do we want it?

A full history would help us think. It would help us get flexible. It would help us be intelligent.

Here, this sentence from an article on global warming in this morning's paper: "Perhaps never before has a single species had such a dramatic impact on the fabric of the biosphere."

It's referring to humans. You see variations on this sentence everywhere today. Perversely they damn us for our excess as they praise us for our dominance. That they're untrue in every part passes unnoticed since the early history required to set the record straight is all but unknown. The plain fact, of course, is that anaerobic bacteria—among whom the idea of "species" loses force—*created* the biosphere that we know by polluting *their* atmosphere—at the time a hydrogen-rich reducing gas—into the oxidizing atmosphere we've known since. Had these bacteria not rendered the world unfit for *themselves* to live in, *we* wouldn't be here to tinker with it at all.

More useful, then, to have said something like "Not always do individual life-forms have the dramatic impact on the fabric of our biosphere that humans are having," and then to have gone on to say a little about *what has happened in the past*, before launching the dreary catalogue of contemporary loss and destruction. Yet from fear such acknowledgments might reduce the enormity of our effect on the biosphere—and so the urgency of addressing our behavior—they are almost never made. There is no doubt about the enormity or the urgency, but there is doubt that we will be able to do anything intelligent about it without understanding what's going on. Absent history this is hard to do. It's like trying to think about "native rights" while celebrating Columbus's "discovery" of America. Under such a burden of distortion it's hard to make sense.[1]

Here, another example, this one from a recent *National Geographic* map supplement, "Australia under Siege." A time line emerges from a darkening void. The line gains focus 80,000 years before the present. Sixty-five thousand years before the present the ancestors of Australian Aborigines show up. During the ensuing 50,000 years a handful of animals go extinct. Two hundred years ago Europeans show up. During the ensuing 200 years there are *extinctions galore*. According to an adjacent caption, "The recent growth of Australia's population has contributed to a mammalian extinction rate higher than anywhere else in the world."

The implication is that extinctions are novelties, somehow attendant on humans, especially on European colonists. Indeed Europeans *have* wreaked havoc in Australia, but the fact that 99.99% of the unfathomable number of species to have ever existed has gone extinct is not only ignored, it is effectively denied by the way the time line . . . *fades out* before the past can be said to have really begun. During the Permian alone nearly 95% of the life-forms then extant went extinct.

Again, to acknowledge precedents for ongoing extinction events seems to diminish this terror, and so the urgency of the tasks we confront. While neither of these is doubtful, it may again be doubted that a meaningful confrontation can occur in a fog of delusion. This way of *thinking* our situation, certainly this way of *talking* our situation, isn't working. As its hysteria has risen, cars in the United States have only gotten bigger.

We *need* history. It will not change the need to act, but it will change the way we do so. It may even make action possible.

Look: you're involved with a couple of people, perhaps they're co-workers, or neighbors, or members of a group you're in that's trying to do something, and every time you, I don't know, initiate something involving them it doesn't go the way you expected. And somebody says, "You know they used to be married and went through a terrible divorce, don't you?"

Knowing this doesn't mean you're going to be able to get things to work in the future, but it does give you a perspective that thickens your understanding, and this at least makes the world a little more friendly, a little less a place you'll never understand. This is an awful, trivial example, couched in the outcome-oriented pragmatism so popular in our contemporary corporate culture, but it does, I think, make the case for history about as simply as it can be made.

Some other examples: you've moved into your dream house and you wake up in the middle of the night because all hell's breaking loose but it turns out to be just a lot of noise from the nearby university; or you wake up and all hell *is* breaking loose, the house *is* folding inward and beginning to move; or you wake up and mud is pouring into the bedroom, it's rising toward the ceiling, and it's lifting your bed with it.

In the first case—which could occur in many places, but let's set it in Raleigh, North Carolina—State's football team has just had its third victory in a row, reversing a recent history of lackluster teams and giving the university its first national ranking in a long, long time. Fans are partying in the street. They are whooping and hollering and honking car horns. There is nothing random about this. Collegiate sports celebrations take a form that has evolved over time. It is part of the history of celebrations. It is part of the history of college life. That it is happening here is part of the entwined histories of the college and the town. The kids aren't feral. They're part of the story.

Do you go back to sleep? Do you get up and join the party? Do you move?

Let's situate the second example in North Carolina too, on one of the barrier islands of its Outer Banks. All summer long heat has been building up in the tropical waters of the eastern and central Atlantic. In the history of the atmosphere—for the atmosphere too has a history (it has a beginning in time, it has changed, it has matured since it's been around)—storms (disturbances, a vast suite of large-scale horizontal eddy transport behaviors) have evolved to move heat around the planet. Among these are hurricanes. In its heat redistribution activity the hurricane in question *reshapes* one of the emergent sandbars that real estate speculators have been able to sell you as continent (i.e., as a potential home site), and in so doing the hurricane takes your house with it. There's a 4-billion-plus-year history of atmospheric circulation involved here, a corresponding history of continent assembly, reassembly, structuring, and restructuring, and ocean forming, a subtext of which involves a comparatively recent history of barrier "islands" in ceaseless motion, to say nothing of the essentially contemporary history of their promotion as sites for "second" or "vacation" homes. This in turn implies a social, an economic, history (it is a history in which most people own no home at all). Again, nothing random's happening. There's a compelling story (there are compelling stor*ies*). The only thing that doesn't make sense are all the houses on the sandbars.

Actually even those make sense. There's a story here too. It turns out you and I are in it, that without (perhaps) realizing it we're paying the insurance bills for the cleanup and repair. It's not about *liking* the stories. It's about using them.

The third example is set in Los Angeles. There the Pacific Plate has been sliding past the North American Plate for millions of years. This sliding is an event, *a moment*, in the history of the Earth's crust that has been unfolding now for some 5 billion years. Immediately to the north of Los Angeles the rapidly uplifting and almost vertical San Gabriels—a compressional subtext of the history of the sliding plates—have experienced one of their far-from-rare fires, the chaparral has been incinerated, and there's nothing to anchor the soil. Indeed, the fire has rendered the soil impermeable. Winter storms off the Pacific have pumped unfathomable quantities of water into the mountains. Sluicing off the mountains, the water rolls up everything in its path. This includes millions of tons of rock debris. The debris flows downhill, it picks up cars, and here it is, having smashed its way inside, lifting your bed toward the ceiling. Surprise? Only for those with no sense of history. Much of what is going on today is a surprise solely because we begin our history 5 thousand years ago instead of 5—or 15—*billion* years ago.[2]

In fact, Los Angeles spends 100 million dollars *a year* to fight debris flows. Most of the city is built on top of an alluvial fan of such debris. Up toward the mountains the fan is 900-feet thick. This is a fat story. We can tell it. It makes sense. It's about the entwining of tectonic histories, climate histories, ecological histories, transportation, settlement, and land use histories. It's about the selling of Los Angeles as paradise (it's about advertising).

The Loss of Paradise

As examples of the value of history, these stories may not be much better than the first one about the divorced couple. These too are ultimately almost tacky examples of *a* role history can play, still more or less couched in an outcome-oriented pragmatism, in the sort of talk scientists indulge in at the end of their interviews on National Public Radio (NPR) when Scott Simon asks, "What's it good for?"

In the face of the widening temporal horizon, however, in the face of the growing list of players, it gets harder to ask that question. It gets harder to think pragmatism in the face of an evolving Earth, in the face of 5 *billion* years. Something gives, something else begins to emerge, something like a relaxation of hostilities. Whereas the stories are all tense with . . . hostility, are permeated with a sense of loss which, since each one of them is about the loss of paradise, is only to be expected.

No surprise, then, either that all the stories take a common form. In each *one is jolted from sleep by something breaking in from the outside*. In the guise of the rapist in the night this is a dominant myth. In the guise of an attack on the country by a "rogue" nation—or by Martians or by an asteroid (by "the Other")—it is *the* dominant myth.[3]

This myth is the narrative form of irresponsibility. The agent of change—the threat—always comes from . . . *somewhere else*.

Implied is the drawing of a line, the making of a distinction. This line severs a here from a there, an inside from an outside, an us from a them. *Inside* things are cool (you're asleep in bed). *Inside* things are also, as we shall see, pure and innocent, self-present and authentic (your bed is the Garden of Eden). Therefore things *inside* are *outside history* or, if in history, then in the penumbra of a *pre*history ("I was sleeping when . . . "). In a word, *inside* things are *natural*. The stories are about the moment when an external perturbation—something from the outside, from over there—inaugurates history (" . . . I was awakened . . . "), the moment history is precipitated by an aggression external to the system (" . . . by someone breaking into my room . . . "). Since once your sleep has been violently broken you are always alert to its being broken again, the historical is an anxious world. It is self-conscious. It is alienated. It is corrupt. In a word, it is *cultural*.

The natural order of things—your continued sleep, as it were—is disturbed, broken: by the allurements of the snake in Genesis, by the revelation of the marriage in the story about the couple, by the noise of revelry in Raleigh, by hurricane-force winds on the Outer Banks, by a debris flow in Los Angeles, but under whatever name, by the rapist, by Soviet bombs, by Islamic terrorists, by aliens. In order to hold this line (this fiction), the actual state of things cannot be acknowledged. That is, the history *behind* the state of things, the prequel, the back-story, cannot be told. So in Genesis it is not acknowledged that Adam and Eve are late additions to a long-running drama starring Lucifer and Elohim (at least as Christians for the past 2,000 years have been reading Genesis in the light of the New Testament). It is not acknowledged that the couple had once been married. In the house shown you by the realtor the coterminous development of the subdivision and the university is not acknowledged. When you buy your property on the Outer Banks nothing is acknowledged—if known, if understood—about the role of hurricanes in the ecology of these emergent sandbars. Historical memory is short in Los Angeles, if the city can be said to have one at all. That Los Angeles is built on detritus the San Gabriels have been shedding and continue to shed on the city at a rate, on average, of 7 tons a year from *every acre* of the mountains is simply . . . *impossible to grasp*.

In fact it is less the snake that intrudes in the Garden than *Adam and Eve*; not the couple's marital history that is revealed, but *your ignorance*; not that the revelry is exceptional, but that the larger life (of which both it and you are integral parts) built your house; not that the hurricane is a freak, but that housing on emergent sandbars is; not that the debris flow invades the bedroom, but that *the bedroom invades the mountains*.

It is another way of seeing things.

Yet this other way no more than inverts the way we saw things originally. No *relationship* has changed. The arrows have switched direction, but it's still a blame game. There is still an us and a them, still a here and a there, still an inside and an outside, still a time outside of time and a time of history. Except that in this flipped version the San Gabriels are the bedroom and *we* are the rapists bulldozing our way through the window at night. In this flipped version *longing for the past* is replaced by *guilt about the present.*[4] Alternatively, inside and outside are flipped. In a version like this—which is the dream of those committed to progress—*here* is already historical. The timeless paradise of *there* lies in a future on the other side of tomorrow. In this version, longing for the past is replaced by *hope for the future.*[5]

To maintain the structure while inverting the terms is like looking at things from the same place but standing on your head. The only difference is you get a rush when you stand back up.

Neither intruder—neither the intruding debris flow (into the bedroom), nor the intruding bedroom (into the mountains)—is "real." *Both* are no more than spawn of our line of distinction, the one that split, *and so created,* inside and outside, nature and culture, prehistory and history. As the terms flip sides, all that remains constant is the form. Soon enough, *it* becomes the . . . *real.*

In the stories it is easy to see how inside and outside, nature and culture, prehistory and history arise from each other or, more precisely, are no more than different ways of *naming* the same thing. Here is how we draw the line: we assume that things *start* when *we* show up. What it comes to is that *we cannot imagine the scene apart from ourselves.* We literally cannot see how it would be organized without *us* to give it purpose. Everything that took place prior to our showing up, everything *outside* our system of intentions, is excluded, is pushed to the other side of the line. It is this that potentializes the intruder effect by establishing a domain from which the intruder can materialize, a domain out of which the intruder can . . . break into the room. The ego-, the ethno-, the chronocentricity here is invisible to us. We can't see it because it is an aspect of our seeing, it is a function of our gaze: the field of the here is established in—and *by*—our presence. Given the domain of the here, history only commences once the intruder, invisible *but implicit in the temporal deeps of the taken-for-granted world,* is actualized, is precipitated. That is, history commences when the past enters our consciousness or, to put it more concretely, when we become aware that the structure of the given world privileges some futures over others.

Here: you are shown a house. It's on a seemingly quiet street in a seemingly quiet neighborhood. Attention is not drawn to the number of houses around you that are rented to students when school is in session.

No allusion is made to the increase in traffic students occasion, or to the students who will park in front of your house. Certainly no mention is made of the (perhaps less than annual) street parties. Indeed, the realtor may not know of their existence. Then one night you're jolted from bed. It turns out the main drag is on the other side of the block behind you, the noise tonight is *insane*, you can't imagine what's happening. You stagger to the window. A guy in a letter jacket is urinating on your lawn. The next day there are empty beer cans in the gutter.

It feels like an intrusion. It feels like an *attack*. But the feeling arises as a trick of perspective, one entirely dependent on the datum. Had you been shown the house *during* the school year (with students walking past on the way to their cars), perhaps even on a Sunday following a blowout (beer cans scattered around, toilet paper in the trees), the effect would have been different. The party would have been a *known quantity*. Though still perhaps you would not have *liked* this situation (and maybe you wouldn't have bought the house), the party would nonetheless have been a part of things, included, *inside*. It would have had the to-be-expected quality of the Christmas vacuum, the students all gone home (plenty of the homeowners too), business suspended, the traffic evaporated, everything . . . *dead*. The fullness of the party would have been no more surprising than this emptiness, no less "natural," if of another order. It would have been like Halloween, or the Neighborhood Art Walk, spikes of a kind (*attacks* for the unwelcoming), but nonetheless *in* the scheme of things. But because you started the clock during the summer recess, *that* sense of ordered calm became the datum, *this* was the way the world *is*, and the party appearing suddenly, unexpectedly, seemed like an attack, precisely because uncalled for, and so . . . *unnatural*.

So in Genesis everything's cool until the snake slithers in. Had the story opened with Lucifer's rebellion in Heaven—that is, with the backstory—there would be an entirely different sense of things. God's disappointment in humankind would have had a precedent. The expulsion from the Garden would have been no more than an incident in an ongoing story (God had already thrown the bad angels out of Heaven). The expulsion wouldn't have seemed . . . the beginning of history.

Were people looking at houses on the Outer Banks shown aerial photographs charting the recent migration of the islands (i.e., the motion of those emergent sandbars), they would have a better sense of the entanglement, *of the interiority*, of hurricanes in the ecology of the islands. Perhaps people might still move there, *choosing* to dwell in so dynamic an environment. Such people would scarcely be surprised when the land disappeared from beneath their feet, they might even stand in the surf arms outstretched to welcome the destroying storm.

As a few do who live below the San Gabriels, embracing, if not quite the debris flows themselves, then the whole dynamic plate-crunching process of it all. They accept the risk and do not sue the county when the debris basins fail on the slopes above them.

But such are rare. Most don't have a clue. When the mountains fall on them they do sue the city. After one awesomely destructive debris flow a Forest Service hydrologist going door to door asked residents whether they remembered the debris flows of nine years earlier. Almost no one did. Most hadn't even lived there then. Realtors and residents along the active front are certain that "It can't happen here." As last year's hurricane recedes into the past, it recedes into myth. Each year residents are startled anew by partying students. The couple's divorce slips into . . . *ancient history*.

The Meaninglessness of the Ahistorical World

Without a past everything's always brand-new. *Everything*, the Earth, the whole shebang. No future is privileged. All are equally likely.

What this means is that at the level of daily living nothing *means* anything. This boulder in Los Angeles: It just *is*. It's like a stage prop. Merely *being* here—that is, not having *moved* here (or not having *been* moved here)—it can't work as a sign pointing to the forces *swirling around us* that while supporting daily life are also capable of moving rocks the size of houses.[6] The rocks are just here. Or they were put here by men from outer space (such a world is literally alien). Or it's all done by magic (it's all done with smoke and mirrors). Or nobody knows.

Waves wash up and down a beach. They too might as well be stage props rocked by a stagehand. There is little in the froth that tickles your feet to suggest the forces that keep them in motion. Building a sand castle on the beach, you're surprised to come back and find it washed away. *Where did the water come from?* But figuring out about the tides is not enough to immediately point to the tug of the Moon's gravitational field, it scarcely begins to suggest the forces involved. And with this you would have but the beginning of a sketch of the things that water can do in the grip of the Sun.

Houses just *are* (they don't imply livelihoods, patterns of socialization, craft traditions).

People have no pasts to drag around (it's just the way people are), they have no historically constructed expectations to wrestle with (it's just the way things are).

The snake was only doing . . . what came naturally.

Nothing in the world means anything. To the high school students I once taught in Worcester, Massachusetts, all the buildings along Main Street just *were*. They'd walked by them for years without noticing the dates carved in the crowns of the older buildings or into the cornerstones of the newer ones. Or they'd noticed them, but read them as . . . *names*. There was no sign of the city's past in its ancient Mechanics Hall, no sign of the city's ongoing transformation in its rapidly emptying storefronts, no sign of a kind of future in the new bank tower rising like a tourmaline crystal in the city's midst. Like the boulder, these things just were, and the kids and their parents were like that too, not given or shaped in any way by the tides of human migration, or the weight of ethnic stratification, or the winds of social change that swirled around them. Some people had money. Some didn't. That was that.

It's like not being able to read. And then you take a walk with a geographer and this freeway is revealed as the route of a former interurban railway that was laid along an older road that connected ranches (long since absorbed into the larger urban fabric) built at the mouths of canyons where, thanks to the mountains (rising along the fault) that wrung the precipitation out of the moisture off the Pacific, *some* water trickled onto the plain; and the scene acquires a structure given in a process that has unfolded over time, the place has a history in a system of histories, and things have meaning, they can be read, they can be understood; and *no*, it doesn't *decide* anything (though it may suggest which futures are privileged), but the understanding is thickened and this makes the world a little more friendly.

If only because it makes it a little more like you, if only because it implies that there are processes active in the world's integration (it's not done by magic). And you can look at that boulder and look up at the distant mountains and out to the Pacific Ocean and think, *"Holy shit!!"*

World History (Global Change)

It's not necessarily easy to see this, though. And the more subtle or convoluted, the more distant or large scale, the harder it is. You can walk on the yielding sand, and that *you're* making the footprints trailing behind you is no stretch at all. But to see in the sand a "footprint" of the eroding continent is not so easy. It's even harder to connect the everyday things you do, the shitting, the coffee drinking, the music listening, the automobile driving to the quality of the water in the river downstream, the loss of rain forest in the tropics, the movement of capital around the world, the character of the air . . . *in your lungs*. As close as you are to it, you may not be able at all to connect the nature of your gaze to your coconstruction of the social order.

It's not that we lack models. We have models in abundance. The idea, the threat, maybe even the promise of comprehensive, large-scale—even *global*—change lies deep in our collective imaginations.[7] It's just that the models incorporate the same ahistorical models we've been looking at: they're all about being awakened from sleep, they all incorporate the story of the loss of Eden. In each a line has been drawn. In each a clock has been started way too late. In each there's an intrusion, an attack, some kind of invasion. After the story of the expulsion from the Garden of Eden, Genesis relates the tale of Noah's God who floods the Earth out of anger at his creation: "And the flood was forty days upon the earth . . . and the high hills that were under the whole heaven were covered . . . and every living substance was destroyed which was upon the face of the ground, both man and cattle, and the creeping things, and the fowl of the heavens."[8] Another (not unrelated) vision of global change is unfurled in the eschatological account of St. John that concludes the New Testament: "The first angel sounded, and there followed hail and fire mingled with blood . . . and the third part of the trees were burnt up . . . and a third part of the creatures which were in the sea, and had life, died . . . and many men died of the waters because they were made bitter," and so on through a deadly catalogue of precisely *environmental* disasters.

It is these two accounts—and the extensive cataclysmic, apocalyptic, and millenarian literatures they structure—that provide the framework within which the contemporary imagination unfolds its understanding of the reality, its understanding of the threat, its understanding of the promise of global change. Because this framework incorporates the story of the loss of Eden it is not one within which useful thinking is possible.

In fact, the essential difference between these accounts and the stories about the victory celebration, the hurricane, and the debris flow is that here each component has been dramatized: in place of an inside and an outside (in place of a bedroom and a street), there is an apocalyptic conflict between Good and Evil; in place of an awakening from sleep (however rude), there is a retributive cataclysm (but still . . . *"sleepers awake!"*); and in place of an anxious, alienated, historical world (the one we live in), there is a . . . *new millennium*. The inversion here occurs because these accounts are already set in, *already take for granted,* the anxious alienated historical world that is the experienced form of the loss of Eden. Indeed these accounts are about reversing that loss. If our original stories were about the moment an external perturbation inaugurates history, and so replaces nature with culture, these accounts are about the moment an external perturbation *concludes* history, and so restores culture to nature.

It's the same line, but everything's been flipped and exaggerated.

These visions of a decadent historical world destroyed by its creator in repentance or anger, which for centuries organized the struggles of many to understand and deal with conquest, slaughter, famine, and death, were complicated—muddied (inverted again)—in 19th-century Europe by the development of a vision of the world undone not by an angry god, but by . . . the scientific progress, the technological hubris of its inhabitants (i.e., "pride goeth before destruction and a haughty spirit before a fall"). The cornerstone of this literature was Mary Shelley's *Frankenstein* (1818), the subtitle of which, "a Modern Prometheus," gives the game away; but it was her *The Last Man* (1826) that sketched the devastated landscape of urban ruins reverting to nature, with its dwindling band of survivors (in this case from the plague), that has become the touchstone of contemporary visions of global change: a reversion to nature anything but millenarian.[9]

Jules Verne was also attracted to this theme. After the string of novels that made him an exemplar of the promise of scientific progress, Verne began writing books that reverted to the dark apocalyptic strain he had explored in his very first novel (the only recently published *Paris in the 20th Century*): *Propeller Island* foresaw the devastation of the indigenous populations of Polynesia; *The Ice Sphinx* predicted the extinction of whales by overhunting, as *The Village in the Treetops* predicted the extinction of elephants; *The Will of an Eccentric* anticipated the pollution of the oceans by oil. In *The Purchase of the North Pole*, a sequel to Verne's highly popular *From the Earth to the Moon*, the protagonists propose to exploit the mineral wealth beneath the North Pole by using a giant cannon to change the tilt of the Earth and so melt the ice.

Despite their prophetic (and often subtle) character, Verne's novels slighted humankind's *general historical* complicity. The villains may have been human but they were never allowed to represent human*kind*, and in this way Verne helped shape the form of the narrative of irresponsibility we retell today: that of a master villain whose domination of the world is (just barely) thwarted by a Superman, a James Bond, an Austin Powers. In these stories we're barely roused before we're returned to sleep, Good and Evil alike careening in from outside (Superman from the planet Krypton!) to battle in our presence. All we have to do—*all we're able to do*—is watch.[10]

Verne was not alone. Other writers in an similarly prophetic vein pursued equally dystopian visions, notably Ignatius Donnelly whose *Caesar's Column: A Story of the Twentieth Century* (1891) anticipated the autodestruction of New York (in 1988—the survivors set up camp in . . . Uganda!); and the H. G. Wells of *The Time Machine* (1895) and *The Invisible Man* (1897). If few couched the environmental consequences of industrialization, urbanization, and globalization in terms as embracing as

Verne's, writers of many persuasions—Dickens, Hardy, the later Twain, Spengler, London, and Conrad—observed environmental change, albeit local, in terms no less bleak. In their work the sun shone only on childhood or on the vanishing countryside (or on both, as in the popular Victorian childhood fantasies of Kenneth Grahame and others), while cities became increasingly the sites of fog, gloom, anonymity, alienation, destitution, repression, and oppression, approximating in Hardy, for example, the urban whore of the Apocalypse of St. John. This historicization of the city against the enduringly timeless paradise of the countryside was scarcely new either (see Theocritus's *Idyls*, Vergil's *Bucolics* and *Georgics*), but the work of these novelists restored the theme to the heart of popular consciousness.

Nevertheless, it remained possible to mount counterfutures of equally compelling power, in Wells's utopian writing, for instance, that looked forward to the promises of technology and socialism (especially on the radio and in the movies); but subsequent to World War I, the Great Depression, World War II, and the Gulag, these became more and more difficult to sustain. Didactic classics like Aldous Huxley's *Brave New World* (1932), George Orwell's *Nineteen Eighty-Four* (1949), and Anthony Burgess's *A Clockwork Orange* (1962) capture the increasingly apocalyptic tone. Images of brighter futures survived for a while in off-planet fantasies (*Amazing Stories*, whose publisher is responsible for the term *science fiction*, first appeared in 1926), and to a degree in superhero comics (*Superman* debuted in 1938), or they were absorbed into historical or ahistorical fantasy. Yet here too the tone darkened. J. R. R. Tolkien's *The Hobbit* (1937) is a sunny tale about the destruction of a dragon by the least of a little people. Its subtitle, "There and Back Again," makes its ahistoric character clear.[11] His *Lord of the Rings* (1954–1955), which "ordinary readers" overwhelmingly regard as the greatest book in English of the last century, concludes its Manichaean conflict with the end of the world as it had been known. In *its* historical world, as in ours, there is no going back.[12]

The often optimistic images of future metropoli that had ushered in the century—with toga-clad men and women navigating sleek vehicles along highways swooping among gleaming spires—survive, but only in automobile advertisements. Elsewhere they have gradually given way to the overcrowded, dirty, violent, nightmare conurbations of John Carpenter's *Escape from New York* (1981), Ridley Scott's widely influential *Blade Runner* (1982), William Gibson's *Neuromancer* (1984), Terry Gilliam's *Brazil* (1985), and Andy and Larry Wachowski's *Matrix* trilogy (1999–2003).

Decisive in this revisioning was the bombing of Hiroshima and Nagasaki. These local catastrophes were immediately and everywhere read as harbingers of a nuclear Armageddon, of global change at the hands of man on a scale and of a completeness heretofore the exclusive province of the gods. As Americans locked themselves into a real-life Manichaean struggle with the Soviet Union, they began to dig fallout shelters, practice air raid drills, and otherwise indulge the delusional fantasy of civil defense. That is, as if in a nuclear version of the Flood, they began to construct a nationwide underground Ark. In coping with the imminence of catastrophic global change and "the possible suicide of the human race," the popular imagination mobilized both its most ancient biblical resources and those variants that had matured during the 19th and 20th centuries. It developed powerful images of blasted wastelands and primitivist survival fantasies not infrequently in the form of "radiation romances" that reworked the story of Adam and Eve in a world cleansed by cataclysm.

Unlike the delayed—and ultimately sublimated response of elite writers (with the exception of John Hersey's 1946 *Hiroshima*)—the popular response to the bomb was immediate, florid, and enduring, initially evolving within the increasingly broad domain of science fiction. Important examples include Bernard Wolfe's *Limbo* (1952), John Wyndham's *The Chrysalids* (1955), Pat Frank's *Alas, Babylon* (1959), and Walter Miller's influential *A Canticle for Leibowitz* (also 1959). Films such as *Five* (1951); *Them!* (1954), with its radiation-mutated ants attacking Los Angeles; *The World, the Flesh and the Devil* (1959); *The Day the Earth Caught Fire* (1961), in which United States–Soviet Union atomic testing causes global warming by tilting the Earth's axis (echoes of Verne); and *A Boy and His Dog* (1975) about communities surviving below the Earth's surface; as well as midbrow responses like *On the Beach* (1959) and *Dr. Strangelove* (1964) typify the range. J. G. Ballard generalized the sense of threat in a series of novels in the early 1960s in which he successively drowned (*The Drowned World*), sandblasted (*The Wind from Nowhere*), baked (*The Drought*), and petrified (*The Crystal World*) the world.

The increasingly *fantastic* tone was transformed in 1979 when, 12 days after the release of *The China Syndrome*—James Bridges's film about the meltdown of a nuclear reactor—the accident occurred at Three Mile Island. Work heretofore read as fantasy or satire began to be read as dramatized reality. Soon enough it was written that way. *The Day After*, 1983's made-for-television movie about nuclear holocaust, was seen by an immense audience that took it as a kind of "documentary of the future." Mike Nichols's *Silkwood* (1983), about the radiation pollution associated with weapons production, was viewed as investigative reporting. *Silkwood* proved to be paradigmatic. Whatever its intentions, its adherence to the dominant myth in its post-Verne version enabled most of us to wash our hands secure that, thanks to heroes (or heroines), villains would get their comeuppance, and the world would remain safe for sleeping.

The growing environmental self-consciousness that began to appear in films like these was not independent of the emergence in the 1960s of the "environmental movement." Frequently dated to the appearance in 1962 of Rachel Carson's *Silent Spring*, this was a recovery of the idea of *nature* as *paradise*, that is, a recovery of the idea of the *changeless* nature of harmony and balance in place of the *voracious* nature of Mary Shelley or J. G. Ballard. It owed more to Walt Disney than to John Muir and Henry David Thoreau. The fact that *Bambi* (1942) was the highest grossing film of the 1940s encouraged Disney to think about films featuring live animals as well as animated ones. He launched his True-Life Adventure Series with *Seal Island* (1948), which not only played to large audiences and won an Academy Award, but returned far more on investment than expensive animation. Films like *The Living Desert* (1953), *The Vanishing Prairie* (1954), *Secrets of Life* (1956), *The White Wilderness* (1958), and *Islands of the Sea* (1960), among others, exerted a profound effect on popular attitudes toward nature; were recycled in Disney's theme parks and on Disney's hugely popular television show; and paved the way for an increasingly popular genre of television nature shows that has included *Nature, Wild Kingdom, Wild America, Untamed World, Animal Safari, National Geographic Explorer*, and the Jacques Cousteau series.

All this reinvigorated—and sharpened—the Edenic idea of an ahistorical nature subverted by sinful humans, so that Carson's *Silent Spring* detonated in a public that had come to care for nature, or at least for Disney's version of it, in a way that was both novel and yet rooted in the first book of the Bible. This was a public that folded the threat from pesticides into the looming threat of nuclear Armageddon to produce a radically generalized concern for the future of nature as well as for mankind, a concern it conceptualized as it always had, by turning it into an invasion epic in which the corporate rapist is heroically repulsed by the environmental activist. *Erin Brockovich* (2000) is a recent example. The public's hands, as ever, remain clean.

During the 1960s this "environmental" anxiety even became the subject of popular song, culminating in Marvin Gaye's socially conscious theme album *What's Going On* (1971). Its "Mercy, Mercy Me (The Ecology)" remains the most frequently played song about the changing environment: "Woo ah, mercy, mercy me / Ah things ain't what they used to be, no, no / Where did all the blue skies go / Poison in the wind that blows from north and south and east / . . . oil wasted on the ocean and upon our seas, fish full of mercury / . . . radiation underground and in the sky / Animals and birds who live nearby are dying / . . . What about this overcrowded land / How much more abuse from man can she stand." As topical as it was, the echoes of The Apocalypse of St. John in this work of a preacher's son are hard to miss.

When you're used to thinking *across* distinctions, it's hard to start thinking *through* connections. It's no great wonder that ecology proved, and continues to prove, an intractable subject for popular art, especially popular narrative art.[13] Despite sporadic efforts like Ernest Callenbach's *Ecotopia* (1975) and, at the very time that phrases like "ozone depletion" and "global warming" were seeping into popular consciousness via the news media, the theme languished. Earth Day, inaugurated in 1970, went uncelebrated during the Reagan–Thatcher 1980s. Films like *The Road Warrior* (1981) and *Mad Max: Beyond Thunderdome* (1985) reverted to post-Hiroshima blast–wasteland survival fantasies, updated with a postpunk sensibility and a novel concern for resource depletion brought into popular consciousness by the oil crisis of the early 1970s (though in the post-*China Syndrome* era even *The Road Warrior* carried a less fantastic and more prophetic weight).

Such survival fantasies seemed indeed to prophesy the world of the late 1980s. In 1988, environmental crises dominated the news. Among others, medical waste washed onto the beaches of the Northeast, huge numbers of Atlantic seals died from unknown causes, the worst drought in years set the West ablaze (a third of Yellowstone National Park burned, the photos were everywhere), and George Bush "drowned" Michael Dukakis's bid for the presidency in the polluted waters of Boston Harbor.[14] When in March of the following year the *Exxon Valdez* spilled 10 million gallons of crude oil into Alaska's Prince William Sound—luridly fulfilling Verne's ugliest scenario—and later that year it was concluded that changing environments, principally rain forest destruction, were the essential promoters of the AIDS, Ebola, and Marburg infections, the popular imagination was seized by the reality and imminence of catastrophic global change. When Earth Day was revived in 1990, 200 million people around the world turned out to celebrate it. Marketers responded to the air of crisis and unprecedented display of public concern by pouring billions of dollars into "environmentally friendly" products, packaging, and advertising; and Hollywood, in 1990 alone, announced no fewer than 14 major projects with "ecological themes."

All this did raise the level of popular awareness of the importance of "the environment," but without a *history* of the environment all it did was reinforce the framework within which change had traditionally been conceptualized. There was still no acceptance of the presence in our daily lives of forces that could move boulders, shift islands, and mount victory celebrations, no sense of the ways in which our lives were interwoven by processes set in motion eons ago, or of the ways in which *every*thing we did—including breathe—contributed to their operation. Popular awareness remained focused on the exceptional and the extraordinary; and so, on heroes and villains, on those asleep in their beds and those breaking into their bedrooms.

Characteristically, among the proposals announced by Hollywood in 1990 were a CBS television series, *E.A.R.T.H. Force*, recycling the cold war *A-Team* as ecowarriors; an enviro-action film from Sylvester Stallone in which his action hero Rambo would battle earth exploiters instead of communists; and *Captain Planet and the Planeteers*, a cartoon series from Ted Turner featuring a caped enviro-crusading superhero. The lack of fit between the systemic, global, polycausal, and often invisible incremental change characteristic of global warming or species loss, and the time-worn apocalyptic, cataclysmic, and millenarian framework into which it has by default been crammed, has had the consequence of trivializing and sentimentalizing its causes, principally by isolating them to the explicitly criminal behavior of a heinous few (in other words, to the Devil), no more than supplanting cold war spies and drug dealers with environmental criminals. "We'll be seeing more and more *James Bond-* or *Lethal Weapon*-type films in which criminals are doing anti-environmental things," is how one Hollywood producer put it. The past decade has more than proved him right.

Although the cataclysm in Kevin Costner's *Waterworld* (1995) was an unexplained, global-warming-induced, Noah-like flood—breathlessly animated in the film's opening minutes—the film otherwise reverted to the bomb-inspired blastland fantasies of *A Boy and His Dog* or the later *Road Warrior*, played out on water instead of land (though admittedly with a new sensitivity to microissues of resource use as, scarcely minutes into the film, the Costner character uses his recycled urine); precisely as *Medicine Man* (1992), in which Sean Connery scours the rain forest for a cancer cure, reverted to the "radiation romances" of *Five* or *On the Beach*. But far more typical of environmental films was *Naked Gun 2½* (also 1992)—still perhaps the most profitable environmental film ever made—in which the hapless Lieutenant Drebin defeated an unholy alliance of oil, coal, and nuclear lobbies (S.P.I.L.L., S.M.O.K.E., and K.A.B.O.O.M., respectively) to make the world safe for solar energy; and Steven Seagal's *On Deadly Ground* (1994), an old-fashioned western in which grazing or water rights have been supplanted by the environmental consequences of oil extraction.[15] The documentary trailer that concluded Seagal's film, in which Seagal appropriately laid the problem at the door of oil *consumers*, only underscored the inability of the film itself—and indeed film in general—to describe, much less work through, a problem caused not by something *outside the system*, but one *arising precisely from within it* (i.e., it doesn't break into the room—*it's in bed with us*).

In movies, comic books, and the omnipresent television cartoons the now-pervasive issue has been even further distilled to the classical megalomania of evil criminals, business tycoons, or politicians with conventional world-takeover plans (i.e., the Devil)—see *Austin Powers: The Spy Who Shagged Me* (1999) and *Mission Impossible II* (2000)—only now they're armed with powerful biological or other anti-environment weapons. Or the threat has been . . . *displaced* . . . to that of children equipped with uncontrollable psychokinetic powers capable, as in Katushiro Otomo's *Akira* (1985–1995), of cataclysmic environmental change. The best of these comics, for example, Frank Miller's and Dave Gibbons's *Give Me Liberty* (1990), are more subtle in their analysis than anything film or television have had to offer, but remain shackled by the dramatic imperatives of the apocalyptic model to portray global change in exclusive (indeed in moral) terms—black and white, all at once or not at all—with causation lodged in a single, more-or-less evil individual (or cabal), effectively isolated, essentially an outsider; and the solution—which in this model is to say . . . *the savior*—in a more-or-less heroic "other," no less outside the system, thus absolving the rest of us, as in almost all popular art, of complicity *in* or responsibility *for*, the state of the world.

Or for that matter the block we live on, the house we inhabit, or the bed we sleep in.

Rethinking the Narrative

Obviously these models don't work. That is, they don't model the world we live in. Not one of them. Not even close.

There are seven *billion* of us humans, but you'd never know this from the stories. In the stories there are only two of us, one with the key to the satellite laser cannon (or the mutant virus or the nuclear device), the other desperate to get his or (very rarely) her hands on it. The enormous labor of millennia involved in developing the laser and the rocket and the research labs and the steel mills and the glass factories and the railroads and the mineral exploration and—

To say nothing of the ongoing labor required to maintain and operate these systems, and reproduce the labor and—

The activities required to feed everybody . . . et cetera . . . all this is backgrounded, crammed into a back-story, a prehistory. It's all stage.

What this does is cut the hero and the villain off from their pasts. In doing so it cuts them off from the rest of us who are in this with them. Isolating them does heighten the drama but it trivializes everything else the rest of us are doing. Compared with what they're doing, what does it matter what we're doing? It becomes easy to justify the extra bucket of popcorn, the Ford Expedition, the AC on high.

It's not hard to knock the movies, but the movies are indistinguishable from the magazines, the newspapers, the textbooks in colleges and high schools. Take that sentence we looked at, the one that said "Never before [humans] has a single species had such a dramatic impact on the fabric of the biosphere." How does that sentence work? Doesn't it really work the same way as that climactic scene in the movies? Isn't it about heightening the drama? *Never before,* which is to say, *for the first time. . . .*

In either case the phrase draws a line, it demarcates a threshold that we humans are crossing. Needless to say, it's the threshold of the Garden of Eden, *we* are taking the step *now,* now is the moment. . . .

Implied are forces for good and evil. The evil forces will accelerate this dramatic change. The good forces will slow it, maybe even keep us from taking the fatal step. The enormous labor of the billions of years involved in developing metabolism, the cell, its nucleus, multicellular organisms—

To say nothing of the soils they constructed and the ecosystems and the atmosphere and the biosphere and—

The activities required to keep the whole thing up and running (while the output of solar energy rises and falls) . . . et cetera . . . all this is backgrounded, crammed into a back-story, a prehistory. It's all stage.

What this does is cut us humans off from *our* past. In doing so it cuts us off from the rest of life among which humans are an integral but tiny part, here for only the most minute portion of time, and probably not for long. Isolating us does heighten the drama, but it trivializes the rest of existence. Compared to what humans are doing, what does the rest of it matter?

One price of this delusional self-aggrandizement is panic. It's hard to think when everything depends on you, when you have to act . . . *now!*

What if you knew we weren't crossing any kind of threshold, that there wasn't even a threshold to cross, that humans weren't the first species to have such a dramatic impact on the biosphere, that there was never any Garden of Eden in the first place, that there never had been a paradise of harmony and balance, that from the get-go it had been a story of disharmony and imbalance driving ceaseless change, it had been a story of continuous destruction and renewal? What if you knew that this world we're in, and which we are anything but alone in (and that is not "ours"), were but the most recent scene in a drama that's been going on a long, long, long time, and that will—*whatever we do*—keep on going on for a long, long, long time?

I think we could relax a little. I think we could relax a little and that if we did relax we could think more clearly.

But because implicit assumptions about a world of nature from which we are excluded by definition are coiled up in phrases like "never before," and because we construct our situation as a fall from the grace of a natural order, whenever we try to think about how better to behave it is always in terms of a return, a recovery, in terms of preserving, conserving.

But the world we are changing from is the very world that brought us to where we are. Going back would only reset the mechanism. It would only return us to this very spot.

Not that any kind of return is *possible*, but thinking this way blocks us from even trying to imagine futures that are both sane *and* unlike any we've ever lived, that is, futures in which 4% of the world's population *does not* consume 25% of its fuel resources, and yet are just as cool and fun to live in.

If instead of being stuffed with hero sagas and fables of return our minds were stocked with images of the Archean Eon, "the Age of Prokaryotes," when the stromatolites ran into the oceans and microbial mats bloomed across the mudflats; or of the Ediacaran Epoch with its algal lawns and soft-bodied animals ploughing the seafloor sediments and fern-like sea pens waving in the currents; or of the Mesozoic Era with its dinosaurs, with *Criorhynchus* snatching a fish from the jaws of *Elasmosaurus;* or of the Pleistocene, only yesterday geologically speaking when, in what is now the United States, 14 major episodes of glaciation ran sheets of ice thousands of feet thick over what are now Boston and Cleveland and Chicago and Milwaukee—if, as I say, *these* were the scenes from history with which our minds were stocked (*if these were the scenes from our childhood*), we might be in a better position to imagine how . . . *unlike* today tomorrow could be.

We might sit up for a second and think about what we're doing, for real.

As long as we keep starting the clock now, and fret about our fall from a mythical grace, our minds will remain forever frozen. If we're to deal with our grotesque situation, our minds will need to be well warmed up. They're going to need to be *very* flexible.

Yet so much around us seems aimed at keeping us stiff. That *National Geographic* map supplement "Australia under Siege," for instance, that's all it does. *Under siege,* which is to say, *under attack* . . . and by whom? By us, colonial invaders of a Garden of Eden, which the *National Geographic* only 20 years ago on another map supplement called a "Land of Living Fossils." Then Australia was home to the world's only black swan, the pied goose, and the "once endangered" Cape Barren goose "rescued by conservationists." Now, not 20 years later . . . *everything's* endangered. But making us out to be villains only gets our back hairs up. It only makes us defensive. It locks us into place. The sense of urgency the title wants to imply only makes us throw up our arms and say, "What the hell? It's all down the tubes already anyway!"

What if you knew that the history of the Earth was marked by *continuous* extinction? What if you could rattle off the ages of the great mass extinctions (surely more important than the presidencies of Chester Arthur or Gerald Ford)? The great Permian extinction, that of the Cretaceous (when all tetrapods over 25 pounds disappeared), those of the Ordovician, the Devonian, the Triassic, and the Pleistocene (maybe caused by the appearance of that formidable hunter *Homo sapiens* with his lethal stone spearpoints)? What if you could see in your mind's eye the endless parade of flora and fauna that had flourished and vanished? I'm not saying you'd dismiss what we're doing in Australia—and have long since done in New York and the Netherlands and the Nile Valley—that's absolutely not the point; but you might think about it differently, *would have to think about it differently*, could not think, "OhmyGod, what are we doing?," but instead would have to think about the causes of extinctions and their consequences. Thinking it through might give us grounds for taking *useful* action instead of panicking because the show's already over. Thinking it through might let us understand that we're *all* involved, not just a handful of heroic conservationists.

History matters.

- What if you'd known the couple had been divorced? (The question is, *How were the signs overlooked?*)
- What if your knowledge of urban history had encouraged you to ask the realtor what population the subdivision had been built for? (What if your memories of school had let you recall the rallies?)
- What if thinking historically had prompted you to ask about the history of the beach, the island community, the island? (What if you used history as a crap detector?)
- What if your understanding of geology as a historical process allowed you to see the San Gabriels as young? (What if your knowledge of processes made you wonder how that boulder had gotten there?)
- What if you knew that 10,000 years ago humans *invented* gardening? (How would you read Genesis then?)
- What if you'd studied the history of flooding and understood how paving watersheds and channelizing streams worsens floods? (What if you knew your local history as well as you knew that of the globe?)
- What if you understood what it means to say there are seven billion of us, going on eight, each of us breathing and eating and shitting, not just Union Carbide no matter how heinous a corporation. (In any case, we're all completely complicit in corporate behavior.)

It's not about cataclysms. There will be no apocalypse. Certainly there will be no millennium. It's not about balance and harmony either, and never has been.

It's complicated is what it is. *It's dynamic.* Change is polycausal and it's incremental.

It's time to reflect these things in the stories we tell. It's time to tell a new story, one that does without distinctions between an inside and an outside, between nature and culture, between prehistory and history. It's time to tell a story that doesn't just have room for heroes and villains, but instead has room for the rest of us.

All the rest of us, humans and the rest of the animals, the animals and the plants, the plants and the fungi, the fungi and the protista, the protista and the monera.

It's time to tell the whole story instead of a little part of the story. Knowing the story doesn't mean we're going to be able to get things to turn out the way we want, but then . . . *that's one of the things the story is about.*

Nor is it that nobody knows this story. It's that the story is so fragmented among biology and history and anthropology and chemistry and economics and astronomy that we don't think about it as a story. But we need to think about it as a story. A simple story. One we can recite as easily as a nursery rhyme. We need to get this story deep enough inside us that it will always be at hand, helping us think everything we do. We need to get this story deep enough inside us that we will be able to read it in the present, seeing in everything around us the strands in time that tie us all together.

Of course this story *is* just another story, but it is time to try a new one.

Note

Nearly 500 notes accompany this history of the land. These notes fulfill the usual role of letting you in on where I read the things that, in the text proper, I'm passing off as facts about the world. But the notes play two other roles as well.

The first of these is . . . to second-guess the text. What in the text wants to pass itself off as assured—as fact—appears in the notes in a different guise. It is less self-confident there, it is contradicted by other "facts," it is embroiled in controversy, it has a history of its own. These doubts and hesitations comprise a shadow text. I *do* want you to take away a strong simple story (this from the text), but I also want you to understand this story as . . . a work in progress, one we continue to write together, one whose unfolding is a function of the inadequacies in the story we've told so far (this from the notes).

I want you to feel that the story is authoritative enough to stand on, but also that it is not definitive. With newspapers everyday reporting new "facts" that have to be integrated into the story, the story's got to be flexible enough to fit in all these facts, yet strong enough not to collapse. This way the history of the land turns out to be one of those stories that gets better in the telling. The notes are about how this "getting better" happens.

Then, at the same time, I want the notes to be a guide to some wonderful reading. I'm a geographer, not a cosmologist, astronomer, astrophysicist, earth scientist, atmospheric scientist, tectonicist, oceanographer, chemist, molecular biologist, microbiologist, biologist, botanist, geologist, paleontologist, primatologist, anthropologist, sociologist, historian, or specialist in any of the cross-breeding and proliferating subdisciplines (e.g., panbiogeography). Because I'm none of these, I've had to read a lot to put this story together. My prejudices are simple: I prefer to read books, and I prefer to read them in my Adirondack chair in the backyard, or propped up on pillows in bed.

Much science, especially breaking science, is not available in book form. Such science has to be read in specialist journals in university libraries, is meant to be read by other specialists, and is rarely written in anything much resembling English even when it's *in* English.

I have rarely cited this literature, though it *is* widely cited in the literatures I *do* cite. One reason that I don't have to cite this specialist literature often is because it's increasingly translated into popular English, increasingly by the primary authors themselves. They do this in the pages of such popular magazines as *Scientific American*. But these scientists are also increasingly inclined to write the material up in books. In both cases the writing is much more readable than it is in primary publications like *Nature* and *Science*. (But still, if you're interested in any of this stuff, you should read these magazines too.) Some of the books these scientists are writing are really terrific, the scientist-authors actually forging insights as they struggle to find the words to say what they mean. Some of these scientists find out they're good writers, and their books can make exciting reading. Almost anything by Lynn Margulis is worth reading. Ditto Humberto Maturana and Francisco Varela. Harold Morowitz, Christian DeDuve, Mark McMenamin, and Tim Ingold are all exciting scientist-writers. Vladimir Vernadsky's *The Biosphere* is a classic. William Schopf has at his fingertips *everything* we know about the Proterozoic. Peter Westbroek ties everything together. Jack Harlan forces you to see food production anew—*forces you!* Carl Sauer never had a dull thought in his life. Neither did Paul Feyerabend. These are just a few that come to mind.

By no stretch of the imagination have I read everything, nor do I want to pretend I have; but I have read a lot, and I have read widely, and I wear my enthusiasms on my sleeve. Sometimes these *are* for abstruse articles bristling with mathematical expressions (Andrei Linde's little four-page gem, "The Universe: Inflation Out of Chaos," is a great example), but far, far more often they are genial or angry or crazily enthusiastic or exhaustively compendious *books* by the scientists who did the work (Linde's *book* is *Inflation and Quantum Cosmology*).

For example, if you're interested in the debate between those who think modern humans came *straight* out of Africa and those who think we evolved all over the place from humans who had *much earlier* come out of Africa, the best things you can read are Chris Stringer and Robin McKie's *African Exodus* and Milford Wolpoff and Rachel Caspari's *Race and Human Evolution*. Both these books—by the scientists most deeply involved in the debate—are chatty and informal without being the slightest bit condescending. In fact they demand a lot of hard thinking, not least because the two points of view are so . . . viciously opposed.

In other cases I recommend books by nonspecialists. I can't say enough in praise of John McPhee's books about American geology and the rise of plate tectonics. As collected in his *Annals of the Former World* they are, I think, a masterpiece.

I could go on, but my point is there are two texts here, the text that follows and its shadow in the notes. You can read them separately or flip back and forth. What would I do? Read the text straight through for the strong, simple story. Then go back and pull it apart with the notes.

And then put it back together again.

And read. Read and write and talk.

3

The Beginning of History

THE LAND (AND THIS BOOK AND YOU AND ME)
IS MADE OF MATTER

So. The Big Bang. The Big

B A N G !

Before this . . . nothing. More accurately, *who knows?*

Another universe collapsing? Crunching itself into the singularity that exploded as ours? Some other structure, utterly inconceivable?

Hard to say, because at the singularity, the laws of physics that we know break down—lose their grip—and without them we have literally no guide into events at this incomprehensible remove.

But for all intents and purposes: *nothing*.

And so, out of nothing . . . everything.

There's an irony for me that this new story we're telling—and it is new, it dates only back to 1922 when Alexander Friedmann concluded from Albert Einstein's general theory of relativity that the universe could be expanding—is like the old one we used to tell.

More accurately again: *still* tell, for it's probably only a small part of the population of even a country like the United States that buys into this new story, the story of the Big Bang.[1] Pretty much everyone else buys into the older story, or some older story, in which a demiurge—say, God—brings the universe out of nothingness into being.[2]

Buying Into

"Buy into"—what do I mean by that?

Certainly something other than "believe." Or perhaps I'm just trying to finesse the issue.[3] "Accept as true or real" my dictionary gives me for *believe,* but I have a hard time making sense of these words in the context of a story, which we're rewriting all the time, about improbable events that happened billions and billions of years ago.[4] "Fashions change very rapidly in this area," says T. Padmanabhan in his recent book on the large-scale structure of the evolving universe. "The half-life for the 'best' models is about four years. Hence I do not expect much of the details of the models discussed in this book to be relevant after, say, five years."[5]

Of course, he goes on: "But the physical principles on which these models are designed are likely to remain useful for a much longer time." This raises two points. *Useful,* first of all. If the story is useful—that is, if it helps us to move forward in our lives—then it's worth telling.

It has to work, though. By which I mean it has to *do* work.

And then second, *principles,* which perhaps can be usefully thought of as distillations of the most useful stories, distillations which, despite many efforts to undo them, have retained their utility. Continuous utility builds confidence. I guess "buy into" means "have confidence in." I buy into the ideas that the Earth is round, that it spins on its axis, that it revolves around the Sun. I have confidence that it's made out of atoms, that these can be converted into devastating quantities of energy. (I've seen a lot of pictures of Hiroshima.)

I'm trying to negotiate tricky ground here. Robert Boyle was one of the founders of the modern scientific method. One of the things he thought hardest about was facts: what they are, how you establish them. Boyle's idea was that matters of fact would be established by the aggregation of individuals' beliefs. A community, an intellectual collective, had mutually to assure themselves, and then others, that belief in an empirical experience was warranted.[6] The multiplication of the witnessing experience was fundamental to this warrant: *lots* of credible witnesses had to see it, and then *agree* on what they'd seen. But this made facts preeminently social constructions: they arose out of looking and agreeing on what had been seen (facts weren't . . . *revealed*).

The deal is, the rest of us have to trust the witnesses and their reporting. With my own eyes I've never seen the earth round, and I have seen what look like the Sun rising and setting; but still I buy into in the sphericity of the Earth and its rotation.[7] I know personally no witnesses of the bombing of Hiroshima, but I accept it as a matter of fact. I guess this means I'm sure I can find someone who did witness it, or records I'm willing to trust *left* by witnesses, or that I could find indubitable traces were I to go to Hiroshima. Buying into this, I buy into the power of the bomb, and I accept it as one among (very) many pieces of evidence attesting to the existence and nature of atoms.

Certainly *no one* witnessed the Big Bang. The Big Bang is a conclusion drawn from other matters of fact that *have been* witnessed by a lot of people. How this conclusion has been reached is the subject of the rest of this chapter, but it's probably important to acknowledge again that the Big Bang is still a pretty new idea, a pretty novel way of making sense of the matters of fact that have been witnessed. Paul Feyerabend put it like this:

> I shall argue that our entire universe from the mythical Big Bang via the emergence of hydrogen and helium, galaxies, fixed stars, planetary systems, viruses, bacteria, fleas, dogs to the Glorious Arrival of Western man is an *artifact* constructed by generations of *scientist-artisans* from a partly yielding, partly resisting material of unknown properties.[8]

Yet this is not to devalue the artifact. It is important to acknowledge that the story science tells—and that I'm convinced we have to get by heart—*is* a story, no matter how well connected it is to scientific matters of fact. It's important to acknowledge that it's a story *written by people*. Historians of science and sociologists of scientific knowledge don't call it a "story." They call it a "construction," a "social construction." But a narrative social construction is a "story." That's what a "story" is.

Why is it important to acknowledge it's a story? First of all, because *it is*.[9] But second, because the power of this story arises precisely from its self-acknowledged and ongoing social construction. It's a story without an end that we keep on working on. It builds on itself, and it keeps getting better and better.

Buying into ideas like these has all sorts of consequences. Richard Feynman tells this story:

> As science developed and measurements were made more accurate, the tests of Newton's Law became more stringent, and the first careful tests involved the moons of Jupiter. By accurate observation of the way they went around over long periods of time one could check that everything was according to Newton, and it turned out to be not the case. The moons of Jupiter appeared to get sometimes eight minutes ahead of time and sometimes eight minutes behind time. . . . It was noticed that they were ahead of schedule when Jupiter was close to the earth and behind schedule when it was far away, a rather odd circumstance. Mr. Roemer, having confidence in the Law of Gravitation, came to the interesting conclusion that it takes light some time to travel from the moons of Jupiter to the earth. . . . When Jupiter is near us it takes less time for

the light to come, and when Jupiter is farther from us it takes longer time. . . . In this way he was able to determine the velocity of light. This was the first demonstration that light was not an instantaneously propagating material.

I bring this particular matter to your attention because it illustrates that when a law is right it can be used to find another one. If we have confidence in a law, then if something appears to be wrong it can suggest to us another phenomenon. If we had not known the Law of Gravitation we would have taken much longer to find the speed of light, because we would not have known what to expect of Jupiter's satellites. This process has developed into an avalanche of discoveries, each new discovery permits the tools for much more discovery, and this is the beginning of the avalanche which has gone on now for 400 years in a continuous process, and we are still avalanching along at high speed. [10]

I hope it's not necessary to observe that this is a physicist's—an insider adherent's—view, not a historian's or a sociologist's. Even discounted, it's heady stuff. It's so heady Albert Einstein warned us to be on guard against it: ⌐"Concepts," he wrote, "which have proved useful for ordering things easily assume so great an authority over us, that we forget their terrestrial origin and accept them as unalterable facts."⌐Instead of an inducement to toss them, however, this was an invitation to understand and appreciate their foundations, to probe them, as Einstein had probed Newton's Laws of Gravity: "It is therefore not just an idle game to exercise our ability to analyze familiar concepts, and to demonstrate the conditions on which their justification and usefulness depend, and the ways in which these developed little by little. . . . "[11]

Little by little: which is to say, in history and, I would add, in place (*some*place anyway), *on the land;* not any old where, but . . . *there* . . . as a *function* of the land (as it were), as it might be said Du Fu's "Spending the Night at the Palace Chancellery in Spring" was a function of the land around Ch'ang-an in the Wei River Valley where the chancellery was located.

Which would be not quite right, which would be to miss the real point I was trying to make. For "Spending the Night at the Palace Chancellery in Spring" was *not* a function of the land around Ch'ang-an, it was a function *of the universe* (as we shall see), of at least (as we can imagine) *a planetary history* involving crustal plates in motion, the evolution of animals, of *Homo sapiens*, the domestication of millet (among other things), and the development of a bureaucracy, burgeoning already 12, 13 hundred years ago.

Science no more grows out of the æther than poems do, anymore than this story I'm trying to tell does. Would I be telling this story if I

hadn't ended up in a landscape architecture program at a land grant university in the United States at the end of the 20th century? *This* story?
Not likely.

And so, as Feynman says, *yes*, avalanching along, but not, as he says, for 400 years, as much as I too admire Tycho Brahe, out there on his island, in the cold, night after night, plotting the positions of the planets; for the story doesn't begin with Tycho Brahe, or with Copernicus, or Aristotle, or the Babylonians, or the peoples of the Indus Valley, or the Chinese, or with the protoscientists de Santillana was on about,[12] it begins—where else?—with the Big Bang.[13]

The First Hundred Thousandth of a Second

Pow! Bang! Descriptions of the actual beginning don't get a lot more sophisticated than this.

And after it everything happened extremely rapidly.

Here's what I mean by rapidly: "The first 10^{-43} second is inaccessible to our current theories," says Joseph Silk, speaking of the Planck epoch.[14] Silk is one of the creators of what these days is confidently called the "standard model." He has no problem referring to a million quadrillion quadrillionth of a second as an "epoch." The current theories to which he refers are quantum theory and the theory of relativity. The theory of relativity deals with gravity; quantum theory deals with the electromagnetic force, with the "weak" nuclear force, and with the "strong" nuclear force.

They're useful principles.

The *Schwarzschild radius*—a measure of the size below which a given mass will collapse to a black hole (and which is derived from the theory of relativity)—and the *Compton wavelength*—a measure of the quantum uncertainty in the location of a particle (and which is derived from quantum mechanics)—set effective limits on what these theories can describe. Setting the Compton wavelength equal to the Schwarzschild radius defines a mass scale—the Planck mass (10^{-5} grams)—with associated density (10^{94} grams per cubic centimeter), energy (10^{19} billion electron volts [GeV]), length (10^{-33} centimeter), and time (10^{-43} second) scales.[15]

Neither the theory of relativity nor quantum theory lets us "see" anything more dense than this, anything hotter, anything smaller, anything less long lived, or anything earlier. Neither the theory of relativity nor quantum theory, that is, lets us see *anything at all* of the universe during its initial Planck epoch, anything at all of the universe when it was—presumably (we can't know)—even smaller, hotter, and more dense.[16]

Neither the theory of relativity nor quantum theory lets us see anything of the beginning, and as Wittgenstein says, "Whereof one cannot speak, thereof one must be silent."[17]

And so time begins—*history* certainly begins—at the end of this Planck epoch, at the Planck instant.[18] That's our starting point. That's our datum.

That is, on our side of the first 10^{-43} second.[19]

It was hot then—in the beginning—hot and dense, hot and dense and . . .

It was exploding. Violently. It was expanding ferociously. It was—

Everyone says it was a fireball. Fireball? *Fireball?* Words like that aren't even metaphors. They don't begin to suggest what was going on. Look: there wasn't any matter around. For that matter, there wasn't any space. The universe—the *stuff* of the universe—wasn't expanding *into* space. It wasn't happening *in* time.

It *was* space and time.

The universe *is* space–time and so, in the beginning, it was space–time that had to come into being. This is hard. It's *not* easy to get. Space–time was coming into being, and it was coming into being in . . . *nothing*, into *nothingness*. Can you imagine that, space–time coming into being . . . into *nothingness?* You know, it's not like there was empty space and it endured in time until the Big Bang brought matter and energy into it. Before the Big Bang there was nothing at all, no space, no time, no energy, no matter. The Big Bang didn't happen in *empty* space. It happened in no-space. It happened in nothing-at-all.[20]

With the space–time came the stuff of the universe. Both of them, the stuff of the universe and space–time, all *four* of them, space and time and matter and energy, came into being together. Really they're just different ways of seeing, of saying, the same thing, for space–time is a consequence of mass–energy; or, vice versa, mass–energy is a consequence of space–time: space–time–mass–energy, all or none (*we* tease it apart and name the pieces).[21]

It came whole, though in the beginning it was all energy, or all what we would come to *think of* as energy, as radiation, photons, and neutrinos, only the most elementary constituents of matter, electrons, electrons and quarks. But this is because it was so hot and dense that, what in a cooler, more rarefied, older universe, we would come to experience as the electromagnetic and nuclear forces, were indistinguishable.[22] Because of this there was no distinction between strongly and weakly interacting particles. All of them just whizzed around—nearly at the speed of light—smashing into each other with incredible energy, just busting each other up. Not until things cooled down enough for the strong force to express itself could what we would come to think of as matter begin to be distinguished from what we would come to think of as energy,[23] for it's the strong force that binds quarks up into nucleons. Only from that point forward could particles with mass—that is, matter—be precipitated out of the stuff of the expanding and cooling (but still mind-bogglingly hot) universe.[24] When the strong force chilled from the mix the epoch of grandunification was over. It was the end of that historical period when the three quantum forces were one. It happened about 10^{-35} second after the beginning.

Which is when things got really crazy.

Up to this point space was mostly in what is known as a "false" vacuum state.[25] This already may be more than you want to know, but it's important, because once particles with mass were "precipitated" out of the cooling universe, this "false" vacuum acted as a repulsive force to exponentially increase the rate at which the universe was expanding.[26] I mean, it was expanding already—it was blowing up—but now it . . . *blew up!* It inflated. In no time it was 10^{50} times bigger than it had been, and it didn't really stop expanding at this exponential rate until the "true" vacuum was established at about 10^{-33} second.[27] Paradoxically it was this inflation that's responsible for most of the mass, most of the matter, in the universe. The *virtual* particles and antiparticles ceaselessly spawned in the "false" vacuum—even today—were so rapidly separated from each other by the violent expansion of space that they were prevented from recombining and annihilating each other. Thus, in a manner of speaking, they were made "real" and flooded the universe with matter.[28] For this reason the inflationary epoch is also known as the epoch of baryon genesis, though until things got considerably cooler there weren't any actual protons or neutrons, just a kind of thick quark soup.

Baryons are fundamental particles of matter: protons, neutrons. . . .

At about 10^{-11} second the weak and the electromagnetic forces separated; and then in a sort of phase transition—almost like water freezing into ice—protons and neutrons began to materialize. This was about 10^{-5} second—almost a number you can imagine, a 100,000th of a second—after the . . . beginning.

There, that wasn't so hard to say, was it?

Start? After the start?

Is that easier?

This Story Connects with Other Stories

And *why* is it that I prefer this cockamamie tale of virtual particles materializing out of a vacuum, and a primordial fireball that explodes into existence, only to cool into the continuously expanding present, to

the almost comparatively sane, "In the beginning God created.... "?
Richard Feynman tells another story about Newton:

> It is easy to figure out how far the moon falls in one second, because
> you know the size of the orbit, you know the moon takes a month to go
> around the earth, and if you figure out how far it goes in one second
> you can figure out how far the circle of the moon's orbit has fallen be-
> low the straight line that it would have been in if it did not go the way it
> does. This distance is one twentieth of an inch. The moon is sixty times
> as far away from the earth's center as we are; we are 4,000 miles away
> from the center, and the moon is 240,000 miles away from the center, so
> if the law of inverse square is right, an object at the earth's surface
> should fall in one second by $\frac{1}{20}$ inch × 3,600 (the square of 60) because
> the force in getting out there to the moon has been weakened by 60 ×
> 60 by the inverse square law. $\frac{1}{20}$ inch × 3,600 is about 16 feet, and it was
> already known from Galileo's measurements that things fall in one
> second on the earth's surface by 16 feet. So this meant that Newton was
> on the right track, there was no going back now, because a new fact
> which was completely independent previously, the period of the moon's
> orbit and its distance from the earth, was connected to another fact,
> how long it takes something to fall in one second at the earth's sur-
> face.[29]

Which is what I like about the Big Bang, the way the parts all connect
up. Most people don't tell the story the way I'm trying to, as a simple
chronological narrative. Well, no one does. I've never read one anyway.
They always start—I mean, the story is *so* incredible—with some simple fact
of daily life, the way the pitch of a siren rises coming toward you, for example, and drops after it's passed. This is known as the *Doppler effect* and it
holds for light as well as sound. The "pitch" of light rises, that is, its wavelength shortens, from a source coming toward you (it shifts toward the
blue), and drops from a source moving away (it's redshifted).[30] In 1929
Edwin Hubble realized that light from distant galaxies was redshifted, that
is, the galaxies were all moving away from us, and moving faster the farther away they were. This is to say the universe was expanding.[31]
So there was that.
And then it was true that if Newton was right about gravity, the universe had either to be expanding or falling in on itself. Even Newton
knew this, though he was able to put himself—and us—off with mumbo-
jumbo about an infinite universe.[32] But it was clear as daylight in Einstein's generalization of gravity in his general theory of relativity, though
in 1917 Einstein too put everyone off with a cosmic constant he threw in
just to keep things from blowing up.

I mean, the stars *looked* like they were standing still.

And then in 1922 Alexander Friedmann found the right solution to Einstein's original equations and the universe had to be expanding,[33] and in 1927 the Abbé Georges Lemaître independently reached the same conclusion.[34]

It was the *galaxies* that were moving.

The point I'm getting at is that before Hubble figured it out about the redshifts, Einstein and Friedmann and Lemaître—reasoning from the period of the Moon's orbit and how long it takes something to fall at the Earth's surface (reasoning from gravity)—understood that the universe had to be expanding. They even predicted the redshifts.

So once Hubble found the redshifts it meant we were on the right track, there was no going back now. . . .

To reach the Big Bang all you do is run the tape backward.

This is how people tell this story, as though they were afraid you wouldn't believe it otherwise. They want you to know how they were *forced* to postulate the Big Bang, how *inevitable* it was if this simple other thing were true.[35]

The evidence just kept piling up.

There was evidence from the evolution of stars, and from the relative abundance of radioactive isotopes, and then there was the cosmic microwave background radiation. A lot of people start the story here, in 1965, with Arno Penzias and Robert Wilson in Holmdel, New Jersey, with the pigeon shit in their radio antenna, picking up this 3.5°K hum—this 7.35-centimeter microwave static—no matter which way they turned it. What could the hum be, they wondered? Penzias learned in a phone conversation with a friend that a young theorist at Princeton had recently given a talk in which he predicted there should be background radiation left over from the Big Bang just about this hot.[36] Why? Because "if there had not been an intense background of radiation present during the first few minutes of the universe, nuclear reactions would have proceeded so rapidly that a large fraction of the hydrogen present would have been 'cooked' into heavier elements, in contradiction of the fact that about three-quarters of the present universe is hydrogen."[37]

We're talking about nucleosynthesis now. The universe is about 3 minutes old and it's cooled off to the point where every proton isn't immediately smashed into smithereens.[38] Some of these protons—these hydrogen nuclei (they're the same thing)—attracted neutrons. A proton with a neutron makes a nucleus of heavy hydrogen (i.e., deuterium). This wasn't a novel reaction, but in the past—yes! all 180 seconds of it!—the deuterium would have been broken right up, whereas now in the "cooler" universe it survived. A fraction of this deuterium attracted another neutron. This is tritium. Right off the tritium would absorb another proton. This is helium. Helium has a tightly bound nucleus, so it holds together, even in this still relatively energetic environment.

Helium has a mass of 4 (2 protons and 2 neutrons). Because there's no stable element with a mass of 5, the process—this is fusion—slackens off here. It's different in the interior of stars. It's dense enough there and there's enough helium for the reactions to go on and build up the rest of the heavy elements. I mean, that's all that's going on inside stars—they're heavy element factories—whereas early in the history of the universe stars had yet to be formed and there was so much free radiation still flying around that nuclei heavier than helium were blasted apart as fast as they formed anyway. This is the radiation that, having cooled for 15 billion years, Penzias and Wilson tuned into.[39]

Do you see how simple the components are? The observable proportion of hydrogen? Radio noise you can tune into on your own TV (it contributes about 1 percent of the snow)? The redshifting of the spectra from distant galaxies?

And you say, what could cause this redshifting?

Hmmm? Well, maybe the galaxies are all speeding away from each other?

But if that were true, then at one time they all would have been crunched up together.

Ohhhh . . . I get it! The galaxies are all spreading apart now because they were *thrown* apart by some sort of primordial explosion.

But if that were true, then there should be a sort of background radiation left over from the early explosion of the universe with a temperature of . . . and you do all these calculations based on what we've been able to figure out about nucleosynthesis and the measured abundance of various elements . . . with a temperature of . . . about 3°K.

And there is.

So. . . .

From Atoms: Galaxies, Stars, the Sun

So the universe went on expanding and cooling. It had cooled enough 300,000 years after the Big Bang that photons were no longer able to knock electrons out of the box and hydrogen could assume its atomic form (i.e., with an electron). Atomic helium too (maybe even a little lithium).

With the electrons bound up like this—that is, *out of the way*—photons could travel in straight lines (no careening electrons to bounce off of), they could propagate without scattering. We say that matter and radiation had decoupled, but the effect was to make the universe transparent for the first time.

Had there been eyes, for the first time there would have been light to excite them.[40]

A few hundred million years later there might even have been something to see, vast clouds of hydrogen collapsing into pancakes of great density, protogalaxies beginning to coalesce in the thinning pancake, galaxies evolving out of the fragments.[41] Or right off, dwarf galaxies forming, clustering up over the billions of years into groups and clusters and superclusters as the density fluctuations—present from the beginning (well, more precisely, since 10^{-33} second)—were amplified by gravitational instabilities (small local excesses over homogeneity), their slightly stronger gravitational fields *pulling in* regions of average or lesser density, the process continuing for millions of years, for hundreds of millions of years, until, a billion years after the Big Bang the universe was filled with protogalaxies, maybe even galaxies, and by implication . . . stars, trillions and trillions of stars.[42]

Stars being born and soon enough stars dying, supernovas flooding the universe with the detritus of their lives, heavy elements flying everywhere, the great clouds of interstellar gas, of interstellar dust, continuously enriched, new stars forming *everyday* (the process continuing as I type), one day, 5 billion years ago, the Sun. At first it was no more than a rotating stellar nebula, a clump, a sheet, a veil of gas and dust, drifting along the Milky Way, moving around, tugged this way and that by the awesome tidal forces of the galaxy, beginning to rotate, and very gradually (under the influence of its own gravity) beginning to contract.

The contraction caused the cloud to spin faster, as—simple fact of daily life—*you* spin faster with your arms closer to your body. A center began to condense from the cloud, surrounded by a disk (the basic model is 200 years old, it's Laplace's[43]). As the center contracted, its internal pressure rose. It got hotter, but also denser. It shrank faster, its temperature climbed (its temperature zoomed), until hydrogen started fusing into helium.[44]

We've been here. The birth of a star is a sort of local little Big Bang running in reverse, saved from collapsing completely by the radiation pressure of the fusion (it's hydrogen bombs, enormous ones, going off all the time, beautifully controlled . . . by gravity).[45]

The radiation boiled to the surface.

Suddenly the Sun blazed.

Fiat lux.

Why So Much Physics in a History of the Land?

Because it's not physics. It's history, early history, the *earliest* history, the foundation, the without which nothing.

And there really wasn't much physics anyway, practically none, not even the inverse square law. Instead, *assuming* gravity and the rest of the laws—what I've been calling "principles"—I tried to tell the story of a universe that unfolded into the one we live in today, but from the first million quadrillion quadrillionth of a second. It's not easy to talk about, but if we all keep trying, it *will* get easier.

I know how uncool it is to say this, but what happened was incredibly exciting. *And we're part of it.* That's what I really want to say. We're part of it. It's not something that happened, I don't know, *someplace else.* It happened right here, in our universe, and at the end . . . the Sun bursts into light.

It's awesome.

And the way as the universe cooled the different forces kicked in, and the way each in turn changed the course of this thing—the universe—that was expanding and cooling, it's an incredible story.

Naming the stages is just a way to, I don't know, make what happened easier to feel part of, *the Planck epoch* ending when gravity kicks in, and then the epoch of grandunification. They're like chapters.

I guess I didn't explain why it's called "grandunification." But there's so much to explain, and if we want to make it a story we can chant, we have to keep it simple.

It's called grandunification because then the strong force and the weak force and the electromagnetic force were all unified. "Unified" is to say the forces were indistinguishable, so that the particles each force acts on—that each force *articulates*—were indistinguishable too. And as the universe continues to cool and as each force breaks out, the particles the forces articulate begin . . . to behave distinctly too. They acquire identity, so instead of like, a smooth broth, there's a chunky stew. And since as the temperature drops each force kicks in independently, it's like a sequence of stages, the Planck epoch, or the epoch of supergrandunification (because during the Planck epoch even gravity was indistinguishable from the other forces), and then the epoch of grandunification—

The deal is, because the early universe was so hot and dense, all the forces and particles were like one indistinguishable mess. It was like crude oil, and what happened was like cracking oil, only backward. With crude oil all the things we know as the various grades of oil, and gasoline, and kerosene, and propane, et cetera are indistinguishable—are unified—at the temperature and pressure they're at in the earth. They're a single goopy mess. But when we let it out of the earth and the pressure drops, some components volatilize rapidly; and then when we heat it up, all the other fractions acquires their own identities, in stages, and we can draw each off separately. In the early universe it's the opposite: the forces and particles distinguish themselves as the universe *cools*, the Planck epoch ends when gravity kicks in, the epoch of grandunification ends when the strong force kicks in, the epoch of baryon genesis—

What it's even more like is the way the principles of solid state physics kick in for water when it cools to 32°F. You know how before they kick in you can't skate on the pond and after they kick in you can? Well, before the universe cooled enough for the nuclear forces to kick in there wasn't any of what we think about as matter at all, that is, no water in even its liquid form, and after they kick in there is. Matter, that is, not water. Water comes a *lot* later.

And it's a big deal in *human* history when quarks began to be confined, and protons and neutrons were precipitated from the quark soup, a bigger deal by far than Columbus stumbling onto the Americas, which is something every schoolchild knows. Unless quarks had been confined there wouldn't have been any Americas, much less Columbus.

It's easy to imagine that none of this matters for *humans*, however much it might matter for the universe. The fact is, though, it's the ancient history that really bites. Like you trip and fall and break your hip. We're all still subject to gravity—it's in our face all the time—and gravity's . . . *15 billion years old*. We all endlessly circulate carbon and oxygen and nitrogen and sulfur and the rest of them, and every one of those atoms is individually 5 to 10 billion years old—which means the stuff we're made of, that we can pinch, is 5 to 10 billion years old. The metabolic pathways through which the atoms flow—what we call "life"—the pathways in use today are all 3 to 4 billion years old. Our upright posture, walking, all that—4 to 7 million years old. Talking, toolmaking, making love—hundreds of thousands if not millions of years old. All the really important stuff? It's *all* ancient history. It's only, like the surficial scum that's new, the rhythms we dance to, the words to the most popular songs, hemlines. . . .

And the idea—since we can change these things, but not our atomic structure—that our atomic structure is not anything we need to concern ourselves with overlooks the fact that our atomic structure is precisely what we *can* change. It's why we don't paint our houses with lead-based paint any more. It's what drugs are about. It's how pesticides work. It's what people are talking about when they talk about global warming, us changing the chemical composition . . . of the atmosphere.

It's not my point that the surficial scum isn't important. The rhythms, the words, the clothes—they're where we live. It's just that in living there *at the very same time* we're living in this huge ancient structure that very much keeps us from changing even the scum any way we choose. Undoubtedly, the social structure into which we're born also inhibits us from changing things the way we choose, but far more determinate, far, far more determinate, is gravity, is chemistry, is biology. . . .

The point is to feel all these forces working at the surface. It's what I was trying to get at when I was talking about the boulder in Los Angeles, about the waves in the ocean, not necessarily *specifically* the early universe, but all the history, all the history informing, all the history *present* in the present. And no, there are no tags on the atoms I'm exhaling as I speak that say "Antique: 9 billion years old." As I said, the past doesn't announce itself as the past. It announces itself as the present. But those atoms *have* been around.

There's no understanding ourselves unless we understand what we're made of, how we work, how we ended up the way we are. It may not be necessary to read Stephen Hawking—though it wouldn't hurt—but it is necessary to have read *some* history of the early universe. Indeed, it's useful to have read more than one. This is one reason for the notes, to point out a few of the doors to the literature. I'm not a physicist. I'm not trying to write physics. What I'm trying to do is integrate a bunch of things that we don't think about *as*, but which really *are*, history. The goal is to write a real history, not one that excludes almost everything that matters. Given the way we divide everything up—physics over here, chemistry over there, biology over here, evolution off there, paleontology in another room altogether, and history somewhere else—it's pretty hard. But if we all wandered more in disciplines not our own, it would help.

This has got to be one reason we're so confused, one reason we find it so hard to think straight, because when we look to history to orient ourselves in time, it gives us so distorted a picture, as though the only thing that mattered were the last few generations. We should be seeing the Big Bang in every sunset. We should be feeling it in every one of our bones. We should be listening for the Big Bang in every song.

To hear the universe in every song, that would be . . . something.

4

The Land Is the Functioning Skin of the Planet

And how do *you* remember *your* past?

"Oh . . . that must have been the year we visited Jeremy in Cheney."
Then whatever it was that prompted the recollection is tied more tightly
to the web, or tied to the web for the first time, the web is more capacious,
can help make more—or at least different—sense of things.

This Is History and It's Ours

"Oh . . . that must have been the epoch when gravity and the electronuclear forces were impossible to tell apart, the epoch of supergrandunification."

When things were really hot. Before history started.

The Planck epoch. The first 10^{-43} second.

The one followed by the epoch of grandunification.

By then gravity had distinguished itself—which is to say space–time
existed (and photons and neutrinos)—but the electronuclear forces still
ran together.

The epoch of grandunification lasted from 10^{-43} to 10^{-35} second.

The epoch of grandunification ended when the strong force chilled
free of the electroweak force. This inaugurated the epoch of baryon genesis. Then the repulsive force of the false vacuum blew the universe up like
the first big puff blows a piece of limp rubber up into a balloon. The universe grew 10^{50} times larger.

This inflationary epoch lasted from 10^{-35} to 10^{-33} second.

It left the universe awash in quarks and electrons.

At around 10^{-11} second the weak force chilled free of the electromagnetic force and at 10^{-5} second protons and neutrons began to chill out of the mix.

Nucleosynthesis began 3 minutes later.

Three hundred thousand years after that hydrogen nuclei had attracted enough electrons that matter and energy decoupled and the universe became transparent.

Within a billion years you could have seen galaxies everywhere.

Within 10 billion years you could have seen the Sun break into light.

People talk about where they're from—they say, Africa, Europe. But this is where they're *from*. This is where *you're* from.

Pinch yourself. That stuff: this is where it came from. All of it.

Out of the false vacuum, randomly fluctuating.

It's been here ever since. It's been here 15 billion years, through supernovas and gas clouds, recycled through who knows how many stars (through who knows how many *bodies*).

It *is* the history of the universe, but it's yours too—and mine—and to segregate it from the study of history as though it were . . . *only science* . . . is insane. It's *our* history that opens at the Planck instant, that begins with the epoch of grandunification. *Ours* is the story of an expanding gas, which like any expanding gas cools, and what happens when it does (*we* happen).[1] It's all—the epoch of grandunification and the inflationary epoch, the time when protons and neutrons were fractionated out and the epoch of nucleosynthesis, the beginning of the atomic universe and the formation of galaxies—*simple physics*. Which is to say *we're* simple physics. Which we are.

It's history—the story of all the simple physics working over time *at the same time*—that complicates things.[2]

Atoms Are Children of History

Like the atomic elements: it's history that explains their presence, their relative abundance. Most atoms—*atoms* now, things capable of forming molecules (i.e., protons *and* neutrons *with* electron shells around them)—come from the fusion within stars of hydrogen into helium, helium into carbon (*et cetera*). Every atom on Earth—every atom in your body (pinch it again)—was built up this way inside stars that blew themselves to smithereens long before the first birth contractions of the Sun. Atoms are nuclear slag (they're . . . *stellar shit*) spread through the galactic medium—

the dust and gas—by those terminal spasms of stellar evolution we know as supernovas.[3]

Stars that formed in clouds of gas and dust that had been enriched in this way by supernovas had more heavy elements in them than stars that formed earlier in more impoverished media. Older stars have a *hundredth*, a *thousandth*, of a percent of the oxygen, of the iron, of the Sun. Younger stars, tykes of say 10 million years, have *twice* the Sun's metal content. This is to say the atomic history of our star, that is, *the Sun*, is a function of the history of the Milky Way—that is, at which point in the *Milky Way's* history the Sun burst into light—whose composition was what *it* was because the history of the universe had taken the course *it* had.[4]

So Is the Earth

Since the Earth condensed from the same cloud of gas and dust as the Sun, it's in the same historical position. The same thing happened to the rest of the dust in the disk around the Sun that happened to the Sun, except there was less to work with, so that under contraction it never built up enough pressure to raise the temperature enough to initiate fusion. Consequently the bodies that formed—we call them "planets"— were cold, *comparatively* cold, and compact, spinning. Basically they accreted into clumps—simple fact of daily life—by collision, the way rain drops do.[5] The dust particles just kept running into each other. When they did, gravity and electrostatic forces held them together. As they grew, the clumps, the pebbles, fell—were pulled—to the midplane of the disk. There they clumped up still more and formed planetesimals. The planetesimals then accreted like the particles did. They just kept smashing into each other and gravity glumped them up into protoplanets. Finally the protoplanets—they were Moon-sized now—started running into each other. Spectacular collisions. Again and again. And . . . if the resulting planets were large enough, they attracted gases and formed atmospheres.[6]

All this took 100 million years, about the same time it took the contracting protosun to reach the 8 million°K it took to get fusion going. So by the time our Sun broke into light its planets were there to receive it.[7]

What? 10 or so billion years after the Big Bang?

About 4.5 billion years ago?

So . . . from 4.6 to 4.5 billion years ago, the condensation of the solar nebula, the evolution of Sol from protostar to star, the accretion of the planets, of the Earth.

Immediately We're Plunged into Geological History

It's the beginning of geological time. Instantly we're in the Precambrian (the Prephanerozoic). More specifically we're in the Archean, early in the Archean. In fact, we're in the Hadean, that is, back when it was hell on Earth.[8]

When I said the new planets were "comparatively cold" I didn't mean they weren't molten. I just meant they weren't hot enough to initiate fusion. In fact the Earth had condensed from gas and dust rich in elements with high condensation temperatures: iron, silicon, magnesium, sulfur. There was plenty of aluminum, calcium, and nickel too. The continuing headslams of planetesimals—the Earth continued to accrete—and the decay of the numerous radioactive elements produced enough heat to melt most of this original material.[9] Between 4.5 and 4.2 billion years ago the molten iron and nickel settled into a core and the mantle bobbed up around it. Think almost anything with a dense liquid center and lighter crust around it: state fair food, think fried cheese, think hushpuppies. Think. . . .

Think about the way the water in whatever it is you're frying spatters into steam as soon as it hits the oil. Coasting toward 4.2 billion years ago a similar devolatilization of the mantle led to the formation of the earliest atmosphere and the earliest hydrosphere, a lot of effervescence, outgassing, a lot of volcanic . . . burping.[10] If the surface were molten, the hydrosphere would have been in vapor phase at this time—I love that, "in vapor phase"—that is, the hydrosphere would have been part of the atmosphere, so . . . there would have been water vapor (H_2O) in any case, and most likely carbon dioxide (CO_2) and nitrogen (N_2) and hydrogen (H_2), with smaller amounts of hydrogen sulfide (H_2S), ammonia (NH_3), and methane (CH_4),[11] most of these still plentiful today in those clouds of interstellar gas where new stars are condensing.[12]

But no oxygen.[13]

Did I mention the lightning? *All* the time.

During the next 200 million years the heavy-duty bombardment slackened off as the new planets swept their environments (comparatively) clean.[14] Temperatures began dropping, and by 4 billion years ago any residual molten stuff had crystallized. As the planet pulled itself together, the mantle continuously reorganized itself (and still does). There were tremendous heavings and settlings, subsidences and swellings. Degassing continued to thicken the atmosphere, but up to this point the Earth would have been too hot for compounds of biological interest to remain stable.[15]

The Cooling Earth Experiments with Compounds of Biological Interest

Between 4.0 and 3.8 billion years ago the upper mantle finally calmed down. It chilled out into a relatively solid lithospheric basement covered everywhere by crust. The atmosphere began raining out too—forming the global ocean—and this multiplication of phase boundaries created increasingly various niches for chemical evolution.[16] Not that chemical evolution required lacustrine pools or estuarine shallows.[17] Atoms are gregarious—most of them anyway. They hook up quite spontaneously, even, as just noted, in high-energy environments like the interstellar medium. There's nothing mysterious about this. Atoms, coughed up by a supernova, jostle around. They bump into each other. They collide. The next thing you know they're bonding, sharing electrons. The simple fact of daily life example is rust. I mean, this oxygen *will* bond with that iron. Spontaneously. You don't need to do anything to promote it. Of course *you* don't need to do anything to promote most chemical reactions, and as Lynn Margulis and Dorian Sagan remind us, "Chemicals do not combine randomly, but in ordered, patterned ways."[18] This is no big deal either. It's the simple consequence of atomic structure and the operation of the electromagnetic force. We're talking about atoms *in intergalactic space* spontaneously organizing themselves into dozens and dozens and dozens of variously complicated molecules: formaldehyde, vinyl cyanide, dimethyl ether, ethyl alcohol. Yeah, even booze. The wags who discovered ethyl alcohol in molecular cloud Sagittarius B2 back in 1974 figured that the cloud was about 1 proof. At that there was hooch enough for 10^{28} bottles.[19]

That was in *outer* space.

What it meant on Earth with its rich variety of niches was the rapid evolution of *networks* of molecules, networks of self-referencing, autocatalyzing macromolecules so complex that they exhibited all the characteristics of life; networks of reactions that produced the very molecules that made the reactions run so they could run again and that at the same time set the boundaries of the spaces in which they ran; molecular networks that . . . were self-producing and self-bounding and, well . . . *that's life*.

"It's a chemical reaction, that's all"—at least that's what Cyd Charisse used to say.[20]

Harold Morowitz says: "The bottom line . . . is that life occurred very early in the history of the Earth. Life formed shortly after conditions were favorable and spread rapidly over the surface of the planet."[21]

We've moved from the Hadean to the Early Archean. We're talking 3.8 billion year ago. Morowitz says:

> I think it is conservative to say that continuous life on Earth formed 3.8 ± 0.2 Ga [billion years] ago. This is not a precise estimate, but it places the event in the late Hadean or early Archean period, suggesting that as soon as the Earth cooled down sufficiently, life formed rapidly on a geologic time scale. A less conservative estimate would be 3.9 ± 0.1 Ga ago—a very different view from the classical perspective involving random chemicals reacting for eons and finally lucking out, resulting in a living cell coming together.[22]

This is another of those stories most people don't tell this way, as a simple chronological narrative. They always start—I mean, people *so* want people to be special, so want *life* to be special, that to (as it seems to them) *reduce* it to no more than . . . *chemistry!* . . . arouses such resistance that, well . . . people telling the story often start it back in 1953. That's when Stanley Miller sort of ripped the veil from the prebiotic by cooking up a bunch of amino acids and other organic compounds—life stuff—in a laboratory flask with nothing in it but a simple reducing gas of CH_4, H_2, H_2O, and NH_3 that (guess what?) was the one his teacher, Harold Urey, on purely geophysical grounds, had hypothesized must have been the atmosphere of the early planet.[23]

This was more than another of those "*We're on the right track, there is no going back now*" moments, though it was surely that. People writing about Miller's experiment use words like *bombshell* and *explosive*. They call it a "dream of a crazy idea,"[24] and say things like "not since Friedrich Wöhler synthesized urea in 1828 had a chemical experiment been hailed as a comparable milestone."[25] It really was the end of something, the end of a mystery. After Miller it was just details. *You* could do it, it was so simple: a source of heat, some simple gases in a flask, a spark discharge device (like the ones in all the horror movies), a way of drawing off the condensing fluid. You too could cook up some amino acids, something, prior to 1953, only God was supposed to be able to do.

It's like *all* the old stories wanted to come back—for here *is* life from a handful of dust!—but again, the new story connects up differently, to different things, Miller's to nucleosynthesis and protein, to the age of rocks and radioactive decay, to the birth of stars, to DNA.

Everybody began cooking up simple reducing gases then, varying the mix and this or that parameter to simulate different assumptions about the prebiotic atmosphere, and *all* kinds of organic compounds showed up.[26] It began to seem that if you cooked it long enough—say millions of years—sooner or later some bounded network of self-replicating macro-molecules was bound to turn up. This is where the idea of the "primeval soup" came from, and maybe it did happen that way, though from the beginning "sooner or later" was never more than a code for "we aren't exactly sure how yet." It's clear today it was right off and nothing random about it.[27]

The Emergence of Protocellular Vesicles

Combining in ordered, patterned ways, atoms formed, right off—oh, among other things—bilayers of glycerophospholipids that spontaneously rolled up into tiny vesicles. Some of these vesicles rolled up around chromophores—one of the other things—endowing their interiors with the energy to drive series of thermodynamic reactions, the likelihood of which was a simple function of those granted by the position of the available elements in the periodic table. Phosphoralated compounds generated keto acids. These ammonified into amino acids and small peptides adsorbed to the phospholipid bilayer where they took on simple catalytic functions.

Among the reaction products were more fatty acids—that is, glycerophospholipids—which squeezed in among those already making up the membrane of the vesicle. This went on until the vesicle grew so large it broke down into two or more vesicles.[28]

Replication!

Replication of a protocellular vesicle anyway. . . .

With nothing in the atmosphere to screen the ultraviolet light, there was plenty of energy. And for sure no predators.

Or maybe it didn't happen this way, maybe it *was* random action in the primeval soup. Maybe Christian DeDuve is right and there was a whole intermediate thioester world in which thioesters formed proto-enzymes that evolved into RNA[29] . . . or as Graham Cairns-Smith has it that biogenesis took place on the surface of clays,[30] or on those of pyrites as Gunter Wächtershäuser believes.[31] But listen to Margulis and Sagan:

> A hydrocarbon chain linked to a group of phosphorus and oxygen atoms manifests an electrical charge on the end bearing the phosphate group and no charge on the other end. The chemical as a whole attracts water on its charged end and repels it on the noncharged end. Such chemicals, phospholipids, tend to line up side by side with each other, the noncharged ends pointing away from the water while the charged ends point down to it. (This is essentially what happens when a drop of oil enters water, instantly forming a film.) These and other types of

lipids tend spontaneously to fold into drops, secluding materials on the inside from those on the outside. They have also been shown to form double layers when waves bring two water surfaces, filmed with lipids, together. When this happens, the charged ends of the sheet of lipid molecules point toward each other, sandwiched between noncharged ends. In this way the first membranes were formed—the first semipermeable boundaries between "inside" and "outside"; the first distinction between self and non-self.[32]

Not that Margulis and Sagan buy into the necessary primacy of the vesicle as do Morowitz ("The first crucial event in the origin of cellular life is the formation of the plasma membrane"[33]), Weissmann ("In the beginning there must have been the membrane"[34]), and Harold ("In the Beginning was the Membrane"[35]), but certainly a vesicle could concentrate a solution, could keep interacting components close together, could let "nutrients" in while keeping water from getting out. "Most scientists," conclude Margulis and Sagan, "feel that lipids combined with proteins to make translucent packages of lifelike material before the beginning of life itself. No life without a membrane of some kind is known."[36]

Anyhow: *life.*

Or protolife.

Some kind of dissipative structure anyway, another of those . . . *fluctuations* . . . without which nothing.[37]

Though nothing random about this one.

The Emergence of Bacteria

We have fossils.

We're still in the Early Archean, which, since it ran from 3.9 to 2.9 billion years ago, isn't saying much. With the 3.5-billion-year-old "prokaryotes *incertae sedis*" of the Apex Basalt of the Warrawoona Group (of Western Australia) we're right in the middle of it, though with the stromatolitic cherty carbonates of the ~2.8-billion-year-old Fortescue Group (also of Western Australia)—from which we have evidence of cyanobacteria—we're coasting out of it, we're coasting on into the Late Archean.[38] The fossils are of prokaryotes, that is, of simple, single-celled organisms—members of the Kingdom Prokaryote—whose genetic material is more or less free-floating in their cytoplasm. That is, they're fossils of bacteria. We call them *pro*karyotes since they appeared before *eu*karyotes, whose genetic material is organized into chromosomes tucked inside nuclei, that is, us and the rest of life on Earth (Protoctista, Fungi, Plantae, and Animalia).[39] The deal with bacteria is that their simplicity (and the fact that different strains sort of free-wheelingly exchange genetic material quite independently of reproduction) makes them endlessly adaptive.[40]

Genetic material: descriptions of adaptations to environmental transformations.

And so—if it makes any sense to speak this way—they took over. "With respect to the history of life on earth, the Archean Eon might well be labeled the Age of Prokaryotes," Steven Stanley says.[41] They evolved all the enzymatic systems, all the major metabolic devices we know. They evolved thousands of them. They had the time. They pushed trillions of tons of gases and soluble compounds around through the air and water, which is to say they got the ecosystem up and running. They were everywhere. They slimed everything, their sheen was on every surface. They painted rock and river bank and mudflat and pond with crazy colors, with chemical colors, acidy and sharp, hallucinogenic oranges and aquamarines, and brilliant reds and greens. They made the world look like a geological map. It looked like one of those *Calvin and Hobbes* planets Calvin used to crash on, with the volcanoes in the background and the scummy water like the Cuyahoga when I was growing up and the fragile crust all shattery, the greenstones and greywackes and the dark mudstones. But mostly the Earth was blue–green and in the air were clouds of spores and in the beds of the rivers there were pebbles and boulders.

They weren't dry and white in the sun though. They were wet and scummy yellow and hot pink and magenta.

Some of the colors came from pigments bacteria had evolved to hook up the energy of the Sun. Others came from pigments they'd evolved to protect themselves from its ultraviolet radiation. The earliest photosynthesizers, anaerobic green and purple sulfur bacteria, came in all kinds of pinks and greens. Cyanobacteria, the earliest oxygenic photosynthesizers, bloomed scums bright and thick like oil paint. Some of the scums were green. Some were blue–green. Others were purple. It depended on the mix of chlorophyll and phycocyanin, of allophycocyanin and phycoerythrin. Because the photosynthesizers ran off the energy of sunlight, they couldn't bury themselves in the mud like other bacteria to hide from the destructive energy of the ultraviolet. Instead they shaded themselves with colorful mats made of the carcasses of bacteria that had died from exposure to the radiation.

Or they learned how to tan.

In that eon the color of living things was a simple indicator of the way they coupled with what was around them.

Histories of Recurrent Interactions Constitute a Coupling

Coupled. To say "coupled" is to acknowledge a history. This is the history of recurrent interactions between *the living thing*, which the self-producing, self-bounding character of its internal interactions foregrounds for us, and *everything else*, which the organization of these interactions reduces to a medium, to an environment, to . . . background.

We've been here before. The self-organization of stars reduced the clouds of gases from which they arose to . . . interstellar media.

It's a question of perspective, one in which *we*, as systems of interactions coupled to a medium, have great interest.

Pinch yourself again.

This time feel between your fingers—*without question*—the 15 billion years that tie you to the epoch of supergrandunification, to the *stuff* of the universe fluctuating randomly up out of the vacuum; but in the *liveliness* of the flesh—in its color and the way it hurts when you pinch it and the way it bounces back, in its integrity and elasticity—feel as well the 3.8 billion years that link you—that link *you*, with your name and your easy chair and all your things about you—to the unbroken chain of interactions that stretches back to these Archean bacteria, and back past them, into the fluctuating here-again-gone-again networks of self-referencing, auto-catalyzing interactions out of which the bacteria pulled themselves together.

Back to when the atoms were trying out this and that, working out the ways they could hook up, glomerating up into this molecule, falling apart into those, taking themselves down this (potentially metabolic) pathway, down that. . . .

Not going anywhere. They weren't *going* anywhere. They were just being themselves.

This *organizational* history—of agglomerating stellar masses, of supernovas and atomic evolution, of molecular evolution and biogenesis—this organizational . . . *adventure*: it too is ours.

There's no break in this chain (*our* forebears were colorful anaerobic prokaryotes), though there are plenty of dead ends (more than 99.99% of all the species that ever existed have died out) and many forks (anaerobic prokaryotes are the forebears of *all* living things).[42] Each fork arises from a novel coupling. There's a change in the hook-up with the environment, and this inaugurates a new phylogeny, a variant history. The different colors of the Archean bacteria arose in the course of different histories as successive generations, in order to preserve their coupling, *drifted* in their adaptation—that is, changed a little here, a little there, in order to maintain an unbroken history of interactions with the medium in which they found themselves.

Initially all bacteria coupled exclusively with the biochemical soup, but as bacteria multiplied they used up the nutrients the Sun had created. Margulis and Sagan put it like this:

> No doubt in the first few million years of life's tenure, each "famine," change of climate, or accumulation of pollution from the microbes' own waste gases always extinguished some and probably sometimes almost all the patches of life on the face of the earth. Life might have fluctuated in tenuous balance with the rate at which the sun could create more nutrients, or might have quickly died out altogether, were it not for a vital trait: the ability of DNA to replicate itself, thus leading to extra copies that could playfully, experimentally change.[43]

A little here, a little there. . . .

Among the metabolic pathways explored this way was one that exploited the reduction of sulfates to sulfides. It's still used by desulfovibrios like *Desulfurococcus halohydrocarbonoclasticus*. As they explored this pathway, Archean desulfovibrios evolved—drifted into—the production of porphyrin rings to use in their electron-transport chains, and this turned out to be an important development for the evolution of photosynthesis.

Coupling Directly with the Sun (Instead of with Things the Sun's Cooked Up in Water)

Photosynthesis let bacteria couple with the energy of the Sun directly. That's the deal with the pigments in cyanobacteria: through the light-absorbing properties of their conjugated double bonds a cyanobacterium can couple with the Sun and use the energy of the Sun's photons to drive the network of internal interactions that continuously constitutes and reconstitutes the cyanobacterium (we too are continuously falling apart, we too are continuously being rebuilt).

Each pigment couples with a different wavelength of light. Chlorophyll, for instance, the one in bacteria, absorbs light in the violet–blue and orange–red ranges. Phycocyanin absorbs widely across the blue–green range, and phycoerythrin absorbs in the yellow–orange range. The pigments complement each other. You can imagine them evolving over time to bring more and more of the spectrum of potentially useful solar energy into the cell. Since color is determined by the absorption spectrum, the mix of pigments determines the color of a bacterium. The color ends up indicating the way it couples with the energy around it.[44]

Of course, we're not just talking about energy when we're talking about coupling. We're talking about *all* the interactions between a living thing and the medium required to sustain it. Here's Humberto Maturana and Francisco Varela:

> Among all possible interactions between systems, there are some that are particularly recurrent or repetitive. For instance, if we look at the membrane of a cell, we note that there is a constant active transport of certain ions (such as sodium or calcium) through the cell, in such a way that in the presence of those ions the cell reacts by embodying them in its metabolic network. This active ionic transport occurs regularly; and we, as observers, can say that the structural coupling of the cells with their medium or environment enables these cells to interact recurrently with the ions they contain. The cellular structural coupling enables these interactions to take place only in certain ions, for if other ions (cesium or lithium, for instance) are introduced into the medium, the structural changes that these ions would unleash in the cell will interrupt its autopoiesis [i.e., its self-production; i.e., they will kill it].
>
> Now, why is it that autopoiesis occurs in each cellular type with the participation of only a certain kind of regular and recurrent interaction and not others? This question can be answered only by referring to the phylogeny or history of the corresponding cellular strain; that is, the type of recurrent structural coupling of each cell is the present state of the history of structural transformations of the phylogeny to which it belongs. In other words, it is a moment in the natural drift of that lineage which results from the conservation of the structural coupling of the previous cells in the lineage. Thus, for the example given before, in the present state of that cellular natural drift the membranes operate by transporting sodium and calcium ions, and not others.[45]

We Too Are Coupled with the Environment

To engage in the interactions required to keep yourself going, so you can engage in the interactions required to keep yourself going (and so on), this is what it means to couple with your environment.

If you think about yourself as a system of interactions, there's nothing abstruse about this, that is, if you think about yourself as a living thing, especially if you float up from the cellular level (where indeed ions of sodium and calcium are streaming back and forth across the membranes of each of your trillions of cells), to the level where the stuff (*food!*) is taken on that ends up in your cells. Now, evidently what ends up in your cells has to get into your body. I mean, it's a truism: you are what you eat.

The thing of it is, eating implies interactions with the medium in which you're living. You have to go to the store, you have to push a cart up and down the aisles, you have to run your credit card through the swipe reader or hand over wads of cash (which through a complementary network of interactions you had to get).

Living system: you.
Medium: stores, aisles, credit cards, carts, cars, roads, houses, people.
Interaction: plucking box of All-Bran from shelf, signing credit card slip.
Coupling: doing this again and again.

Now, Maturana and Varela refer to a *structural coupling* whenever there is a history of recurrent interactions leading to a *structural congruence* between two (or more) systems. That is, the coupling is structural when the structure of each of the systems—you, on the one hand, the store, say, on the other—move toward a kind of congruence: the store sells what you buy, you buy what the store sells.

This is in fact the case: you and the store *are* structurally coupled.

Less innocently: you are what your grocery store allows you to be; your grocery store is what you allow it to be.[46]

I know, I know: you only choose from what's available, *but*—and this is the nut—what's available is *not independent* of what you choose. This is most obvious at your refrigerator *which you stock* but, though the effect of your choice is diffused as we move through the system from refrigerator to store (where you might know the manager), to distributor, to wholesaler, to . . . importer, to . . . farmer, it remains nonetheless real.

When Maturana and Varela ask, "Why is it autopoiesis occurs . . . with the participation of only certain regular and recurrent interactions and not others?," what they are asking at this level is, "Why All-Bran?"

And when they answer that "[t]his question can be answered only by referring to the phylogeny or history," they are saying, "Because your mother always gave you All-Bran and when she took you shopping she let you put the box in the cart."

Your choice of All-Bran is a moment in the natural drift of that lineage (the ancestral tree from which you are descended) that results from the conservation of the structural coupling of your mother, your father (your ancestors)—that is, that results from *your* recurrent repetition of the interactions *they* had with their medium.

When one day instead of All-Bran you reach for . . . Barbara's Shredded Oats ("May Reduce the Risk of Heart Disease") you're . . . drifting . . . and your children will conserve *their* coupling by putting Barbara's Shredded Oats into their carts unless they too . . . drift.

As when back in the days cyanobacteria drifted into . . . photosynthesis.

And what this is to say is that we're living things, metabolizing networks of ongoing interactions. In order to keep *on* going, we have to be caught up in encysting networks of ongoing interactions, shopping, but also breathing (and so littering and shitting). We call these encysting networks . . . *the environment* and we and our ancestors have maintained a continuous history of coupling with it, without disintegration (or we wouldn't be here) . . . for 3.8 billion years.

Which is a long time by any measure.

During this time the features of this encysting network—of our medium, of the environment—have become inseparable from its history of coupling with us, with us and, well, with the rest of the living things with which we're (like it or not) inseparable. Through the history of these interactions—this *structural coupling*—medium and living system have come to codefine, cospecify, codetermine each other.[47]

Which is to say . . . we're one with the land. It's only our choice of perspective, that of a system of interactions coupled to a medium, that allows us to see the distinction in the first place.

One with the Land

One by one these assertions are easy. We're made out of cells. Cells are chains of molecular reactions. The molecules are composed of atoms. So at that end we sort of dissolve into this big sea of atoms that were born in stars that grew out of (or into) the galaxies into which the universe ripped itself—and continues to rip itself—under the tides of gravity.

While at the other end *we're* the atoms, participating in the chains of molecular reactions called shopping and work composing the body of what we think about as socioeconomic culture in order to satisfy the demands of our cells for the actual atoms provided by food.

Which is to say we're sort of middlemen—middlepersons—continuously transforming the macroworld into the microworld and the microworld—of food and stuff—into the macroworld of the landscape—the land—of highways and supermarkets and fields and warehouses and power generating plants—

And, of course, forests and fjords and deserts, for it's not just the man-made world we're tied into, we're tied into the whole shebang.

Actually, though, there's not a lot of *land* in the story so far, which is because there're not a lot of cratons in it yet. *Cratons* are the cores around which the continents get built. Continent building is an ongoing project in the early Archean, along with the emergence of life. So there isn't a lot of land in the story in the beginning, especially since what we usually think about as land isn't . . . nude crust. Somehow if you think about lava pillows emerging from the sea, they're not land. They're like . . . *proto*land. Land as we know it isn't really a thing so much as a complex process of interactions among the oceans and atmosphere and biosphere as well as all the—

In fact, as we'll see, weather and vegetation may actually be *responsible* for some mountain building. But what I'm wondering is, can you imagine the land without the sky? Because I can't. What I'm saying is, it's not rock. The land is the functioning skin of the planet. It's as much sky as crust, as much rain as sap.

Which may be a sappy way of alluding to structural coupling. With, for a few billion years, bacteria doing most of the coupling.

There would have been flows of elements among the lithosphere, hydrosphere, and atmosphere even without life, driven by mechanical and thermal and chemical gradients (themselves interacting in all kinds of complicated ways), but the biosphere exists *in virtue* of its coupling, and as a consequence it changes the flows, and speeds them up, and concentrates and directs them.

People talk about how hospitable a planet the Earth is, but it's important to understand that it's hospitable *because* it's pervaded by life, by bacterial life for most of its history. It's not hospitable *and so* life emerged here, although life *did* emerge here because it was able to. It's hospitable *because* life emerged here, because life took control here. Life is what makes this a great place for living. Humans tend to be so derisive of bacteria, but it's really their planet. I mean, we're not only their descendants, we're their heirs.

Bacterial life emerged coupled to the flows of elements driven by mechanical, thermal, and chemical gradients and so life emerges . . . *as a form of the flow.* That's what Miller's experiment really showed. He basically mimicked the hydrological cycle in his flask, under the conditions of a prebiotic atmosphere. You had all this hydrogen and carbon dioxide and ammonia and hydrogen sulfide and methane and nitrogen being fixed by lightning and vented volcanically and ionized by the ultraviolet and circulated hydrologically; and the amino acids and nucleotides and lipids and hydrocarbons form as—how else to say this?—chemical eddies on these main fluxes. Can you see that? There're these flows and these eddies along the interflows, and first these eddies are like, just the amino acids et cetera; but then there are eddies off these eddies, and these are the self-referencing, autocatalyzing chains of reactions we'll end up calling "life." That's what I mean when I say "a form of the flow." I guess basically what I want to insist is that life didn't have to, like, impose itself on an existent set of interactions. It didn't couple like an alien shanghaiing the planet's machinery, but, on the contrary, was . . . precoupled, because that's the only way it could have emerged.

Can you see that?

The thing is, these eddies in the eddies, let me call them "lively reactions," are self-directed. What I mean is, the logic of the reactions is not the logic of thermal systems or mechanical systems or even simpler chemical systems. The logic of these lively reactions is . . . *to maintain their structure*—what Maturana and Varela would call their organization—*at the cost of energy*. It's a bio-logic. And their appearance in the mix of flows greatly complicates the existing flows, enriches them, gives them a whole new self-maintaining character; and it's this that has kept the surface of the planet on such an even keel for the last 4 billion years. These chains of reactions can only sustain themselves within a relatively narrow range of temperatures and chemical concentrations and they're grown incredibly sophisticated at maintaining this range of temperatures and chemical concentrations.

The thing about the bio-logic that maintains the skin of this planet as a hospitable place for itself to keep on being itself is that it has zillions of feedback loops working simultaneously at every scale. If intelligence is the ability to take right action—action that's life-sustaining, life-enhancing, life-forwarding—then the biosphere can be said to be intelligent.

Our self-awareness is no more than an epiphenomenon of the traffic of the cells of which we're composed. We make too much of it, as though it gave us some claim, some special ability to act rightly, and there weren't other ways of being intelligent. But we've only been around a few million years and the bacteria have been here for a few billion. I'd say they're pretty smart too. Our kind of self-awareness is just . . . our organismic *style*.

It's a style that foregrounds us too much against the world. We lose sight of the fact that it only arises from the interaction of our cells that arise together as eddies in the great elemental flows of (and between) the lithosphere and the hydrosphere and the atmosphere and, you know, if you want to make much sense of yourself and your situation in the world, you need to be reading Morowitz, and Maturana and Varela, and Margulis and Sagan, as well as the physicists.

In the first place they've written some great books, and in the second place I'm no more a biologist than I was a physicist. I'm not trying to write biology any more than I was trying to write physics. All I'm trying to do is integrate a bunch of things that we don't think about as, but which really are—*really* are—history. Earlier I said matter was the without which nothing, but without this bio-logic and the life-forms that follow it, the Earth would be like Mars, it would be like Venus: it would be a freezer or an oven.

It's life that's kept the temperature equitable by inaugurating a bunch of chemical feedbacks that control the composition of the atmosphere. Stage? Actor? Only if what the play's about is the actor's construction of the stage, and the stage's construction of the actor. And that's *all* the play's about.

5

The Land Lives, Suspended in a Network of Unholy Complication

It didn't use to be apparent that we were one with the land. Of course, it didn't use to be apparent that the hydrosphere, lithosphere, and atmosphere codetermined, cospecified, codefined each other either, that they were all coupled up in a network of interactions of *unholy* complication—*forget* the biosphere.

Or rather you can't forget the biosphere because it turns out that the other systems are coupled *through it,* coupled through it to a degree that's ridiculous, to a degree that would have been completely laughable only a few years ago, when I was coming up, for instance, and not just coming up through elementary school either (where Mrs. Clark lectured about the Torrid, Temperate, and Frigid Zones), but even in college, where, at Western Reserve in the 1960s, each of these circulatory systems was, to the extent *any* was thought of as a *system . . . on its own.*[1] There were the oceans and there was the solid earth and the wind blew and the rain fell. Yes, the air rising over the mountains cooled, and the water it had picked up passing over the oceans rained out, and, yes, this water rushing off the mountains carried bits of the mountains back to the sea where they settled to the bottom, and, yes, sedimentary rocks later arose . . . *somehow.*

Things . . . *touched.* They sort of pushed each other around. But there was none of the sense of interpenetration, of multiple causation, of feedback, of mutual interdependencies, of . . . *the structural coupling* that is the essential characteristic of our situation as we understand it today. As recently as the 1960s it was possible to believe that you could look *out* at a landscape without being *in* it, to imagine that in looking you weren't touching, that by not touching you weren't implicated, in other words, that the grocery store would exist . . . like a fact of nature (hollow laughter) whether you shopped there or not. We believed lions and tigers were forever, and that the lives of the sons could replicate those of the fathers— *god*, it was an age of innocence!

But the thing about the land is that it turns out to be not some simple story about rain carving valleys and life perduring, it turns out to be this practically obscene tale about trees . . . *making* rain and rain . . . *raising mountains* and . . . *bio*minerals—

*Bio*minerals!

The very name obliterates a boundary, leaks the organic into the inorganic—or vice versa!—suggests a complication, a messiness, a confusion. I happen to be thinking of the coccoliths of *Emiliania huxleyi*, those minuscule perforated umbrellas of $CaCO_3$—of calcium carbonate—those crazy calcite crystals extruded as scales by this unicellular planktonic alga so minute 5,000 of them lined up don't make an inch. Peter Westbroek says:

> This tiny organism plays a role in earth dynamics far out of proportion to its modest size. Figure 6.4 [not reproduced here] is another representation of *Emiliania*, seen not through an electron microscope but from a satellite in space. One can easily recognize the northern part of Scotland and the islands of the Outer Hebrides. The vast cloud in the ocean is *Emiliania*. Under appropriate conditions, these cells can grow explosively, forming giant blooms thousands of square miles in size. Such transitory but recurring and gigantic accumulations occur in all the world's oceans. *Emiliania* is believed to be the most productive calcite-producing species on earth.
>
> The cells and coccoliths are so small that, given the turbulent motion in the seawater near the surface, they might remain suspended forever if left on their own, even long after their death. But when a bloom of *Emiliania* dies off, the cells clump together so that they sink rapidly through the water. Or they are eaten by small animals, especially by a tiny swimming crustacean, the copepod. These creatures digest the soft parts of the algae, wrap the coccoliths in a slimy substance, and excrete them as fecal pellets. The pellets are biological bullets, loaded with coccoliths and other small particles, that shoot down, reaching the ocean floor in only a few weeks. That is how coccoliths accumulate; they are produced and transported by biological forces.
>
> Enormous carpets of coccoliths, larger in area than all the continents together, cover the floors of the world's oceans. These terrains represent the largest calcium carbonate sinks on earth. The tiny

coccolith-generating alga that we can observe with an electron microscope is geological dynamite, a formidable force helping to conduct vast fluxes of calcium and carbonate toward the ocean bottoms.

Recently it has been pointed out that *Emiliania* is a potent producer of a smelly sulfur-containing gas, dimethyl sulfide. Blooms such as those in figure 6.4 are thought to be powerful sources of this gas. The fumes rise from the sea high up into the atmosphere and are oxidized by solar radiation into sulfuric acid. These minute acid droplets drift around in the air, forming ideal nucleation sites for the condensation of water. *Emiliania* is one of several algae held responsible for the formation of clouds over the oceans. Such clouds reflect a considerable proportion of the solar radiation, and they may have a cooling effect on climate.[2]

Creating clouds of water vapor over the oceans, creating clouds of algae *in* the oceans, transducing solar energy, manufacturing minerals, feeding animals, modulating the climate (not least of all thanks to the carbon it sucks from the atmosphere to spread over the ocean floors), *Emiliania,* this barely visible self-producing, self-bounding, macromolecular system of interactions, structurally couples *through* the interactions that comprise it: hydrosphere, lithosphere, atmosphere.

And what's most remarkable is nothing's remarkable about it at all.

Every life-form does it. After harboring the elements it assimilates for a while, every life-form releases them—having used them—in forms other than it took them on. It cobbles them up metabolically (as having taken on CO_2 and H_2O, plants release O_2); or gives rise to them during decomposition (as the decomposing bacteria at work on dead plant tissue release CH_4); or creates them in the course of some cultural activity (as the meltdown in Chernobyl released the radioisotope iodine-131). Life-forms release these to air and water, soil and rock, to whatever planetary sink is most convenient.[3] In so doing, life-forms, that is . . . *the biosphere,* both accelerates and modulates the flow of elements through the three nested spheres on which it's superimposed, or . . .

. . . *on which it superimposes itself.*[4]

Fourteen Layers, Three Terrestrial Spheres

In any case, three terrestrial spheres, three shells.[5] More precisely: *three phases*, the atoms—cobbled up inside the stars that splashed apart as supernovas—fractionating out here, at least in this moment in the drift of these systems, into solid, liquid, and vapor *phases*. Each phase in turn has settled out in layers, like a parfait, like one of those rainbowed cocktails, a

Broadway smile or a pousse cafe, with the cocktails themselves stacked three-deep: innermost, at the bottom, (1) *the solid inner core*, then (2) *the fluid outer core*, next (3) *the lower mantle* (the mesosphere), then (4) *the upper mantle* (the asthenosphere), and finally (5) *the crust* (or lithosphere proper) with the oceanic crust and continents. All that's the solid phase, the sphere of stone, the stone-sphere, the *lithosphere*. Next comes a thick layer of (6) *uniform deep waters* (most of the oceans), on top of which rides (7) *the thermocline*, and then (8) *the upper wind-mixed layer* of the ocean. All that's the liquid phase, the watery phase, the water-sphere, the *hydrosphere*. Over land and water floats (9) *the troposphere*, where evaporation, condensation, and precipitation happen; (10) *the stratosphere* (with its layer of ozone); (11) *the mesosphere* (this one in vapor phase); (12) *the thermosphere*; (13) *the exosphere*; and last and largest, (14) *the magnetosphere*, with its tail stretched the length of 30 Earths by the charged particles of the solar wind streaming from the sun. All that's the vapor phase, the vapor-sphere, the *atmosphere*. Three nested spheres, 14 layers, *and each layer has layers*: it's as though you'd unraveled the periodic table of elements, tacked Ni and Fe to the core and then just read your way up the chart, nickel and iron at the bottom, alumina and silicates above them in the mantle, nitrogen and oxygen in the atmosphere, the *very* lightest elements, helium, hydrogen, working their way into the exosphere and . . . *just blowing away.*

More or less, always accepting there're no compounds on the periodic table, notwithstanding which ours remains a story told by atomic mass and gravity.

And by radioactivity and solar energy, for despite the discontinuities and pauses—the Mohorovicic and Gutenberg discontinuities, for instance, the thermopause, and the thermocline—there *is* a churning of the layers within each phase; and then, the phases too are coupled, radically along the interphases by the life-forms that arose there.

Each phase exhibits a planet-scaled structure, its own planet-scaled system of fluctuations. Like the life-forms that couple them (and are coupled *to* them), these systems of tectonic drift, of ocean currents, of atmospheric circulation, can also be thought of as dissipative structures, as eddy forms which, even as matter and energy stream through them, maintain themselves.[6]

They do this every bit as well as we do.

What's at stake remains the dissipation of heat—for the plot of our story remains that of a cooling gas—in the course of which great convection cells arise (with their attendant panoply) in water and rock and air alike. Those in the air we recognize as winds (the westerlies, the trades) and in high-elevation currents like the jet streams. Those in water we recognize as currents (the Gulf Stream, the Peru Current) and in flows like the thermohaline circulation (the "conveyor belt," the "grand oceanic conveyor"). Those in rock we recognize as earthquakes and volcanoes (Etna, Popocatepetl) and the motions of the great tectonic plates with all their . . . *Sturm und Drang.*

What we call *the land*–the solid earth with its eternal verities, a season for every purpose, summer and all the rest of it (the kids partying after their football victory, the neighborhood art walk, the Christmas vacuum)–is but *our* moment in the natural drift of these systems. In *this* moment, in the *present* state of this drift, the continents are distributed more or less as they appear in any decent atlas, the jet stream is where it shows up on the weather map, and the thermohaline circulation is dumping cold salty water into the abyss of the North Atlantic at a rate somewhat lower than has been common of late.

"Of late"–of course–is relative, and systems far from equilibrium (such as the atmosphere), especially when coupled to other systems equally far from equilibrium (say, the circulation of the oceans), can find themselves moving from one "stable" state to another without a lot of forcing, into and out of ice ages, for instance. . . . [7]

The land hasn't been in the shape it's in for long. It's not likely to *be* in this shape long either.

Atmospheric Circulation

The deal is, because the Earth is a sphere, the equator, where the Sun is most frequently overhead, and so where the greatest number of photons strikes per unit area (i.e., where the Sun is most intense), gets hotter than the poles (to which the Sun's rays are essentially tangential). It's this difference in temperature that drives the global circulation of air and water. This is, stripped to its essentials, a simple thermal system.

Think about having your feet up on a radiator in a room in the wintertime. Unless the heat being pumped into the radiator by the steam passing through it were in turn passed into the air and circulated through the room, the radiator would get hotter and hotter–way too hot to leave your feet on–while the further reaches of the room would get colder and colder. Instead, the radiator heats the air near it, which rises, diffuses through the room, cools, sinks along the walls, and finally returns along the floor to replace the air rising above the radiator.[8] In a room we call this moving air a "draft," but on the Earth a "wind," where a similar circulation is occurring. Low-latitude, that is, *warm*, air and water move toward the poles, while higher latitude, that is, *cold*, air and water flow toward the equator. More or less. It's not as straightforward as it might be. For one thing, all the layering gets in the way—all sorts of friction and eddy motions—and for another thing, the movement's not so simple (there are turbulences everywhere). The insolation's not so simply distributed either. One reason for this is that the tilt of the Earth effectively smears the equator into a band that's 47°s wide, into the *tropics* (i.e., into Mrs. Clark's old Torrid Zone), which seasonally bounces the line of greatest insolation from the Tropic of Capricorn to the Tropic of Cancer and back again.[9] Another reason is that there are broad differences in cloud cover, in albedo generally. The Amazon in January—it *is* a rain forest—and monsoonal Southeast Asia in July receive far less insolation than you'd expect. It's the South Atlantic in January—free of clouds—and the Middle East in July that get all the sunlight.[10]

To speak of only *this* moment in our planetary drift. . . .

The way it works is hot tropical air rises. Hot air molecules are more excited than colder ones and they bounce around more frenetically. So they take up more room. This means there are fewer of them per unit volume, making warmer air relatively light and buoyant. So it rises.[11] We're going to create this huge climate system out of little parts like this, but there's nothing any more reductionistic about this than noticing that Chartres is built out of blocks of stone. *The system's not in the parts, it's in the pattern.*

Anyway, the rising air creates, well, sort of a vacuum—anyhow, low pressure—that sucks air in from higher latitudes to take its place. We'll come back to this in a second. The heat's coming from the ground.[12] It's *coming* from the Sun, but it's the long-wave *re*radiation from the Earth that the air absorbs, and dissipates in rising.[13] Once the rising air gets as cool as, and is no more buoyant than, the rest of the air around it, it stops rising. This is the tropopause. We're 6 or so miles high at this point. Up here the air begins to spread out toward the poles to balance the air being sucked in toward the equator along the ground. The air doesn't get far, though, because it keeps getting colder and therefore denser and . . . well, it starts to sink.

Where? Around 30° north and south of the equator. This is the latitude of New Orleans, but also of Cairo in Egypt and of Durban in South Africa. These are the subtropical high-pressure zones. The air falls here like water from an aerial Niagara, though since it's bone-dry this is hardly the metaphor. It *was* wet when we started. Being less dense, warm air can hold a lot of water vapor, but as it cools the water vapor is wrung—it condenses—out of it, so that the rising air along the equator is sort of like a sponge being squeezed: water just pours out of it (tropical thunderstorms, rain forests). By the time the air hits the tropopause its temperature is way, *way* below zero and it's pretty purely oxygen and nitrogen and argon.

But once it starts sinking, compression starts to heat it up again and it sucks up moisture like a dry sponge. The air descending in the subtropical high-pressure zones is dry, warming air, and it is responsible for most of the world's great deserts: the Great Victoria, the Mojave, the Sahara, the Kalahari. When it hits the ground it spreads out. Some of it moves poleward, some slips back toward the equator to complete the circuit, continuing to warm and soak up moisture until, by the time it reaches the equator, it's hot again and sopping, ready to rise and rain.

These currents of air—these *winds*—are deflected by the Coriolis effect of the Earth's rotation: to the right of the direction of movement in the northern hemisphere, to the left in the south. *Go on. Do it.* Ride the wind. *Toward* the equator now, you're heading toward the equator. You're turning right in the northern hemisphere, left in the southern. Which way are you heading? *Yes! West.* So the wind at your back is coming from the . . . east, and therefore these are the tropical easterlies, the trade winds, the trades. The air heading toward the *poles* follows the same rule, so it's deflected in the *opposite* direction (but it's *heading* in the opposite direction). *Go on! Do it again!* Ride the wind *toward the poles.* Turn to the right in the north and the left in the south just like you did before. *Yes.* You're facing east now, the wind at your back is from the west, these are the westerlies, the midlatitude westerlies.

The whole circuit—the rising at the equator, the poleward traveling, the descent around 30°, the return to the equator—constitutes a single reel around a Hadley cell.[14] What a Hadley cell amounts to is a giant doughnut of gyring air girdling the planet. There's one on either side of the equator, *two* doughnuts, that is, between the equator and 30° north and south. Between the doughnuts lies the intertropical convergence zone, the ITC, where the easterlies of the two doughnuts converge. This is where we started. It's where the air rises. It's where it rains.

Poleward of the Hadley cells lie a pair of Ferrel cells.[15] This is another pair of doughnuts girdling the planet, north and south between 30° and 60°. They're what *we* experience as the westerlies. They're driven by frictional coupling with the Hadley cells (equator-side) and the weak cells swirling over the poles (pole-side).[16] Seen this way, the atmosphere consists of four writhing doughnuts—two Hadley cells and outside of these two Ferrel cells—with seething caps over the poles.[17] Starting from the equator there's the low pressure ITC with its air rising (the doldrums); then the Hadley cells with their easterlies; then there's the air descending into the subtropical high-pressure zones (the horse latitudes); then the Ferrel cells with their westerlies; then the low-pressure zones around 60° where the air rises into the ill-formed cells of the poles; and finally—such as they are—the polar easterlies. Again, starting from the equator: there's a low, then the easterlies, then a high, then the westerlies, then another low, then the polar easterlies. It's a lot like a wave moving out from the equator, a kind of peristalsis (something like that), a planet-scaled fluctuation, in any case a dissipative structure.[18]

Actually. . . .

It's not even this tidy. For one thing, the great convection cells are stitched together by eddies that get snapped off as if by the crack of a whip to whirl across the midlatitudes. Along the polar front where the westerlies slam into the polar easterlies there's a profound mixing of pole-bound tropical and equator-bound polar air. The terrific temperature gradient across this front fans a thermal wind near the tropopause, the jet stream.[19] Due to the coupling of winds aloft and alow—it's one system—the characteristic undulations of the jet stream spawn great storm systems which the westerlies sweep across the surface. These cyclones—cyclones and anticyclones (what you read about in the paper as warm and cold fronts)—are the most effective features of the whole system for getting heat out of the tropics.[20] However they may strike *us*—as we settle in for an evil night at Chicago's O'Hare (so much for *that* trip!)—such eddies, often hundreds and hundreds of miles wide, and hurricanes, and tornadoes, and other midlatitude traveling pressure systems (however tiny), aren't serendipities. They aren't accidents. They certainly aren't freaks. Just because we haven't figured out the beat—doubtless in some fractal time signature—doesn't mean there's not a drummer. They're features of the atmospheric circulation that are as characteristic—as reliable, as lawful—as . . . the "steady" trades.[21]

They're not the only thing that's wandering around though. The westerlies themselves are drifting north and south, the whole gyring girdle is, the whole Ferrel cell. Both Ferrel cells are. So are the Hadleys. The whole thing, the whole shebang's being dragged, reluctantly (with a lot of ripping resistance), as the *effective* equator, the *meteorological* equator—where the sun's rays are hottest, where the angle of their attack is most direct—slips from the Tropic of Capricorn to the Tropic of Cancer and back again. It does this because the tilt of the Earth continuously changes the way the Earth presents itself—its aspect—to the Sun that it orbits. Not that it wobbles: the Earth doesn't wobble, or it doesn't wobble much.[22] It's precisely the *constancy*, in fact, with which it points toward the North Star as it slips around the Sun that causes what *we* recognize as, well, the seasons.

Here: make a fist of one hand. That's the Sun. Make a pointer of the other. This is the Earth. Now, keeping your forefinger (the North Pole) pointed at a corner of the ceiling (the North Star), move your *Earth* hand around your *Sun* hand. Note that when the Earth hand is between the Sun and your eyes, the North Pole—in fact, the whole northern hemisphere—is tilted toward the Sun. When it's in this position were you to run a direct line from the Sun to the Earth, it would hit the Earth 23½° north of the equator at the Tropic of Cancer. When that happens it's the first day of summer in the northern hemisphere: it's June 21st. Now move your Earth hand 180° around your Sun hand.

C'mon. Do it! Now *the base of your thumb*, the South Pole, is pointing toward your Sun hand, and a direct line from the Sun to the Earth would hit the Earth 23½° *south* of the equator, at the Tropic of Capricorn. This is the first day of summer in the southern hemisphere: it's December 21st. During the rest of the year—and how to put this?—the Earth is (relatively) . . . *sideways* to the Sun—look at your hands—and on March and September 21st direct lines from the Sun to the Earth run straight into the equator. We then say "the Sun is overhead at the equator," meaning by this that it is overhead at noon (solar noon), and then the insolation's greatest there. And that's the point, of course: where the Sun is apparently overhead is where it's hottest. And since it's the dissipation of the "excess" heat along this meteorological equator that drives the circulation system, when the equator migrates, it drags the rest of the system with it, the whole kit and caboodle, ITC, easterlies, subtropical highs, midlatitude traveling pressure systems . . . so Italy, say, or Southern California, can enjoy the warm, dry weather that comes from having the subtropical high parked over it during the summer *and* the wetter, wilder, winter weather the westerlies bring when the Sun . . . "moves" south.

Our Moment in the Ongoing Drift of These Systems

By repeating the phrase "*our* moment in the ongoing drift of these systems" what I'm trying to do is suggest their evolutionary character without getting into it. Until recently we thought about climate as a given. *Climate just was.* And it was the way it was forever and ever and would be that way forever and ever. It was stage. It had no history. It never changed. *We* had history. Ours was a play that changed and evolved, but the climate was the stage we played it on.[23]

It's true that we knew about the ice ages, and that before them there had been swamps with dinosaurs. But we haven't known these all that long. Five or six generations? It's still not clear it's sunk in. We've always thought about them as . . . *prehistoric.* They've always been radically discontinuous with our present. It's as if between them and us the world had been reinvented. It might as well have been in another world. In talking to people I rarely get the sense that they're aware—no, *that they feel in their bones*—that 150 million years ago right here, *right here*, the world was a very different place, or that it was a *very* different place 1.5 million years ago. Because if they did, they'd be more open to the possibility that things could be totally different now. They'd look out more often and say, "You know, the world wasn't always like this. So obviously things don't *have* to be this way—"

This is the sort of thing you want people to say when they're thinking about social life, and I'm talking about the climate; but I guess what I'm trying to get at is that if people felt the world *had ever* been different in any way whatsoever, they'd feel freer to imagine it different in other ways as well; that if they *really* felt that where they were standing had one day been buried—the shopping mall and the freeway and the new jail, all of it—under a mile of ice—*a mile!*—well, then maybe they could imagine scraping it all off with bulldozers too, they'd know the world wasn't given once and for all—you know, *given*: period.

If people really understood this I want to think there'd be a lot more change, a lot more revolutionary change, shorter lag times. Most people are so uptight about these things: *this* is the world and it's the only world there could ever be. If people could *see* that ice towering over them, I think the arrogance of their convictions would be reduced.

But who can imagine a sheet of ice 1 mile high, 2 miles high? Besides, if they took geology they learned the ice age *ended* with the Pleistocene. Calling the present the Holocene just reinforces the sense that we're living in a *new age* with, I don't know, *different rules*, like after the Flood. All the bad stuff happened . . . in an earlier world.

That we're actually *in* an interglacial period with 30 or 40 or 50 more episodes of glaciation to look forward to in the near geological future, that the next episode of glaciation could be inaugurated "early" by our spewing tons and tons of lovely carbon dioxide into the atmosphere—

That the next glacial episode could come fast, that the "weird" weather we've been having could be related to the cars people drive—point this out and most people just go into denial, or they get hysterical.

"All hysteria arises from a deficiency of historical perspective," Mark Harris says someplace. If you can see not only that glaciers have come and gone, but ice ages too, that the climate has a history just like we do, that everything is connected and nothing is fixed, then it seems to me you have to be more relaxed; and being more relaxed you have to be more attentive too. It would become easier to pay attention to how things change, to the gearing, to the interconnections, and, accepting that things change, be less panicked, be less upset by it. Panic helps push people into denial and resistance, helps kill their curiosity and attention. "How will I deal with it? What will we do?" Accepting that people have been here, have dealt with change again and again and again—shit, if *they* could do it, so can *we*. *No?* But when you think you're the first, the only one who's ever tried it. . . .

It's scary.

It's very scary. And history is so little a part of our lives, and so much of it is dismissed as prehistory—

But there is no prehistory. It's all one everlasting story.

It's just so hard to see.

But it's equally hard to see the Earth move—at least while you're on it—and the apparent motion of the Sun across the year is so gradual as to all but not happen. But not being able to see these movements doesn't change the fact that the simplest everyday things, like the length of your shadow at noon, or will there be light enough after dinner for croquet, or the way the wind lies and should it rain and how much, depend on ludicrously petty details of our relationship with a star 93 million miles away, on its surface temperature, on how fast we're running around it, on our rate of rotation, on the degree of the tilt of our axis of rotation with respect to the plane of our orbit, on intimate stuff like that, stuff you wouldn't think you'd need to know unless you wanted to buy her lingerie, or buy him lingerie.

What I mean is, you can work with someone perfectly well without knowing his or her waist size, but as soon as you want to buy him lingerie, or buy her lingerie, you can't do without it. The day was, knowing the history of the planet was superfluous. That day is long gone. It's not superfluous anymore. With our numbers and our agriculture and our fossil fuel consumption we've become so heavily involved in the structure of the couplings connecting everything that knowledge of its history is—

It's the only real clue we've got.

And what we're doing may *not* be "bad," not even for us, but wouldn't it be nice to have some handle on it, not to be going blindly into that dark night? Wouldn't it be nice to have some idea about what's going on, all of us, so that we can take intelligent action?

If that's what we're going to do.

And how do you do that unless you understand how the things you're caught up in work?

All of which means that books about the weather are relevant to your life. Look, we *live* in the weather! We walk outside and there it is. We open our mouths and there the atmosphere is . . . in our lungs, *inside our bodies*, with every breath. I can't imagine what could be more interesting or important than atmospheric science. And the thing that's freaky is how new our understanding is, how recently we've begun to put the pieces together, especially the long-term dynamics of the atmosphere, to understand how thoroughly it interacts with everything else, especially with the dynamics of the oceans. Almost everything we know about the chemistry of the atmosphere we've learned in the last 30 years. Think about the sea change that knowing about the New World caused in European minds. What we're learning about the atmosphere, about the whole dynamics of the . . . of this skin of the Earth, is so much more radical than that—

And yet most of us don't even have the elements of it down, the major currents of air and water. They should be taught like the continents.

Not that all that many of us know the names of all the continents.

Oceanic Circulation

The thing is, the hydrospheric circulation is every bit as nuts, every bit as complicated, every bit as dynamic, in the short term and the long, as the circulation of the air: cycles inside of cycles inside of cycles, running and evolving, in three dimensions, in three *spatial* dimensions. Don't even get me *started* on the other factors: temperature, density, salinity, Ekman transport, tides, waves, El Niño, North Atlantic Deep Water, Antarctic Bottom Water, authigenesis. . . .

It's up to the same thing, the hydrosphere is. It's another dissipative structure, another planet-scaled system of fluctuations shunting excess heat from the tropics to the poles.[24] So it's the same story of warm currents curling poleward and cold currents dragging their asses toward the equator. Surface currents—the North Atlantic Drift, the Gulf Stream, the Kuroshio—are driven by the surface winds. The deeper ocean currents are driven by density differences that are caused by differences in temperature and salinity.

Among the surface currents, the easterlies drive the great equatorial currents immediately to the north and south of the equator: the North and South Equatorial Currents, from Africa to the Americas in the Atlantic, and from the Americas to Asia and Australia in the Pacific. If it weren't for the imposition of the landmasses, the easterlies would drive the water on around the globe, just as the westerlies whip the Antarctic Circumpolar West Wind Drift along what the whaling captains used to call "the roaring forties."

Where the continents get in the way of these zonal currents great oceanic gyres are set in motion, clockwise in the northern Atlantic and Pacific, counterclockwise in the southern Atlantic, southern Pacific, and Indian Oceans. These gyres center themselves around 30° north and south, that is, under the subtropical highs, and they have much more intense currents on their western than on their eastern sides.[25] These more intense currents—the Gulf Stream, the Kuroshio, the Brazil, and the East Australian Currents—are carrying heat toward the poles, anywhere from 25% to 40% of that required for the Earth to maintain its thermal equilibrium.[26] They're powerful, consistent, swift streams, hard to recognize in the sluggish cold currents that return the water to the equator—the Canaries, the California, the Benguela, and the Peru Currents—to say nothing of the weak unreliable Equatorial Counter Current that flows eastward between the North and South Equatorial Currents in the intertropical convergence zone.[27]

Although this surface circulation redistributes a lot of heat, it's only indirectly driven by it, unlike the atmospheric circulation where—to reiterate—long-wave reradiation of solar energy from the Earth beats the air it heats into the convectional system of Hadley and Ferrel cells.[28] This it does, you'll recall, by making the air more buoyant and so forcing to rise what gravity pulls down again as soon as it's cooled. It's the *friction* of the winds thus set in motion over the oceans that drives the currents.[29] Minute turbulences of air molecules jouncing along the surface of the water stir up and, what? . . . *drag along with them* . . . bunches of water molecules that amount to, again, what? . . . *a thin surface layer of water.* The second this happens, however, the Coriolis force yanks on the water, deflecting it from the path of the air some 45°. If the easterlies, say, really did blow from the east, this would send the water off toward the northwest, but since the easterlies actually blow from the north- and southeast, the currents they drive—the North and South Equatorial Currents—actually roll pretty much due west. The layer of water below this top layer also starts to move, because the layers too are frictionally coupled. The lower layer is *also* deflected by the Coriolis effect—this time from the layer of *water* above it—and so on for each lower layer, though both speed and deflection decline as you descend this *Ekman spiral.*[30] At the bottom, at what is known as the "Ekman depth," the water flows in the opposite direction to that of the surface current, though at a minute fraction of its speed.[31] In this turbulent, wind-stirred upper layer of the ocean, the heat, gases, and particles absorbed at the surface are mixed, in say a year's time, to a depth of some 600 feet or so.[32] This churning helps moderate seasonal fluctuations of atmospheric temperatures, for the marked capacity of the oceans to store heat for long periods—thanks precisely to this mixing—stands in decided contrast to the more evanescent heat storage of the continents, where only the upper few yards exchange any heat at all with the atmosphere.[33]

At the Ekman depth we're at the bottom of this wind-stirred layer, and so on top of the thermocline, a second layer of equivalent depth, through which the temperature of the water—all but decoupled from the surface—decreases rapidly. At the bottom of the thermocline the water's not much above freezing. As the temperature's dropping, the salinity's increasing. The water's cold and salty down here, that is . . . *it's dense.* Most of the world's water lies in a third layer—it can be miles deep—down below the thermocline.[34] It's of an all but uniformly low temperature and high salinity, the "all but" differences driving the world-girdling thermohaline circulation. This is a by-now familiar story of colder and more saline—that is, denser—water sinking, just like colder air. I mean, the whole thing *is* driven by gravity on the one hand and sunlight on the other. The cold, salty water sinks until it reaches water just as dense as it is, where it begins to spread out horizontally. As it sinks, it pulls warmer water in behind it. And so on. You know the deal.[35]

The thermohaline circulation carries water, heat, salt, and other chemicals around the globe with—and need I add this?—momentous consequences. It doesn't really matter where but it's convenient to start in the North Atlantic where cold, dry Arctic air—this is the third time we've run across air while talking about the water (they're completely coupled)—removes heat and promotes evaporation from the North Atlantic Current (the northern extension of the Gulf Stream). The evaporation makes the water more salty. The loss of heat makes it colder. Both make the water more dense. Every winter it sinks.[36] It pools here for a while in the Greenland and Norwegian basins but eventually slops over the ridges connecting Greenland and Scotland and tumbles down into the Atlantic abyss. As it careens south it pulls in water from the Labrador Sea and the Mediterranean. We're talking about a lot of water moving along a mile or more beneath the surface, "the flow of a hundred Amazon rivers," is how one puts it[37]; "more than 20 times the flows of all the world's rivers," another says[38]; while a third describes the North Atlantic Deep Water as "80 Amazon Rivers marching along the ocean floor, toward the equator and across it."[39] In any case . . . a lot of water.

It pours down the coast of South America and out into the Antarctic Circumpolar Current—"800 Amazons of water"—where . . . *anything is possible.* Colder incoming water may push it to the surface. It typically cools again, sinks, and . . . who knows how many times *this* may happen? Or salients may break off and work their way back north, warmer now, so lighter, higher, maybe only half a mile deep, into the Atlantic, into the Indian, but mostly into the Pacific Ocean where, as it drifts northward, tropical heat is gradually mixed down into it, making it still more buoyant. A lot of what's going on is the redistribution of salt—dissolved minerals—and much is left here in the Pacific which, because it is larger, is more dilute than the Atlantic.[40] This redistribution was easier in the days before North and South America were jammed up together—for the land and water too are coupled—and it now takes this great oceanic conveyor to bring these salts to the Pacific. Getting continuously closer to the surface now, losing salt, gaining heat, the water's whipped toward the south, where it works its way through the seas of Indonesia (through the Seas of Molucca, Halmahera, and Arafura, the Sulu Sea, the Banda, the Timor, Flores, and Bali, the Java, the Savu, the Ceram, and Celebes Seas), crosses the Indian Ocean, curls down along the coast of Mozambique and . . . in eddy motions . . . *struggles* around the Cape back into the Atlantic. It flies north now to rejoin the torrent of the Gulf Stream and so returns to the Arctic, to the west wind screaming off the ice pack, the great freeze, the long descent. . . .

It's not a short trip—it can take years—and some masses of water may actually make it, but most, like us ("How *did* I end up in this godforsaken hole?") don't manage to. Most masses of water are shanghaied, get caught up in this or that local gyre, rejoin the main flow—if they do—only after an eddy (or two or three) around this salient or that. But superimposed on the movements of individual molecules of water is this vast flow—with implications for the pattern of climate worldwide—of salt out of the North Atlantic and heat into it.[41]

For while it's the winds that drive the currents, it's the heat flux carried by the currents that supports the temperature differences that drive the winds.[42]

Arrggghhhh!!

Yes, I know, it's enough to make you want to run off screaming into the afternoon, though it's no more than to say that the hydrosphere and the atmosphere are structurally coupled (which they are). The thing is, while the gross features of the atmospheric circulation—the Hadley and Ferrel cells, for instance—and the layering and some movements of the seawater—the Equatorial Currents probably—have been churning away for billions of years, things like the monsoon and the structure of the great high and low pressure systems and *most* of the currents running today and the route of the thermohaline conveyor that has everyone so excited are . . . very much . . . *this* moment in the drift of these systems, a moment entirely dependent *on the distribution of land and water*, a distribution subject to its own evolutionary history and driven by . . . well, not by the Sun.

Lithic Circulation

"Granite is one of the most commonly seen rocks on the surface of the earth," writes Minoru Ozima, who continues:

and is a stone with which we are all well acquainted. Let us suppose that we put a fragment of granite in a small container which is then completely sealed. We will assume that the container is made of an ideal thermally insulated material, and that heat can neither escape from within nor enter from outside. What changes will occur in the granite inside the box?

If the contents of the box are examined after one or two years, probably no changes at all would be observed. However, if examined after the passage of several hundreds or thousands of years, a careful observer would no doubt realize that the temperature within the container was rising very slightly. After a few hundred thousand years, the rise in temperature would be apparent to any observer. If the calculation described later is carried out, it is clear that the granite in the sealed container would melt completely after several tens of millions of years owing to the rise in temperature.

Where is this energy hidden in the seemingly commonplace granite? Granite contains extremely minute quantities of uranium and thorium, and these radioactive elements are the source of this energy. . . . In the long run, the amount of heat generated by the material which forms the earth is extremely slight, being on an average less than one-twenty-millionth of a calorie per year per gram. However an enormous amount of heat has accumulated throughout the earth's history. Repeated volcanic eruptions, formation of the magma within the earth which causes these eruptions, earthquakes, mountain-building and tectonic movements—the driving force behind these events in the colorful drama of the earth's evolution is none other than the energy released by the nuclear disintegration of these radioactive elements.[43]

And it's more complicated than this—Minoru wrote this in 1981—but the movements driven by these (and other) forces are none other than those we have become so familiar with in the ocean and in the air: convection currents, this time churning through the mantle. They rise below the midocean ridges like air along the equator. Like the air over the equator, upwelling magma breaches the surface and spreads out. It's crust now. It spreads to the left and the right, like the air at the tropopause spreading north and south.[44] It settles out down the slopes of the subsurface ridges, cooling, solidifying, moving, pushed by fresh magma from below (pulled by the cooling magma sinking back into the mantle), year after year away from the ridge, widening the ocean a little bit at a time.[45] Like the air that descends at subtropical highs, like the water that drops into the seas of the north, the creeping crust continues to cool, crawls into a trench, dives, sinks again, once again is mantle. Millions of years later it will work its way up again, spurt, dribble, foam forth, crust out, crawl, dive. Around and around.[46]

Spreading centers along the ridges, *zones of subduction* where it dives: scarring the crust, starring it, crazing it, scoring it. There are seven major plates: the Eurasian, the North American, the South American, the African, the Indo-Australian, the Pacific, and the Antarctic. There are many minor ones: the Juan de Fuca, the Cocos, the Nazca, the Caribbean, the Arabian, and the Philippine. There are some splinters, slivers. All are rigid, to a first approximation.

"Floating slowly over a basalt sea . . . " is how James Lovelock put it.[47]

At the spreading centers the plates move apart from each other, they're divergent. These are constructive boundaries. We're talking about the great midocean ridges.

At subduction zones where one plate dives under another, the plates are convergent, or consuming. These are destructive boundaries. Here we're talking about the great submarine trenches, the Mariana, the Ryukyu, the Puerto Rican. . . .

Where crust is neither eaten up nor spit out, the boundaries are conservative—*hoo boy!*—transform fault boundaries. We're talking about the San Andreas Fault, the Mendocino, the Falkland. . . .

Oceanic crust is basaltic. It's continuously welling up and creeping along the ocean floor and sinking, cycling back into the mantle. Sediments it's picked up as it's crept along the ocean—like those carpets of coccoliths—get scraped off when it dives beneath the overriding plate. As it sinks, it begins to melt, and where it does volcanoes punctuate the surface like periods. This oceanic crust is continuously consumed. In fact there's none around more than 190 million years old. That's less than the last 4% of the Earth's history (less than the last hour of a day). It's young.

Continental crust is old. The oldest continental crust is right at 4 billion years, maybe even 4.2. It's granitic. It's a lot lighter than basalt and, as Lovelock said, it more or less floats on top of it.[48] Instead of being consumed, it gets deformed—folded, compressed—by the jouncing and rubbing of the plates. It gets eroded too, worn down, carried away, especially where it gets squunched up into mountains, as where the continental block carried by the Indo-Australian Plate is ramming into that on the Eurasian Plate and squunching up the Himalayas. Then it gets washed into the oceans where it forms some of the sedimentary material that gets scraped off with the coccoliths when the ocean floors starts to dive for the mantle. Some of this sedimentary material gets taken down with the basement rock—it's not all scraped off—to reappear millions of year later, hard for even its mother to recognize, in the magma welling up from the deep.

As magma wells up and seafloors dive and plates are driven this way and that, the lighter continents on top of them get carried . . . hither and yon. Yeah, the solid earth, root metaphor for stability, our foundation, the *rock* . . . floating like a bar of soap in a bathtub. Arctic terrain migrates to the tropics. Or vice versa. Oceanic gyres stretch out into circumglobal drifts. Rain stops falling here. It starts falling there. Changes in this part of the system get fed back into that, get damped down, amplified. It's another of those *who-would-of-thunk-it* stories. The feedbacks can get nuts. People like Nicholas Pinter and Mark Brandon are writing articles with titles like "How Erosion Builds Mountains"—savor *that* paradox—and talking about how "erosional processes can be viewed as 'sucking' crust into mountain ranges and upward toward the surface."[49]

They do it by tearing mountains down.

Listen: it's straightforward. You just have to give up thinking about any of these things as independent. They're not. They're all coupled. *They're all structurally coupled.* They're all drifting—evolving—together. The thing about mountains is, they're light, they're buoyant. "Thicker or hotter crust rises upward, forming mountains, because the crust is essentially floating on the mantle under it, and crust that is either thicker or hotter (less dense) floats higher."[50] I know it sounds ridiculous, but they're like icebergs, which as everyone in the world now knows may float in water but can still rip holes in hulls, only instead of like icebergs—having 90% of their mass below water—mountains have 80–85% of their mass below . . . *the upper crust.* As much as they've got sticking up into the air, they've got *massive* roots in the lower crust and mantle.[51] Now, erosion's like a guy on an iceberg trying to dig down to the waterline. He picks and scrapes, but every time he tosses a shovelful of ice in the water, the iceberg's buoyancy lifts it just that much more out of it. It's just like that with mountains: *the more erosion tears them down, the more isostasy builds them up.*[52]

Isostasy: the buoyancy of the Earth's crust as it floats on the denser, fluid-like mantle below it.[53]

The thing is, if it's not too anthropomorphic to put it like this, the mountains seem to be sort of doing this themselves. What I'm getting at is the fact that they sort of . . . *make their own climate.* For example, there's something called "orographic precipitation." Air that's forced over mountains rises, rising air cools, cooling air dumps water, and so if the air's wet it rains. So the windward sides of mountain ranges tend to be wet, especially if the winds hit the mountains, as in northern California, say (or along the coasts of Oregon, Washington, and British Columbia) after crossing an ocean; or, as along the northeastern edge of the Andean Altiplano, after crossing a rain forest; or, as along the southern edge of the Himalayas, "subjected to the large moisture fluxes associated with the Asiatic monsoon."[54] Even this isn't simple though. Wetter conditions *are*

more erosive (and so build mountains). Yet they also promote vegetation—
lots of it in the sites just cited—and vegetation serves to "armor" the surface (which slows erosion down). Here elevation makes a difference: as
temperatures decline with altitude, so does vegetation (and this speeds
the process up).

I know, it's ridiculously complicated—which is part of the point—but
what do we have in the mix so far? Tectonic and isostatic properties of the
lithosphere; prevailing winds and orographic precipitation of the atmosphere; evaporation of the hydrosphere; with the biosphere tossing in the
vegetation, and not just the vegetation armoring the slopes, but that transpiring the moisture assaulting the Andes (*trees* . . . making rain). It's the
interaction among these that produces, hell . . . that *is* the mountain. The
mountain *is* this moment in their interaction, not something else. It's not
like it exists apart from them, it's not like it's . . . *Platonic lithosphere*. It's a
chunk of floating plate, shoved up into the business of the atmosphere,
sucked up into it. The atmosphere rides up over it, rolls around it, storms
and rains where it otherwise might not. It drags the ocean (or does the
ocean leap into the air?) to cry over—to lash—the rock, to seep into it,
freeze, expand, crack, to splinter off chunks and chips and slivers. Suddenly roots are everywhere. Root *hairs*. And soil. A trillion nematodes. A
jillion bacteria.

Suspended in the Web of the Flows

Okay, the atmosphere sucks rock up into it—isn't *that* an image?—
where the hydrosphere that's in vapor phase condenses on it as rain. The
rain oozes into surface cracks where one day it freezes, and because freezing water expands, it chips up the surface of the rock into rubble that the
root hairs of the biosphere are able to work up into soil, in the process
transpiring the moisture back to the atmosphere and so, thanks to the excess rubble washed downslope and out to sea by the excess water, en-
abling the ever-so-slightly lighter rock mass to rise still higher.

When you run through a sequence of coupled processes like these,
isotasy, orographic precipitation, the physical weathering of rock by frost
action, the development of a C soil horizon, and so on through the development of the rest of the soil, presupposing plant life, et cetera, and
sooner or later fluvial erosion processes and mass wasting and so further
isotasy, it's easy to lose sight of the fact that there isn't any mountain.

This is to say: no mountain *thing*.

It's a process. *We're* processes. So's this book.

All are processes, which is to say, dissipative structures, eddies, spinning at different temporal scales, flows, mountains and humans alike, endlessly, ceaselessly changing, the mountain, you, me, and the book, which will deteriorate, rot, return to the elemental carbon, oxygen, and hydrogen it's made of, reentering the flows, returning to the flows it never left. All that's "real" are the shapes of the eddies that we call mountain and me and you and *Five Billion Years of Global Change*; and I'm adequately anxious about the "real." I'm least of anything an ontologist.

Then how to think about the mountain, you, me, this book, if not as things? I've been saying eddy motion. *Momentary crystallization?*

To the extent that we are momentary crystallizations, we are so by virtue of our suspension in a web of flows that it behooves us to understand, at least as long as we want to keep them flowing in a way that's inclusive of us, that continues to crystallize *us* out of the flux.

History is the story of these flows. They *have* a history, a beginning; they undergo a series of transformations. The history is the story of this cooling gas and what happens as it cools (we happen). In this chapter I haven't been making a point of the history but unavoidably it slips in. Like when I pointed out that although the gross features of the atmospheric circulation—talking about the Hadley and Ferrel cells, and the layering and some of the motions of the oceans—have been churning away in situ for billions of years, other features like the monsoon and the structure of the great high and low pressure systems and most of our currents and the route of the oceanic conveyor, they're very much *our* moment in the drift of these systems, one wholly dependent on the history of plate tectonics—

I also alluded to history when I quoted Ozima about the heat that's been accumulating in the mantle. It's almost impossible, really, to talk about the circulation of the mantle *without* speaking historically. Whereas there's an illusion that you *can* talk about the circulation of air and water purely structurally.

To do this, though, you have to *ignore* the drift of the mantle, the mantle and the biosphere, and, as we'll see, changes in Earth–Sun relationships too. But it's hard to talk about the history of the biosphere without *some* systematics of air and water, so there's this Catch-22. In the end you just have to plunge in somewhere. It's like I said before, each of these systems is basically simple physics. Sunlight heats air which expands, rises, cools, falls, producing surface winds. That's simple. The disintegration of radioactive elements and so on heats the mantle which expands, rises, cools, falls, resulting in continental drift. That's simple. But now these two simple systems interact—they become coupled—and it's not simple anymore, and you can no longer describe either in purely structural terms. You have to enter the historic term, because due to the coupling the systems change over time.

Because the *principles*, the physical laws, stay the same. They're *why* the structures change. As I said, history is the story of all the simple physics working at the same time over time.

To produce the world we live in.

To produce *us*. It's hard, but it's simple too. I weigh 160 pounds. Time was I weighed 16 pounds. Not long ago either. Caught up in this everyday transformation is my entire history, the interventions of all the simple physics interacting in complicated ways that grew me and gave me the identity I have. And the same for the Earth. Five billion years ago it didn't even exist and today it's this hulking water planet, and caught up in that transformation is the history that gives it its identity. And my history is a function of, a subset of, an eddy off of that history. So my identity is a function of its identity. I can't believe we all don't think this is every bit as interesting as the childhoods we don't seem to ever be able to get out of our heads.

What I'm saying is, this whole history doesn't just produce the world we live in, it produces the *us* who are doing the living too, it produces me, it produces you. I can't repeat it too often: this isn't about something long ago and far away. It's about the here-and-now.

It's about the past of the here-and-now.

Which I think we're maybe ready to get into. . . .

6

Emergent Land Turning Green

THE COEVOLUTION OF THE CONTINENTS
AND ATMOSPHERIC OXYGEN

So here it is and it's a mess, this coupled system of systems cospecifying, codetermining, codefining each other, hydrosphere, lithosphere, and atmosphere caught up in networks of interactions of *unholy* complication. It's the winds driving the currents, but the heat flux carried by the currents supporting the temperature differences that are driving the winds. It's the mountains making the air rise and so rain, but the rain eroding the mountains and so making the mountains rise. It *could* be the temperature rising enough to melt enough Arctic ice to flood the North Atlantic with enough fresh water to reduce its salinity enough to keep the water from sinking, and so shutting down the thermohaline circulation, that is, among other things, the Gulf Stream (Antarctic water would be lapping on the beaches of Iceland), so plunging Europe into a . . . well, let's just put it like this: it would be a whole new world.[1]

Or maybe this won't happen. Maybe enough dimethyl sulfide and sulfur dioxide will form enough nucleation sites to condense enough clouds to reflect enough sunlight to keep the ice from melting,[2] though it sure looks like the thermohaline circulation has shut off often enough in the past, every 1,000, every 1,500 years, for hundreds of millennia at least, Heinrich events[3] embedded in Dansgaard–Oeschger events[4] entrained in Milankovitch cycles,[5] though—to tell the truth—it's not easy to know much about the dynamics of these past interactions. The records are not easy to get (they're in cores drilled from Greenland ice sheets, in cores drilled from ocean bottoms, in cores drilled from glaciers 20,000-feet-high in the Andes). They're hard to read (unbelievably tedious analyses, sorting and scanning of microscopic particles and quantities, work with the single hair of a brush). They're even harder to interpret (as though the significance of the changes in the ratios of oxygen isotopes were unequivocal).

It's not easy to know what happened in the past, even in the past for which we have cores of ice and mud. And we don't have ice and mud for much of the past. Ice sheets melt. Glaciers calve. Ocean floor sediments get scraped off when the crust dives for the mantle, get metamorphosed, further deranged. How the atmosphere, the lithosphere, and the hydrosphere interacted, hell . . . how they *acted* earlier is difficult to say, back, you know, at the dawning of the Age of Prokaryotes. Certainly it wasn't the world *we* live in. It wasn't the world I've been describing. Back then there was no Gulf Stream, no North Atlantic Drift. There was no Atlantic for that matter, no Americas, no Europe, no Africa, no *continents* as we know them. There *was* an atmosphere and Hadley cells and so there were easterlies, but it is not easy to know much else. There *was* an ocean but, as I say, probably no large continents yet and so hard to say about the currents. There was salinity stratification, though, and some version of thermohaline circulation.[6]

But Wait . . . Pinch Yourself Again

The continents . . . but *wait, wait, wait, wait.* Before we get into that, a brief, a momentary, a really quick refresher:

> "When I was one I was just begun,
> When I was two—"

No, no, no! That's Christopher Robin. That's the story of our individual lives. It's *the history of the whole* we're trying to keep in mind:

First the Planck epoch, the epoch of supergrandunification, when the four fundamental forces were still unified, when they were . . . supergrandunified: all the time prior to the first 10^{-43} second.

Then the Planck instant, *at* 10^{-43} second, when gravity chilled out of the mix to inaugurate the epoch of grandunification, when the remaining forces were merely . . . grandunified.

At 10^{-35} second the strong force split to inaugurate baryon genesis and the epoch of inflation: poof! The universe . . . blew up. Electrons and quarks were . . . everywhere.

At 10^{-11} second the weak force disentangled itself from the electro-magnetic force.

Around 10^{-5} second protons and neutrons began to chill out of the mix.

Three minutes later nucleosynthesis began.

Three hundred thousand years after that hydrogen nuclei had attracted so many electrons that matter and energy were effectively decoupled. Light could propagate. It was like lifting a veil. The universe became transparent.

Within a billion years there were galaxies everywhere. There were stars and the atoms they made. Soon supernovas were spreading atoms around.

Nine billion years later Sol condensed from a cloud of interstellar gas and dust.

At the same time the Earth began to clump up in the accretion disk.

A hundred million years later the Sun burst into light.

What? 10 or so billion years after the Big Bang?

About 4.5 billion years ago?

Yuh. . . .

It took another 300 millions years for Earth's molten core to settle and the crust to bob up around it.

Near the end of this time the mantle devolatilized. The atmosphere formed. Right off it began to curl into action. There were Hadley cells, and some version of easterlies, but little else is certain.

We're right at 4.2 billion years ago.

Over the next 200 million years the heavy-duty bombardment of the Earth by stuff from the accretion disk slackened off and by the end of this time any remaining surficial magma had crystallized.

That happened about 4 billion years ago.

Over the next 200 million years the upper mantle chilled out into a solid lithospheric basement covered everywhere by crust. The ocean rained out of the atmosphere and immediately stratified. Right off it coupled with the atmosphere and churned into motion.

It was pimpled by small, steep-sided felsic protocontinents.[7]

Along the multiplying interphases networks of self-referencing, auto-catalyzing macromolecules pulled themselves together inside vesicles.

They left fossils.

There are the 3.5-billion-year-old "prokaryotes *incertae sedis*" of the Apex Basalt of the Warrawonna Group in Western Australia. There are the nearby stromatolitic cherty carbonates of the ~2.8-billion-year-old Fortescue Group with their evidence of cyanobacteria.

It's the Late Archean—

You know, *all* these chunks of time have names, most of them anyway. It may be rote, but the rote has a role. Its roll call makes a spine of time with which to articulate the body of history.[8]

Shall we learn it? Some of it we know:

The Planck Epoch, the first 10^{-43} second.[9]

The Epoch of Grandunification, 10^{-43} to 10^{-35} second.

The Epoch of Baryon Genesis, the Epoch of Inflation, 10^{-35} to 10^{-33} second.[10]

The Epoch of Unification, 10^{-33} to 10^{-11} second.

The Epoch of Quarks, 10^{-11} to 10^{-5} second.

The Hadronic Epoch, the Epoch of Confinement, the Era of Protons, 10^{-5} to 10^{-3} second.[11]

The Leptonic Epoch, the Era of Electrons, and on into the Epoch of Nucleosynthesis, 10^{-3} second through the first 3 minutes.[12]

The Photon Epoch, 3 minutes to 300,000 years.[13]

The Protogalactic Epoch, 300,000 through the first billion years.

The Epoch of Galaxies and Stars, from a billion years to the present.

Some 4.6 billion years ago: the Condensation of the Solar Nebula, the Accretion of the Terrestrial Sphere.

This is the dawning of geological time. We're in the Precambrian here (the Prephanerozoic). More specifically we're in the Archean. In fact we're in the Hadean (when things were hell). It's 4.5 billion years ago. The Hadean lasts 600 million years. It runs from 4.5 to 3.9 billion years ago.[14]

The Hadean is followed by the Early Archean (3.9 to 2.9 billion years ago) and the Late Archean (2.9 to 2.5 billion years ago). The Hadean, Early Archean, and Late Archean periods together, running from 4.5 to 2.5 billion years ago, make up the Archean Era. This is the Age of Prokaryotes. It's half of Earth's history. We don't know much about it.

It's like one day you woke up and realized you'd had a childhood, a youth, about which you knew next to nothing. No wonder we've invented stories.

After the Archean comes the Proterozoic, altogether from 2.5 to 0.57 billion years ago. It too is divided into three periods, the Early (or Aphebian), 2.5 to 1.6 billion years ago; the Middle (or Riphean), 1.6 to 0.9 billion years ago; and the Late (or Vendian), 0.9 to 0.57 billion years ago. At this point we give up talking about "billions of years ago," and start talking about "hundreds of millions of years ago."

The Archean, along with the Proterozoic (about which we don't know a whole lot more than we do about the Archean) together make up the Precambrian. This is the common name but since it doesn't make a lot of sense, it's increasingly called the Prephanerozoic. The Cambrian, after all, is no more than the first period in the Paleozoic Era of the Phanerozoic Eon.

Eons into eras into periods.

Earth history, then, following the accretion of the terrestrial sphere, into the Prephanerozoic and the Phanerozoic *Eons*, into, that is, the eon *before* visible life (so to speak) and the eon *of* visible life. And then the eons into eras, the Prephanerozoic into the Archean and Proterozoic Eras; the Phanerozoic into the Paleozoic, Mesozoic, and Cenozoic Eras, and so, since the accretion of the terrestrial sphere, all of Earth history into the Archean, Proterozoic, Paleozoic, Mesozoic, and Cenozoic *Eras*. And then, again, the eras into periods, the Archean into the Hadean, Early Archean, and Late Archean Periods; the Proterozoic into the Early Proterozoic (or Aphebian), Middle Proterozoic (or Riphean), and Late Proterozoic (or Vendian) Periods; the Paleozoic into the Cambrian, Ordovician, Silurian, Devonian, Carboniferous, and Permian Periods; the Mesozoic into the Triassic, Jurassic, and Cretaceous Periods; and the Cenozoic into the Tertiary and Quaternary Periods; and so, since the accretion of the terrestrial sphere, all of Earth history into the Hadean, Early Archean, Late Archean, Early Proterozoic (or Aphebian), Middle Proterozoic (or Riphean), Late Proterozoic (or Vendian), Cambrian, Ordovician, Silurian, Devonian, Carboniferous, Permian, Triassic, Jurassic, Cretaceous, Tertiary, and Quaternary *Periods*. And then the periods into epochs, into series, into stages, and into ages. We'll get into some of this, lightly.

Which is to say, the celestial epochs (the Planck Epoch, the Epoch of Grandunification ...), and then the terrestrial periods (the Hadean Period, the Early Archean Period). They need to be strung into a rhyme like the ABCs: Planck, Grandunification, Baryon Genesis ... Riphean, Vendian, Cambrian ... something like that. . . .

Anyhow, the continents. . . .

Early Cratonization and the Coevolution of Terrestrial Systems

There probably weren't any continents during the Archean, not what we think about as continents. More precisely, there's no evidence that any existed. It's quite possible that the mantle was episodically layered. That is, an upper layer of convection cells may have overlaid a lower one. It's possible that under these warmer conditions plate tectonics wouldn't have worked. In fact things might not have gotten cool enough, oh, say within 50°C of the current mantle temperature, until 1.5 billion years ago, that is, in the Middle Proterozoic (or Riphean). Don't forget, throughout its history the Earth is continuously cooling off. The present, essentially unlayered, state of the *mantle* may have evolved only within the last billion years, that is, since the Late Proterozoic (or Vendian), though "the nature of the tectonic regime during earlier times in the Earth's history still is quite unclear."[15]

To put it mildly. . . .

Radiometric data do show that during the Early Proterozoic (or Aphebian) large bodies of magma were intruded into cratonic rocks, into, that is, those all but undeformed masses, *cratons*, that make up the cores of the continents we live on, and this presupposes a late Archean origin for cratons of at least moderate proportions. So during the Early Archean, there were "small, steep-sided felsic protocontinents . . . separated by numerous marine basins that accumulated lava and volcanic sediments;"[16] and then during the Late Archean, there was gradual, episodic "cratonization," beginning perhaps at the tail end of the Early Archean, ~3.0 billion years ago, with a large craton already in what is today South Africa; and within the next few hundred million years there were a number of large continents in existence.[17]

Don't you just love it the way we toss around these chunks of time. One minute we're dividing seconds up into nearly infinitesimal slivers, the next we're casually muttering about the ensuing several hundred million years. . . .

I can barely remember what it was like before I knew how to ride a bicycle.

The thing is, during these Archean years when cratonization is beginning to transform the surface of the planet, it's also pushing and shoving the bejesus out of the atmospheric and oceanic circulations. I mean the continents aren't, like, striding onto a stage set by the oceans and atmosphere. There is no stage. All three are doing improv.

What I'm trying to say is the *whole shebang* is evolving.

It's hard to see this sometimes. You can read whole long books about evolution—good ones too—and never know that when life arose there probably weren't any of what we'd recognize as continents. Or that there was an evolving thermohaline circulation stirring up the oceans. Or that during the entire time the Sun had been evolving too, getting brighter, dimmer, messing with the climate. . . .

The point I'm trying to get to, the point I've been trying to get to for pages and pages now, is that when the prokaryotes organized themselves, they did so *coupled* to a medium, to an environment, with at least this *double* dynamic, this—as it were—*dynamic squared*. That is, they were coupled to both the nested dynamics (diurnal, seasonal, annual, and longer) of the global circulation of air and rock and water, *and* to the dynamics of *their* own evolution. The prokaryotes coupled not just to a changing environment, you know, *with seasons*, but to an environment in which the rhythms of the changes were changing, drifting into and out of ice ages, for example, seasons this season, but next season no season at all, like professional basketball, that kind of—

So that the biosphere that evolved isn't much threatened by change. It . . . came up *with* change, along the very surface of the churning. It's got change built into it (the very point of sex[18]). It's sole invariant is . . . adaptation, congruent with changes in the environment, of which it also happens to be an integral part.

Its own changes cause itself to change.

It *is* the environment.

Talking about Evolution

Okay, so we're talking about evolution at this point, or about what people think about as evolution, though frankly the whole thing's evolving, has been from the beginning, *from the first 10^{-43} second*, continues to be, at all times at all scales, and how you arbitrarily—

Well . . .

I suppose it's to quell the nausea. It's not, *pace* Martha and the Vandellas, that there's nowhere to run, it's that there's nowhere to stand.

Anyway,

e v o l u t i o n

which Lynn Margulis defines "simply as change through time,"[19] that is, like the typefaces I used to type the word, not going anywhere, one just succeeding another, not *heading* someplace, not heading toward some typeface nirvana, *not even in alphabetical order*, Gill Sans Ultra Bold coming after Elephant, Edda coming after Gill Sans Ultra Bold, Algerian after Edda. From what was available, *yes! from what was on my machine*, and evolution works that way too, *from what's given*—gravity and the strong force and the weak force and the electromagnetic force and the particle zoo—*what we've got*.[20]

But *all* of it. That is, all of it evolving. The lithosphere and the atmosphere and the hydrosphere *and* the biosphere. Nothing fixed. All drifting along together. Margulis says, "Evolution, defined simply as change through time, brings into focus the convoluted history of which we are the living legacy. The study of evolution is vast enough to include the cosmos and its stars as well as life, including human life and our bodies and our technologies. Evolution is simply all of history."[21]

The problem is that everyone sees it from his own perspective. This is Minoru Ozima again:

> Let us finally sum up the earth's evolution. It is thought that the earth was born along with the solar system about 4550 million years ago [i.e., 4.55 *billion* years ago]. This age was deduced from the age of meteorites, and was not found from samples from the earth.
>
> The elements composing the earth were probably formed by explosions of supernovae over more than 10,000 million years ago [i.e., 10 billion years ago], long before the birth of the earth. Finally the primitive solar nebulae were formed from these elements, and then condensation occurred and ultimately the earth was formed. It is thought that about 100 million years elapsed between the synthesis of the elements and the formation of the earth. This conclusion was reached from the existence of xenon-129, which indicates the previous existence of iodine-129 on the earth.
>
> It is believed that a few hundred million years after the formation of the earth—that is, more than 4000 million years ago [i.e., 4 billion years ago]—the formation of the earth's core occurred, and at about the same time volatile degassing occurred from the interior of the earth on a large scale and something similar to the present hydrosphere and atmosphere was formed. This was concluded from such factors as the argon isotope ratio within the earth and the existence of the earth's magnetic field. The oxygen within the atmosphere, however, is due to the photosynthesis of green plants, and substantial accumulation did not begin until after the middle of the Precambrian period.
>
> By looking at the isotopic ratios of strontium and lead, it was concluded that the present division into mantle and crust has been formed virtually continuously throughout the whole evolution of the earth.
>
> Crustal activities in the Precambrian period are not yet understood well [but] the oldest known crustal rocks were formed about 3750 million years ago [i.e., 3.75 billion years ago].
>
> Continental drift and ocean floor spreading are main actors in the earth's evolution, though they account for less than 10 percent of the earth's evolution since the Mesozoic era. Two hundred million years ago the Pangea continent, which was virtually one single block of crust, began to separate. Separation of the South American and African continents is believed to have begun about 180 million years ago.[22]

Ozima summarizes this in an elegant chart on his two following pages. There his only allusion to life is the note "Appearance of abundant fossil records" at the beginning of the Cambrian. In contradistinction, Mark Ridley writes that "evolution means change in living things over long time periods," and again that "evolution means change, change in the form and behavior of organisms between generations," adding that "it all depends on external environmental change and on random genetic innovation."[23]

We know that neither Ozima nor Ridley has it right. Indissolvably the evolution of our Earth is the evolution of Ozima's *geo*chemical Earth and Ridley's *bio*chemical Earth, as Ozima's recourse to photosynthesis to explain the composition of the atmosphere and Ridley's recourse to "external environmental change" (to drive the biological evolution required to explain the composition of the biosphere) assure us.

But it's hard to see this all at once. It's impossible from one perspective. If you want to see what's going on, you have to keep changing your point of view.

Here: another thing I got coming up was the fable of the celestial watchmaker, the way all these parts working together points to the hand of a supreme designer. Of course it doesn't, any more than this prose does, which looks to you so inevitable, but for me was a matter of fits and starts and staggering revisions, page after page tossed in the trash and rewritten, unanticipated encounters with this book, with this article Arthur came across and sent me, with that from my miraculous mother, *nothing serendipitous*, no more chance than design, but . . . *a path laid down in walking*.

I look back and . . . there's this whole finished thing. But it didn't come that way. It came like the world, neither whole nor finished, neither by accident nor design, coemergent with the units that comprise it.

The world wasn't *set* for the biosphere. But neither was it some happy accident.[24] "Before the Gaia hypothesis, the conditions of Gaia were thought to be just right for life by a happy accident," writes Lovelock, but *without life* the physical and chemical evolution of the Earth would have soon reached a state *incapable of supporting life*:

> After the Hadean the Earth cooled and the chemistry and the climate became favorable for life. But this state could not persist; the physical environment was evolving through this favorable "window" to a hostile state with the Earth cycling through freezing and thawing and eventually drying out like Mars and Venus.[25]

But the life that appeared in this window—that will appear in every such window wherever one is opened—stepped in, and spread along the percolating interphases to coevolve self-regulating *chemical* pathways as straightforward as the *thermodynamic* pathways already trod by the late Hadean easterlies, by the early Archean thermohaline circulation, by the Proterozoic microcontinents at sea among the open vents.

It's the *tightness* of the coupling that makes it so hard to see. Where are the phase boundaries in the salt spray and the breaking waves? In the fog moving onto the land and the wind-driven sand? In the tumbling rock in the rolling water? In the snow sublimating into air?

The one *is* in the other. You dig down into the soil and bring up a handful of dirt and an earthworm slithers out between your fingers. And those white things are larvae, and those . . . those things are roundworms. You can't see the nematodes, they're too small, thousands and thousands of them—*in your hand*—and the billions, literally billions, of bacteria. And it's moist and those things are breathing and in situ it would be pervaded by roots and fungi, miles and miles of hyphae and hairs.[26] Taste it. It comes in more flavors than Baskin-Robbins ice cream, and more colors than human skin.[27]

What is it?

Oh, yeah, sure: it's soil. But . . . *soil?*[28] Is it lithosphere? I mean, isn't soil ground-up weathered rock? Clay, silt, pebbles, sand? But it's damp too, and so . . . *hydrosphere?* And clearly there's air down here and . . . without doubt . . . *life.*[29] Some say the soil's a biotic construct, and surely without life, no soil, only sand and dust. But this too is niggardly, for no soil without dust or sand either, without water and air.

Or take the air. Take . . . *the air*, not some hypothetical planetary atmosphere, but the Earth's atmosphere, our air, what we breathe. You know, it's just like the soil. It's teeming with life, teeming with it. I'm not thinking about us and the trees, though we walk in the air and our talk is air and the trees build their branches and wave their leaves in the air; or even about the birds and the bees, the bats and the butterflies, though there they are, overhead, beside us—the bees in the spring in the flowering trees, the clouds, the rivers of them—threading their paths, treading their places in the air. No, I'm thinking about the seeds and the spores and the eggs and the bacteria and the viruses that are part of the air,[30] and I'm thinking about the gases that wouldn't be here but for life, the methane and nitrous oxide and the dimethyl sulfide and the oxygen, the oxygen 21% of the atmosphere, more than a fifth of it, none of which would be here but for life. . . .

If the soil's a biotic construct, so's the air.

Life and the Evolution of the Oxygen in the Atmosphere

And when we talk about evolution, this is the sort of thing that really sticks out. Dinosaurs are all very interesting, but they're gone, along with the rest of the 99.99% of the species that have ever existed, and the oxygen is still here, this violently reactive gas that has no business being in any planetary atmosphere. And as I've said, for a long time there wasn't any, the *oxygen holocaust*, as Margulis and Sagan call it, having taken place only 2 billion years ago, after the Archean, sometime in the Early Protero-zoic. By then life had been around almost 2 billion years, evolving with the evolving crust and air and waters. The interphases were teeming with it. We've been here. A lot has changed but it's still that *Calvin and Hobbes* world, flat and damp, "with volcanoes smoking in the background and shallow, brilliantly colored pools abounding and mysterious greenish and brownish patches of scum floating on the waters, stuck to the banks of rivers, tinting the damp soils like fine molds."[31]

They were all anaerobes, these microbes, these prokaryotes, and they'd been evolving frenetically—drifting—as one of their defenses against the bullying of the ultraviolet. They'd invented all kinds of repair technologies. Bacteria can do this. They don't have to reproduce to ex-change genetic information. They can just pass it back and forth, exchang-ing fragments of DNA like greetings (which is how they evolve resistance to antibiotics so rapidly).[32] It's sex—its earliest form—and its ease and speed is the biospheric face of the unending change in the environment to which the bacteria were coupled.[33] Evolving like this, drifting in conser-vation of their adaptation—*keeping in tune*—they multiplied. And multi-plied. And multiplied.

And since they were building themselves up out of oxygen and car-bon and hydrogen, and since hydrogen is so light that any in the atmo-sphere just gradually works its way up into the exosphere from which it . . . blows away, there was a perpetual . . . *hydrogen crisis*. Even the hydro-gen sulfide bubbling up through subsurface vents and out of volcanoes was inadequate to satisfy the ravenous appetites of the photosynthetic bacteria that increasingly dominated the biosphere.

Now, for a long time the hydrogen in water was inaccessible to these microbes because the bonds tying the hydrogen to the oxygen were too strong for the bacteria to break. But then:

> The cyanobacterial ancestors seem to have been mutant sulfur bacteria desperate to continue living as their stores of hydrogen sulfide dwin-dled. There organisms were already photosynthetic, and already had proteins inside them organized into so-called electron transport chains. In some of the blue–green bacteria, mutant DNA which coded for the electron transport chains duplicated. Experts at capturing sunlight in their reaction center to generate ATP, the new DNA led to the construc-tion of a second photosynthetic reaction center. This second reaction center, by using light-generated electron energy from the first center, ab-sorbed light again; but this was higher-energy light, absorbed at shorter

wavelengths, that could split the water molecule into its hydrogen and oxygen constituents. The hydrogen was quickly grabbed and added to carbon dioxide from the air to make organic food chemicals such as sugar.[34]

Of course, what this left over was the oxygen, more and more of it as the cyanobacteria spread and spread, oxygen mixed and whipped around the atmosphere by the churning of the Hadley and the Ferrel cells, beaten down into the oceans, stirred into their wind-stirred layers, carried everywhere.

Cyanobacteria spread. Because all they needed was sunlight and carbon dioxide and water, they could live anywhere. And they still do, on top of your swimming pool, for instance, and all over your shower curtain. Back then, when there wasn't much competition, they covered the globe; and as they did, they spewed oxygen everywhere. The thing about oxygen is that it's completely promiscuous. It will bond with almost anything. It reacts with, it oxidizes, *it burns*: small metabolites, enzymes, proteins, nucleic acids, vitamins, lipids, atmospheric gases, minerals. I mean, you expose iron or sulfur to oxygen and you get hematite or pyrite like . . . *that!*

Which makes it incredibly toxic. And here it was all of a sudden being produced by the carboy by a wildly successful microbe that was making it out of water!

At first all this oxygen was sopped up by elemental sinks—most of our iron ores were laid down at this time (i.e., environmental iron was oxidized into hematite, into magnetite)—but gradually, and probably episodically, oxygen began to build up in the atmosphere. *Hoo-boy!* And people worry about minuscule increases in atmospheric carbon dioxide! Back at the end of the Archean, at the beginning of the Proterozoic, oxygen concentrations kazoomed from one part in a million to one part in . . . *five*, from 0.0001% to nearly 21%, *oxygen*, now, not some hot-house gas, but intensely reactive, wholly toxic oxygen.

Talk about your air pollution!

Oxygen *destroys* the tissue of anaerobic microbes. It was like something from the movies, frothy bubbles, gonzo. It was wholesale carnage.

You know how these microbes are macromolecular networks of interactions producing the parts required to produce the parts required to keep production going? Well, this was a time when they produced a chemical that *disrupted* the chains that produced the parts required. . . .

Shitting where you eat: that's what it amounted to.

The anaerobes could change, hide, or die. Most died. It's worth repeating: *most died*. Some hid and these survive today in the airless—that is, oxygen-free—refuge corners of the world: at the bottoms of the oceans, underneath glaciers, in the rumen of cattle, in my gut and yours.[35] But others defended themselves against this threat as they had against the destructive power of ultraviolet radiation. They mutated like crazy and worked out a way to *use* the gas that was killing them:

> In one of the greatest coups of all time, the cyanobacteria invented a metabolic system that *required* the very substance that had been a deadly poison. Aerobic respiration, the breathing of oxygen, is an ingeniously efficient way of channeling and exploiting the reactivity of oxygen. It is essentially controlled combustion that breaks down organic molecules and yields carbon dioxide, water, and a great deal of energy into the bargain.[36]

A *great* deal of energy, far, far more than fermentation generated, which is how the anaerobes had previously been extracting energy from food.[37]

With high-energy photosynthetic reaction centers, which let them couple *directly* with the energy of the Sun and cleave water to release oxygen, and respiration, which let them consume and exploit the reactivity of the oxygen they released, the new aerobic cyanobacteria exploded around the world into hundreds of different forms and into every conceivable niche.[38]

Some people think this is a story about how *we're* going to transcend *our* pollution of the planet. Others notice that those who caused the oxygen holocaust, the hydrogen-starved anaerobes, they . . . mostly died. It was the new guys they made an oxygen-rich environment for who thrived.

Anyhow it's all different now and what with all the oxygen it wasn't long before a layer of ozone mantled the planet. Ozone is a molecular form of oxygen with three instead of the more usual two atoms, O_3 instead of O_2. It happens when the high energy of short-wave ultraviolet radiation smashes an ordinary O_2 molecule into its individual atoms and one of them hooks up with an O_2 molecule—then *bang!* O_3. The thing about ozone is that it *absorbs* ultraviolet radiation, so once there's enough of it in the upper atmosphere, no more ultraviolet radiation makes it to the Earth's surface.[39] Without its high energy to weld them together, this brought an end to the abiotic synthesis of organic compounds (an end to the primeval soup). Now the only way to build up "organic" molecules was . . . "organically," and soon "new food relationships developed as other bacteria fed off cyanobacterial starch, sugar, small metabolites, and even the fixed carbon and nitrogen of their bodies. But most significant, cyanobacteria's continuing air pollution forced other organisms to acquire the ability to use oxygen too. This set off waves of speciation and the creation of elaborate forms and life cycles among them,"[40] symbiosis, cells moving inside cells, organelles, eukaryotes, protists, plants, fungi, animals, reptiles, conifers, mammals, primates, *Australopithecus*, *Homo erectus* . . . me!

All Evolved Together

Well . . . and *you*, and your mother and . . . the honey bee and the horse and the eel and the billions—literally—of mutations of *E. coli* drifting down your gut this very second. . . .

Because we're *all* equally evolved, we've all reached the same point—the one we're *at*, this one, the present—in the drift of *some* 4-billion-year-old lineage. Each can trace itself back, generation by generation, to, well, I don't suppose there was a *single* ancestral reaction—the world's too big, there were too many Urey–Miller experiments going on, too many ancestral convergences on the same chemical pathways, too much symbiosis—but if not to one, then to a handful, and who knows what happened to their successors in the pruning, in the prunings, carried out by famine (by the exhaustion of the nutrients the ultraviolet had stirred into the biochemical soup before the development of the ozone layer), by changes in landforms, climate (in the beginning they could have been quite local); or the later ones, that pruning of prunings carried out by the rise of oxygen, those by . . . ice ages . . . by. . . .

So, *maybe* one, but probably quite a number, though in any case: enough alike to exchange genes. . . .

Each carrying traces of its past, so that we're all living history books recording the history of living:[41]

The chemical evolution inside stars, or in interstellar gases. . . .

That on the surface of the planet, the upper mantle calming down, the atmosphere raining out, the oceans churning into action. . . .

All the abiotic Urey–Miller compounds whipped up by the ultraviolet, the glycine, the alanine, the glutamic acid, the n-methylalanine, the. . . .

The networks producing them getting more and more complicated, the phospholipid vesicles rolling up around them, concentrating the solutions. . . .

But still running on the compounds whipped up by the ultraviolet, by the corona discharge, by lightning, chemical heterotrophs, harvesting environmental ATP. . . .

Until the *fermentative* heterotrophs worked out the metabolic pathways required to make their own ATP. . . .

And the *auto*trophs, the anaerobic photosynthesizing bacteria, figured out how to harvest the Sun's energy to reduce the CO_2 to make sugar (and then ferment it). . . .

See, *remembering* the fermentation, building on it, forgetting nothing, building on it all, all recorded in the palimpsest of the metabolic pathways. . . .

Figuring out next how to cleave water for its hydrogen, liberating all that oxygen. . . .

Working out the pathway for using the oxygen. . . .

Liberating more oxygen, and more oxygen. . . .

Poisoning its own environment, poisoning off—obliterating—most of life, just wiping it out . . .

Then working out new carbon pathways, learning how to live off the starch, the sugar, off the very bodies of cyanobacteria . . . setting new nitrogen pathways in motion . . . getting the modern global ecosystem up and running, insinuating metabolic pathways into the circulation of every element and so controlling them, damping down accumulations, making good losses . . . keeping the oxygen, you know, plentiful enough to respire but scarce enough to keep the whole thing from going up in flames, a percent or two more and *whooee!* Fireball city!

Dates?

Ozima gave us a 10-billion-year estimate for the age of the terrestrial elements, but if we start with the hadrons and leptons that make them up, we're right back at the beginning, 15 billion years ago, give or take some part of a second. And then 5 billion years after that the intrastellar evolution of the atomic elements, and 5 billion years after that—right at 4.6 billion years ago—the condensation of the solar disk, the accretion of the terrestrial sphere. Four or 5 hundred million years after that, ~4.0 billion years ago, around the time the Earth stopped being bombarded by large objects, the crust calmed down, the atmosphere rained out, and the oceans heaved themselves into action.

So much for the Hadean.

We've seen Morowitz's 3.9 ± 0.1-billion-year-old estimate for the formation of life, and maybe that's a little radical, but prokaryotes were certainly thriving as far back as 3.5 billion years ago in shallow water and intermittently exposed environments along the margins of those steep-sided felsic protocontinents (*we have fossils!*), possibly even on land surfaces and in open ocean waters, often living in elaborate stromatolitic communities, in microbial mats, at sediment–water interfaces;[42] and so, already: anaerobic heterotrophs and autotrophs, which is to say, photosynthesizers, oxygenic phototrophs, perhaps capable even of cleaving water, *maybe* even the beginnings of some kind of respiration.

So much for the Early Archean.

And then, during the Late Archean, beginning around 2.9 billion years ago, the rise of photosynthetic cyanobacteria, extensive cratonization, the formation of continents. . . .

At 2.5 billion years ago: the end of the Archean, the end of the Age of Prokaryotes, the end of an atmosphere without—or with very little—oxygen, the end of . . . the early Earth.[43]

It wasn't like there was a thunderclap, but in the Early Proterozoic oxygen did begin to assume its contemporary place in the atmosphere. With the concomitant extinction of anaerobic prokaryote lineages, there was a parallel radiation of cyanobacteria, a fluorescent speciation—remember, all these guys mutate like crazy—and by 2.2 billion years ago, a third of the way into the Early Proterozoic (which runs close on a billion years itself), eukaryotic cells begin to appear with their membrane-bounded nuclei and mitochondrial respiration centers.[44]

Pinch yourself: *yeah*, those kinds of cells.

This *was* a thunderclap.

And, okay, maybe it didn't happen 2.2 billion years ago when Margulis and Sagan say it did, but again: *we have fossils!* —acritarchs they're called—from 1.4 billion years ago, of dinoflagellates, maybe of green algae, lots of them, so it sure happened earlier than that.[45] Interestingly enough everyone seems agreed on where these came from, *interestingly*, because not 30 years ago most biologists derided the theory of serial endosymbiosis, Margulis's idea that eukaryotes arose from the union of two or more prokaryotes, one of which took up residence—and again, remember how easily these guys exchange genetic material—*inside* the other.[46]

Symbiosis is the living together of different kinds of organisms, like you and the *E. coli* slaving away in your gut (well, in any case *living* in your gut, living well). There are all kinds of symbiotic relationships. Some have a community flavor. The relationships between ants and aphids, flowering plants and pollinating insects, are of this character. Then there's the more obligatory—*more physiological*—tone of the relationship between termites and the mastigotes living in their intestines that break down the cellulose the termites eat. Without the bacteria, the animals die.[47] Finally there are cases like lichen, an association of fungi and algae so close that *its* life is regulated by a single physiological mechanism. But sometimes "cohabitation, long-term living, results in symbio*genesis*: the appearance of *new* bodies, *new* organs, *new* species," says Margulis, who believes that "most evolutionary novelty arose and still arises directly from symbiosis, even though this is not the popular idea of the basis of evolutionary change in most textbooks."[48]

Her idea of where early eukaryotes came from goes something like this: First, a sulfur-reducing bacterium—an anaerobe—that also happened to be heat-loving (a fermenting archaebacterium or thermoacidophil) merged with a swimming bacterium to become a swimming protist.[49] Later an aerobic bacterium moved in to become in time a mitochondrion.[50] This tripartite complex—anaerobic thermoacidophil + swimmer + aerobe—subsequently became capable of wrapping itself around and engulfing food. It's a eukaryote now, protoctista, an alga, an amoeba, ancestral to all fungi and animals, to me and you (pinch yourself again). In a final merger this eukaryote "ingested, but failed to digest bright green photosynthetic bacteria," the bacteria became chloroplasts, and this swimming green alga became ancestral to today's plant cells.[51]

Which would make each of us—yes, you!—a community of communities, every one of your trillions of cells being one of these little bounded endosymbiotic bacterial communities.

Wow!

Anyhow, eukaryotes, as I say, maybe 2.2 billion years ago, but by 1.4 billion years ago for sure, in the Middle Proterozoic in any case, with multicelled algae showing up a few hundred million years after that, possibly 1.4 billion years ago, but certainly 800 million years ago (*we have fossils!*), with protozoa from ~850 million years ago and invertebrate metazoans extant ~650 million year ago, in the Late Proterozoic, in the Vendian.

Metazoans: animals more complex than single-celled protozoans.

And meanwhile: continental glaciation and mountain building in the modern manner, plates smashing into each other, orogeny, subduction. When? Maybe 2.5, at least 2.0, billion years ago, the Wopmay orogeny for sure between 2.1 and 1.8 billion years ago, the Gowganda glaciation a little earlier (a global phenomenon), a lot happening, *a lot going on then:* mountain building, continental glaciation, aerobic bacteria spreading out over the Earth, oxygen really beginning to build up in the atmosphere, ozone beginning to absorb the ultraviolet, the symbiogenesis of eukaryotes, oxidized paleosols. . . .

People talk about the past with its unchanging eternal ways, but the Proterozoic was something else![52]

Another flush of glaciation at its close, several episodes, even in the tropics, on every major continent except today's Antarctica,[53] and at the same time a flush of speciation: Bang! Coelenterates. Bang! Annelids. Bang! Arthropods.[54]

Ediacara![55]

It's not that *Calvin and Hobbes* world anymore. It's still laced with microbes (it's laced with microbes *today*) and there are still stromatolites and microbial mats (there are still stromatolites and microbial mats today), but for one thing it's turning green—it's *turned* green—at least along the interphases. There are cyanobacterial scums on everything, and algal lawns, and in the oceans there are soft-bodied marine animals, all sorts of them, *Dickinsoniae* and *Spriggnae* ploughing the seafloor sediments, the fern-like sea pens, the jellyfish moving through the waters like flying saucers. You've seen the pictures. It's a translucent world and soft and squishy.

There are icebergs, and spores in the air, and the sky is blue.

Are You Getting This?

Are you getting this?

I don't mean the details, the names, the dates. I hope you are. I repeat them enough. They're to hang the point on, but they're not the point. Are you getting the picture? Of the ceaseless slithering change, of the magnitude, of the unfathomable age, of the minuteness of the tiny parts, the bewilderingly rapid actions, the simple chemistry, the majestic unfoldings (the galaxies spinning in their awful silence), our irrelevance, our puff-we're-here (but-not-for-long) transience. . . .

It's like jumping out of an airplane, there's all the rush of the air beating past your ears and then you don't notice it, all you know is the immensity, the wholeness, the huge calm and indifference. You might be falling at terminal velocity but for all you can tell you might be standing still. It makes you laugh. You're not responsible for any of it. You couldn't be. And you laugh at the presumption that you might have thought you were.

You're free.

Really!

(For a minute.)

And then your chute opens and in the sudden silence it's the lavishness that gets you, the spread of the earth, the richness and intricacy and detail, the patchwork of the fields, the roads and the runnels, along the horizon the river—

And then all of a sudden it comes up fast, and you're responsible again, and you're running to keep from falling.

The names change, and the dates always get pushed back, but contemplating the unfolding of the universe from the Planck instant should be like jumping out of an airplane. There's all the rush and the noise of the first couple of seconds, and then the silence and the stillness of the evolving galaxies—those billions of years—and the cooling Earth and an-

other billion years and then the chute opens and there's the detail of protozoa, primates, language, the Hanging Gardens of Babylon, knights in
shining armor. . . .

And then all of sudden, coming up fast, fourscore and 7 years ago,
Vietnam, the L.A. riots, Monica Lewinsky, 9/11, and . . . where shall we
eat tonight?

Not that it matters, not in the long run.

Yes, "it requires both discipline and effort to deal with the fact that every breath we take is part of a cosmic transformation that began with the
'big bang' and will end either in the heat death of the universe or in some
regenerating catastrophe,"[56] but the fact is that in a couple of years no one
will remember Hurricane Andrew and in a couple of generations the Nazi
Holocaust will be as hoary as the St. Bartholomew's Day Massacre. This is
where the exhilaration comes from: *it all gets scraped off when the crust dives
for the mantle*. You can leave no monument (and forget the estate for the
kids), so you might as well relax.[57]

Eat in tonight.

Everything Matters

This is not to say nothing matters. In fact, precisely the opposite: everything matters. But it's a matter of scale. One thing the history of the
land—or jumping from an airplane—has to teach us is a sense of proportion, a sense of what we have a right to feel responsible for, and what we
don't. One thing we don't have the right to feel responsible for is "the ecosystem," not even if you take us all together.

Think about *Emiliania huxleyi*, the planktonic alga that produces all
those coccoliths. It's microscopic, but it can multiply until it covers thousands of square miles of ocean. It transports *tons* of carbon to the
seafloor. And it churns out enough dimethyl sulfide to maybe keep the
climate cool. What gets me is how like it we are. We may not be microscopic—though that's only a matter of perspective—but we sure can multiply. If you're far enough in space to see a cloud of *Emiliania* in the North
Sea, you're also out there far enough to see the clouds of smoke from the
burning over of the fields in Middle America and West Africa—individual
peasant farmers preparing their fields—or at night the light of the gas being flared off in the oil fields of the North Sea and the Middle East.

An ahistorical hysteria would paint these as disasters, but are they?
And if they are, for what? For whom? Was it a disaster when bacteria
started kicking out oxygen?

Although they almost killed themselves off, without the consequent development of the aerobic metabolism that their "disaster" brought about, *we* wouldn't be here to worry about how we were messing things up. Is it a disaster when *Emiliania huxleyi* kicks out all that dimethyl sulfide? We usually don't take it that way, but . . . *what's the difference?*

To think about our situation intelligently we've got to figure this difference out. After all, we're but one of myriad species, the biosphere is a little bigger than we are, just a *little* more resourceful, and it's way resilient. *It's way resilient.* Remember when we used to fret about nuclear Armageddon? We were going to wipe life off the face of the planet? But asteroids slamming into the Earth with the force of a million nuclear warheads haven't been able to wipe life out.[58] *We* never had a chance.

This is not to say a nuclear war between India and Pakistan would be a good idea, or that it's not incredibly stupid to flare off all that natural gas, or that there aren't way better ways to prepare those fields the peasants are burning over. But we've got to start making sense about *why* we think it's stupid. It's sure not because we can hurt the atmosphere. That's the dumbest reason. We can't hurt the atmosphere.

Again, this is not to say we can't affect it. All action is global, and in fact we *have* affected it. Global warming is a reality and this episode—for the world has warmed up and cooled off many times—is one we've had a hand in. But instead of slackening the pace at which we burn fossil fuels, we just go out and burn more. We buy bigger cars. We put fewer people in them. We're nuts! And then we wonder why the weather's turning "crazy." But if our behavior is "hurting" anything—which again is a perspective issue—it's "hurting" us, not the environment. So it's *us* we should be concerning ourselves with, us and the here-and-now. From the environment's, from the Earth's, perspective our behavior may be like being goosed or tickled. The trouble is, the environment chuckles, we die. Or, worse, suffer excruciatingly diminished lives.

Projecting our perspective onto the environment only clouds what's going on. The other day I heard a newscaster promise a story about "how hurricanes help the environment even as they hurt it." *What was he talking about?* Hurricanes can't help or hurt the environment. They *are* the environment.

They're a big part of the heat redistribution machine that's the climate. They're a form of the interface between the air and the water. Everything in their environment—that is, the coastlands and the vegetation and the animals—has evolved, has *co*evolved, with them, is adapted to their appearance, has come to depend on them for this or that flushing or reproductive or other function. Were the hurricanes to stop, all these other things would have to change, adjust to the new state of affairs. It's not hurt or help. It's coevolution. This hurt-or-help mind-set with respect to things like hurricanes assumes there's this, I don't know, thing like god, called Nature, who has plans and goals, and the hurricanes either advance or retard those goals. Which is nonsense. There aren't any goals. There never was any great plan in the sky on which the Outer Banks were indelibly inked. In fact, the barrier islands are one of the best examples I know of land as process, not thing. Idiots go out there and build hotels on what they think is a thing and then when the island that is a very active process wants to move out from underneath them, when it *does* move out from underneath them, the owners think . . . I don't know: they're so stupid I can scarcely imagine how to talk to them.

This is the real point: we can't hurt the environment because ultimately like hurricanes we too are a part of the same heat redistribution machine, an eddy off of bigger eddies. We too are the environment. All *we* can "hurt" are the chances that we're going to be able to continue to live lives so organized that plenty of endorphins get dumped into our intersynaptic clefts.

Certainly we can make the world a miserable place for us to live in, we can make our lives miserable. I was 11 when I first read—it was in Walt Disney's *Our Friend the Atom*—that every breath I took had a 100 million atoms in it that had passed through the lungs of Leonardo da Vinci.[59] Whoa! Think about that. I mean, that's how interconnected we are. We're all inside each other—literally inside each other—and if we're all that wrapped up in each other, and of course I don't mean just us humans but *all* of us, slime molds and soil fungi and redwoods and viruses and bacteria and whales (and for sure the atoms in question have been part of those things too), imagine what the detonation of even a single warhead does?

But if we're going to worry about this, we need to have a decent reason. If we're worrying about it because we're afraid for the rest of the biosphere, we're worrying about it for the wrong reason, because the biosphere has taken care of itself for billions of years. We might kill ourselves in the process, but all this "Save the planet" crap is insane. It's ourselves in our individual lives we have to save. Of course that's what's turning out to be so hard. We need to be especially clear about this. Our responsibility reaches only so far. The biosphere can—it will—take care of itself. This is the biosphere, remember, that nearly took itself off the planet by dosing itself with oxygen. If we're worried about burning oil and coal because it's going to set off another ice age, that's another wrong reason. An ice age? Whoop-di-doo! As though an ice age were a novelty for a planet that had pulled its first ice age off 2.5 billion years ago, and had them regularly ever since.[60]

Why then? Why worry about burning oil and coal? How about . . . because it stinks? How about worrying about burning coal and oil because it stinks, because it offends our noses and our taste buds when it gets inside our mouths, because we have more wonderful things to do than labor for the exchange value for coal and oil to burn.

We're not responsible for much, but we can claim so much for our noses and our hands and our arms and our feet. We can decide what to eat. This is what I'm getting at when I talk about thinking local.

It's fall as I'm writing this. "The sun has been setting earlier," as we say, acknowledging the drift of the Earth in its orbit around the Sun. The evening skies have been getting more and more awesome, and look! the sun on the brick of the apartment building across the street is flaming a red so deep it breaks my heart.

The air's much cleaner now too. I couldn't sit on my porch this summer and enjoy it so bad was the low-level ozone. But with the reduction in sunlight came a reduction in low-level ozone, and with the cleaner air and the lower temperature somehow the noise of the traffic is less irritating. It's nice out here now. Nicer anyway.

It's still hard to hold a conversation. The grit still settles on my papers. A sheet of white paper will be gray in a day.

We're breathing this. And the cars run into each other and they kill people. From my porch I can see a wreath put up to memorialize a girl killed a couple of years back. I didn't know her, but someone did and it's hard for me to pass her death off like the explosion of a supernova or the passing of a butterfly. There was too much screeching of steel and broken glass and too many flashing lights and wailing sirens. And someone keeps replacing the wreath on the telephone pole in front of White-Wall Auto Repair.

It's important to accept our responsibility in *this*. We've created a transportation system that isolates us and privatizes our lives and stinks and is loud and alienating and kills people. There're six reasons right there, not counting the paving of everything, parking lot deserts, the death of mom-and-pop grocery stores, strip malls, the loss of sidewalks, the end of the walking environment, the end of *walking*, the death of downtowns as commercial environments, the deaths of downtowns, fast food "restaurants," suburban shopping malls, the obliteration of streetcar systems, automobile showroom wastelands, junkyards, nonbiodegradable tires (which make such great breeding places for mosquitoes)—

So what's that? 16, 17 reasons—without even thinking about it—for getting out of your car that have nothing to do with global warming and they're all at the scale of running to keep from falling when you touch down after a jump.

It's a scale on which we can operate with some sense—with any sense—of making a difference. It's only with land we're directly messing with that we can make a substantive difference. We can't truck with the land as it has been thrown up, is being thrown up, in the great churning of the terrestrial spheres. That's beyond us. Not that we're not part of these spheres, not that we're not eddies—

And not that we haven't massively intervened in the flows of carbon and in the nitrogen cycle and in the gene pool. But we've got to be able to sense the distinction. We've got to be able to feel the great motions, to know and understand they're going on, now as in the past; to appreciate and accept our position with respect to them, as minor and transient and ephemeral as it is; to sense the drive train, but also the fine and convoluted gearing (and the clutch) that couples us to them, to it. We have to be able to appreciate and accept our autonomy with respect to that gearing, the time and space we might have in which to shape our world.

Which is *the* world.

Unless you can sense the gearing, and the clutch (all that autocatalysis), it all comes to have an overwhelming sense of inevitability, and then you feel powerless and allow yourself to be just swept along—

And of course that's the ultimate state of alienation. We can't access our own autonomy then. There may be nothing we can do to intervene in the tectonic motion of the plates, but we can decide what to put on our dinner plate.

There's no deterministic connection between these plates, not through the descent to agriculture, nor the great drifts in the gene pool of corn, nor the coupling of your parents to their shopping environment, nor the advertising of the great cereal companies, nor the point-of-purchase come-ons in the aisles of the grocery stores. You don't have to fall for Tony the Tiger. What we put on our plate may not be much but it's our own. As is what we plant in our window boxes and along our walkways and what we read and how we get where we're going and what we do with our eyes, what we watch, what we read, what we pay attention to. If we're not exercising autonomy there, hell, we might as well give up, because we've got *nothing* to say about the next incoming asteroid. Trust me on this.[61]

If we're waiting until then to start living right, it's going to be too late. Armageddon! Jesus! What I want to see is a movie where Bruce Willis saves the world by taking a courageous stand on . . . *what he's going to eat.* At least he has a chance there. He doesn't have a chance against the asteroid, or the earthquake, or the hurricane, or the next ice age, or biogenic drift. No one does.

7

The Land Covers Itself
with Plants and Animals
(and the Human Animal
Comes Down from the Trees)

For someone who is supposed to be writing a history of the land I haven't said much about the land so far, but, (1) there's a lot more *to* the land than . . . *land*; and (2) up until now, there hasn't been all that much "land" to write about.

A Brief History of the Continents

I did point out that large bodies of magma were intruded into cratonic rocks during the Early Proterozoic, presupposing a Late Archean origin for cratons of at least moderate proportions. This is to say there was gradual, episodic "cratonization," beginning perhaps at the tail end of the Early Archean, ~3.0 billion years ago, with large cratons already in what today is Canada and South Africa—and within the next few hundred million years a number of large continents came into existence. Ur may have been assembled as early as ~3.0 billion years ago, Arctica maybe 500 million years later, and Baltica and Atlantica maybe 500 million years after that. It seems likely that Arctica and Baltica combined to form Nena by ~1.5 billion years ago, and that Nena, Ur, and Atlantica began to pull together a couple of hundred million years after that to form Rodinia, the first supercontinent, whole, say, a billion or so years ago, more or less co-terminous with the early protozoan radiation.

This is all quite speculative—it's only been a few years since we realized that the continents have been moving around—but there's already a consensus that Rodinia started coming apart maybe 250 million years after it formed, say 750 million years ago, breaking up into Laurasia and East and West Gondwana. These things don't happen overnight. Continents "consist of cratons amalgamated by networks of orogenic belts that contain oceanic, island-arc, and continental-margin rocks plus, locally, fragments of older cratons," where cratons are "blocks of crust sufficiently stable to have provided a basement for the accumulation of shallow-water or subaerial volcanosedimentary suites," so we're talking some painfully slow processes here.[1] Indeed, breaking up is hard to do. In the case of Rodinia it took 150 million years.

Nor is any of this discrete. The assembly of the Gondwanas is taking place during this disassembly of Rodinia, so that by the time Rodinia is a paleomagnetic memory, East and West Gondwana themselves have come together to form Gondwana, what? 550 million years ago, at the inception of the Cambrian—*at long last*—that is, at the inception of the Paleozoic, opening, this is to say, the Phanerozoic, the age of *visible* life.

Which is *sort of* a misnomer, like you couldn't see the stromatolites and microbial mats. I guess the point is that with the assembly of Gondwana you had the metazoan radiation.

One of the things that's hard but essential to hang onto as these landmasses collide and rift and raft around the planet is that they're opening and closing oceans: Iapetus, Tethys, the Atlantic. This means the structures of the surface and thermohaline circulations were in continuous flux, and that means that as these interacted with atmospheric circulation, local environments—where the medusoids, the erniettids, the early coelenterates, the annelids, and the arthropods actually lived—were constantly changing, probably at relatively rapid rates, relative, at least, to the rates at which the continents were being collaged and coming unglued and shifting their locations on the planet. It's important to keep in mind that the *interaction* among global systems means that local change can happen overnight, no matter how slowly the systems themselves may be evolving, and it's important to keep in mind that these systems *are* evolving. When, during the period between 450 and 320 million years ago, Gondwana hooked up with Laurasia to form Pangea, it completed a supercontinent *cycle*, the episodic rifting and disintegration of one supercontinent and its subsequent reassembly in a new configuration. Pangea began fragmenting 160 million years ago. It's parts are still dispersing today even as it seems that a new supercontinent may be assembling (vide the incredible Himalaya-building collision between India and Tibet). What this means is that we're in the *middle* of a supercontinent cycle (we're at the beginning or end of nothing), in a process we have to imagine continuing as long as the Earth is hot enough to keep the plates in motion.

To say again, Ur, Arctica, Baltica, and Atlantica, to give them in the order of their amalgamations—Arctica and Baltica already unified in Nena—fuse to form Rodinia, which splits into Laurasia and East and West Gondwana, the latter two getting together as Gondwana. Gondwana marries Laurasia to form Pangea, which fragments into pieces that assemble as North and South America, Eurasia, Africa, Australia, Antarctica, and India, the latter, as I write, hard at work uniting with Eurasia.[2] It's like a planetary heartbeat, another of these giant global fluctuations, this one distributing and redistributing the land and water masses, so that this year—or this eon—this terrain is on the tropics that some time ago was over a pole. There's that, then, which can be all determining. Gondwana, for instance, really moved, crossing the South Pole several times in its wanderings. Needless to say, Gondwana experienced significant glaciations at these times, in the Late Ordovician, Late Carboniferous, and Early Permian, glaciations doubtless responsible for the simultaneous declines in sea levels around the world.

Not only are such ripple effects global—and so glaciation on Gondwana results in the desiccation of Laurasian estuaries—but they can penetrate *deep* into the coupled systems. This is how Brian Windley describes the consequences of India's plowing into Asia:

> The uplift of the Tibetan plateau since 8 ± 3 Ma profoundly affected the Hadley circulation cells of the atmosphere, the late Miocene strengthening of the Indian monsoon, the drop in the calcite compensation depth in the ocean, the evolution of plants using the relatively efficient C4 pathway for photosynthesis, the downdrawing of CO_2 from the atmosphere, and global cooling leading to the late Cenozoic glaciations. So, we can finally see a connection between continental tectonics, and the circulation of the mantle, hydrosphere and atmosphere—a whole Earth system.[3]

And, we might add, the cryosphere and biosphere too, the C4 pathway being a way plants conserved their coupling under conditions of scarce CO_2,[4] though this is all speculative. There are many competing explanations for the late Cenozoic glaciations, and indeed for all nine major episodes of glaciation that have occurred since what we might refer to as the "continentalization" of the planet—though the possible effects of rapid continental growth,[5] suprapolar continental drift,[6] changes in oceanic and atmospheric circulation associated with suprapolar drift (especially of Gondwana),[7] and changes in moisture regimes caused by shifts in the Gulf Stream associated with the tectonic formation of Panama[8] provide plausible explanations for all of them—for the glaciation of the Pongola Supergroup some 3 billion years ago; for the Huronian glaciation at 2.4 to 2.3 billion years ago; for the later Proterozoic episodes of the Riphean, at 1.0 to 0.9 billion years ago, the Sturtian, at 750 to 700 million years ago, and the transitional episodes of the Vendian, from 625 to 580 million years ago, and the Sinian, from 600 to 500 million years ago; for the Gondwanan (and Pangean) glaciations of 450 to 400, and 350 to 250 million years ago; and for the recent glaciations of the last 15 million years[9]— without the invocation of other mechanisms (say, Milankovitch cycles) except perhaps as triggers.[10] This is not to deny changes in the level of insolation, but to acknowledge that once the biosphere is up and running, Gaian mechanisms exist for smoothing out its effects,[11] except insofar as the continuous redistribution of land and water masses provides a platform for glaciation—well, for example, by rafting continents over the poles.

The Supercontinent Cycle Pumps Life onto the Land

And with the land . . . life. Again, this is not to deny the anciency and richness of oceanic life. In fact, for all its growth, "dry" land will remain more or less free of life during the early Paleozoic, free of trees, free of insects, free of aerobiota, until "higher" plants and arthropods in the Silurian (~410 million years ago) and vertebrates in the late Devonian (~370 million years ago) learned to couple with terrestrial media.[12] Let me admit that "dry" land draws a distinction harder than land and water, especially in a world where shallow seas periodically spread across vast portions of cratons. In fact, it was in these inland and epicontinental seas, in these broads, in these flat lowlands where shallow seas gave way to brackish mudflats which gave way to freshwater skims—much of it alternately desiccated and flooded (and was this land or water?)—that life in its increasingly various forms then most flourished. You've seen the pictures.[13] Even stromatolites were abundant for a while, and by this time in the early Cambrian stromatolites had been around nearly 2 billion years. There's a terrific sense of continuity, of overlap, in the early Cambrian, with these Archean organisms and the microbial mats along the beaches (so easy to forget) and the late Proterozoic Ediacaran biota and the soft-shelled craziness of the novel creatures of the Burgess Shale (the "weird wonders" of

Opabinia, of ferocious *Anomalocaris*, of chordate ancestor *Pikaia*)[14] and the cuplike *Archaeocyathids*, the sponges and the small shelly faunas of the Tommotian (the earliest interval of Cambrian time), even precursor species of Ordovician fauna, early trilobites most likely, early echinoderms.

Again, these crazy names. As we've seen, geologists have divided the Phanerozoic (these most recent 570 million years) into three eras, the Paleozoic (570 to 245 million years ago), the Mesozoic (245 to 65 million years ago), and the Cenozoic (the last 65 million years),[15] then the eras into periods, the periods into epochs, the epochs into stages, and so on. We won't often go there (though in alluding to the Tommotian I already have). The period names come up all the time, so here they are again, this time with dates. The periods of the Paleozoic are: Cambrian (570 to 510 million years ago), Ordovician (510 to 439 million years ago), Silurian (439 to 408), Devonian (408 to 362), Carboniferous (362 to 290), and Permian (290 to 245). Those of the Mesozoic are: Triassic (245 to 208), Jurassic (208 to 146), and Cretaceous (146 to 65). And those of the Cenozoic are: Tertiary (65 to 1.64) and Quaternary (the last 1.64 million years); this last almost compulsively broken into the *epochs* of the Pleistocene (1.64 to 0.01)) and the Holocene (0.01 to the present) and often enough set outside—on *our* side of—the Cenozoic, in a sort of ageological free zone.[16] You don't *need* to know these names—they won't be on the test (there is no test)—but why not? The rote does help spine time. If you can keep the terms of the presidents straight, you can handle these. The things that happened during the Devonian, say (pinch yourself), are every bit as relevant as the things that happened during Gerald Ford's term as president.

Who?

Exactly.

Anyway, neither the stromatolites (for the most part) nor the Ediacarans long survived the grazing and burrowing activities of the new Phanerozoic species,[17] though for a while there they all lived together. It was still a wet world—completely—but less and less translucent and soft and squishy. Algae galore, of course (what most everyone was eating), but many of the new guys had shells, shells and skeletons, a lot of them did—

And what was this about? Shells? Skeletons?

One of the things it was about was changes in sea level. Which is where I was heading when I began that paragraph with "And with land . . . life." It's not, obviously, that life came with the land. It's that with the land—with the continentalization of the planet, with the growth and movement of the continents—the oceans got so much more interesting, so much more complicated, so much richer and various. The epicontinental seas, for instance—which is where life was really blowing up during the Cambrian—the epicontinental seas, the endless mudflats, the tidal pools (think about them as sea or land as you wish) were pretty much controlled by changing Phanerozoic sea levels, which pretty much were driven by the supercontinent cycle.[18] Among other things, water levels rise and fall with the production of crust, and the production of crust is greatest, by a factor of two or so, when supercontinents are breaking up. So also is the flooding of continental shelves, with all that implies for the expansion of niches for all this novel life.[19] Simultaneously implicated in these processes are variations in the composition of seawater, and for that matter, the atmosphere, as the amount of material available for weathering (and burial) rises and falls, with all this implies for the abundance of atmospheric CO_2, and all *it* implies for global temperatures (and all they imply for glaciation).[20] Frankly the data at this point are at once overly rich and too sparse, the chemical pathways (metabolic and other) too numerous, and the competing explanations and models too conflicted,[21] but given the all but unfathomably rich interconnectedness of the parts, it's no stretch to conclude a truly profound relationship between the continued growth and migration of the continents—with the attendant changes in sea levels and the cycling of critical chemical elements—and the explosive development of life-forms that took place in the early Paleozoic.[22]

Again, there are too many models, but among those advanced for the evolution of organisms that use calcium carbonate for hard skeletons and shells *is* a sudden increase in the availability of calcium carbonate in the oceans. On the other hand, against this, is the suggestion that this increase was due to the evolution of swimming organisms whose muscles—like ours—were under calcium control. "Muscles," Margulis and Sagan remind us, "contract when dissolved calcium and ATP are released in precise quantities around them. The calcium must be scrupulously kept in quantities far lower than those of seawater or chemistry takes over and the calcium comes out of solution in a solid form," as in the calcium deposits in the overworked muscles of athletes.[23] Nor are muscles alone in playing with calcium this way. It's an essential element in the metabolism of all nucleated (i.e., eukaryotic) cells, notoriously in the firing of the neurons that were increasingly important as the novel metacellulars grew in complexity. Margulis and Sagan observe that some early creatures (and recall the coccoliths of *Emiliania huxleyi*) *must have* secreted bits and pieces of calcareous armor and protective films that were not fully skeletons:

> Always used by nucleated cells, excess calcium must be excreted or harmlessly stockpiled out of solution. Since Cambrian times organisms have been stockpiling their reserves as calcium phosphate, which takes

such forms as teeth and bones, or as calcium carbonate, as in chalky shells.

Skeletons did not appear out of nowhere during the Cambrian: Ediacaran muscles preceded Cambrian skeletons. The need to continuously respond to calcium surpluses in the cell made it easy for some animals to stockpile calcium salts inside or outside their bodies in dump heaps that eventually became skeletons and body armor. Just as termite nests are largely constructed of insect excrement and saliva, so skeletons and teeth are made of compounds that originally had to be excreted as waste.[24]

Meanwhile, as we've seen, the seas were rising and falling, and they were rising and falling with both the regular periodicity and low amplitude of the tides, and with the irregular periodicity and greater amplitudes that accompanied the changes in (1) the production (and subduction) of crust (i.e., the changes in the volume of the ocean-ridge system), (2) the number of continents (and their rifting and collision), (3) their location with respect to geoid highs and lows (raising and lowering continental masses and so shores and continental shelves), and (4) the extent of their glaciation (if any). Some of these changes may have been cyclic and had regular periods, but periods in the 10s of millions of years;[25] and all of them caused changes in sea level that came to 10s, often 100s, of meters.[26] Now,

Algae dwelled in wet, sunlit shallows. Occasionally these shallows dried up, and those algae that could remain wet on the inside while dry on the outside had the evolutionary edge. They survived and multiplied to become the early plants, low-lying forms without stems or leaves, related to modern-day mosses and liverworts, that could not support their own weight out of water. Algae became land plants by bringing water with them. . . .

Dry land was as hostile an environment for plants as the moon is for us. It was crucial for the gelatinous tissues of algae not to collapse or dry out. Becoming terrestrial meant developing a tough, three-dimensional structure as contrasted to earlier low-lying forms. Using atmospheric oxygen (that old waste product of the cyanobacteria), the early plants evolved a cell wall material called lignin. It is this lignin that, combined with cellulose, gives such strength and flexibility to shrubs and trees. This sturdiness led to the development of a so-called vascular system transporting water up from the roots and food down from the flattened ends of the branches that were early leaves.[27]

Animals too would have found themselves panting as the water evaporated from sunlit shallows suddenly cut off from receding oceans (as it would have seemed to them), drying out on the beach, struggling as the film of water grew thinner and thinner; and certainly those that could, as it were, remain wet on the inside while dry on the outside had the evolutionary edge. There were, of course, incredible advantages to living on the land. There was abundant oxygen for one thing, and a shocking lack of competitors, and by the time animals worked out the mechanics of surviving, a relatively rich plant life to live on. But there were plenty of disadvantages too. For one, there wasn't the buoyancy of water. For another there was the harsh, desiccating sunlight. Whereas plants constructed lignin from waste oxygen, animals constructed skeletons and carapaces from the waste calcium excreted by the muscles that we've seen were developed by Ediacaran swimmers (and eukaryotic cells in general).[28]

Hypersea: The Ocean Life Brings onto the Land with It

Drifting in this way (in the effort to conserve their coupling with the environment), and exploiting their own waste products (very much at hand), plants evolved and animals began to spread onto the land. The way this story is usually told is as a succession of triumphs, literally, the novel life-forms marching through the victory arch to strut and fret upon the stage of life: the Age of Marine Invertebrates (in the Cambrian), followed by the Age of Early Fish (in the Silurian); the Age of Coelacanths and Early Amphibians (in the Devonian) succeeded by the Age of Coal Swamp Forests (in the Carboniferous); the Age of Early Reptiles (in the Permian) surpassed by the Age of Early Mammals (in the Triassic); that of Dinosaurs (in the Jurassic, in the Cretaceous) trumped by the Age of Mammals (in the Tertiary); and all overwhelmed by the great glory of the Age of Man (in the Quaternary, in the Present).[29]

Ah, me . . . except, as David Raup reminds us, the Phanerozoic wasn't like this: "It was not a succession of global dynasties dominated by organisms of every-increasing sophistication."[30] In fact,

> It is often said that the large reptiles dominated the world, both on land and in the oceans, for long spans in the Jurassic and Cretaceous periods. But this is a gross exaggeration. To be sure, some reptiles were the largest animals then living—big dinosaurs on land and ichthyosaurs and mosasaurs in the oceans. But in terms of the global biomass, they were minor players, never having many species or large populations compared with the millions of smaller organisms.[31]

And I know *we* think *we're* the cock of the walk, but I can imagine this story being told from the perspective of corn, from the perspective, that is, of *Zea mays*, by any measure among the most successful organisms in the history of the biosphere, spreading as it has, completely dominating endless square miles of the planet, just . . . *taking in the sun*, having gotten another species—ours—to take responsibility for, oh, among other things, the dispersal and germination of its seeds. Age of Man? Or Age of the Cereal Grains, Age of Rice and Corn?[32] Yes, we write the books—so take that *Zea mays!*—but really, who's to say which species has domesticated which?

Not that it matters: neither could readily get along without the other. True, corn can no longer survive "in the wild," but . . . can you? In your armchair with your feet up on the hassock, book in left hand, cup of tea in right, air conditioner humming or radiator popping? *Survive in the wild?*

Gimme a break. You'd have a hard enough time surviving on the street.

Certainly no chance at all without corn or rice or wheat.[33]

It's complicated, this business of how to characterize the shifting scene. You can't just look out your window in Midtown Manhattan or across the dappled lawns of your subdivision—or out at Hillsborough Street here in Raleigh—and mutter, "Age of Man." There's all that corn, for one thing, and wheat and rice and the bright fields of quinua in the high Andes and 1,700 species of rodents and 900 of bats, and as Margulis and Sagan would remind us, "Beneath our superficial differences we are all of us walking communities of bacteria."[34]

Communities—that's what you don't get in the Age of Dinosaurs talk, that's what you don't get in the pictures. The new kid in town is always foregrounded in the picture and the drama is heightened to rivet our attention. In the picture I'm looking at now, the pterosaur *Criorhynchus* is snatching a fish from the jaws of the pleisiosaur *Elasmosaurus,* crashing surf enlivens the image in the lower right, and . . . you know, there's none of the quotidian quality that pervades our life, and theirs too could we see the next frame, that annoying cloud of insects and all the shellfish, the smaller amphibians and the centipedes and spiders, the mutualism. . . .

I don't *expect* to see the "invisible" life of the bacteria, the protists, the fungi.

So it's one of the things I really like about Mark and Dianna McMenamin's idea of the Hypersea, that it *is* a community, a community of plants, animals, protists, and fungi, a community first and foremost.[35] It's *our* community. It's the community of life on land. The McMenamins call it "Hypersea" because what best characterizes this community, *our* community (I can't say this too often), is the way, sloshing and slopping up onto the land (a rising tide of life), it brought the sea with it, *inside it,* carried the sea and its distinctive solutes up onto and over the surface of the land and into some of the driest environments on earth.[36]

We don't just carry the sea inside us either. In the effort to stay wet inside we continuously exchange it with each other. The fluids we exchange—I'm talking about the body fluids of eukaryotic organisms (about cytoplasm, lymph, blood, the fluids our ancestors preserved, *created*, in the effort to conserve water in the desiccating environments of the land)—we exchange through direct physical connection, through sex, by eating each other, and in other relationships, mutualistic, parasitic, or hyperparasitic.[37] "Like a string of kettle lakes connected by streams only during times of flood," the McMenamins write, "the bodies of land animals are lakes within the Hypersea hydrosystem—connected by the fluids they exchange during moments of contact."[38] Continuously gaining and losing water through the biologically mediated actions of roots, leaves, lungs, and cuticle,[39] Hypersea is largely a story of parasitism (and hyperparasitism) and fungal hyphae, though I can't help noticing the way we've pushed the envelope, draining plant and animal "lakes"—oranges, cow udders—into the waxed paper cartons, into the plastic bottles, we pick up at the grocery store, freezing fluids, or drying them for subsequent reconstitution (powdered milk is one evaporite, maple syrup is another), in this way storing and transporting not just (or even) the H_2O, but the whole fluids of these animals and plants. I know we don't like to think of ourselves as vampires, but isn't that what we're up to when we sink our teeth into an orange and suck the juice out of it?

No, no, no, *no*: calling it "vampirism" gives eating a bad name, *but . . .* except for the oxygen we get from the air (and the cold viruses and respirable quartz and ragweed pollen and dust mites and mold spores and tuberculosis bacilli), the main way we connect not just to Hypersea but to the rest of the cycling elements essential to our metabolic ongoingness—*the way we connect to Gaia*—is by sinking our teeth into whatever it is and ripping it to shreds. Or sucking it up. Gulping too. Chewing. Grinding.

Licking, lots of licking.

And, except for salt (and for the few geophagists, dirt), if what it is we're sinking our teeth into *isn't* alive, it once was.

Each of Us: A Lake in the Hypersea, a Link in the Carbon Cycle

Pleasant to think about or not—but why not? whence the squeamishness?—putting our mouths on them is the main way we have for linking up with the rest of our community.[40] What did *you* have for dinner last? I had a nice linguine with a fresh tomato and olive sauce from a recipe I got from the Williams–Sonoma catalogue. Thereby I hooked up with all kinds of Hypersea "lakes"—those in tomato plants, in garlic, in olive trees, in pepper and basil plants, in onions, in lemon trees, in pepper trees, in wheat, in cows (source of the mozzarella)—*and through them* with the great reservoirs of essential elements circulating through the soils (and percolating through the waters and wafting through the airs) of Italy and sunny California and Florida and Sampson County, North Carolina (where my lawyer, Kyle Hall, grew the garlic), and Indonesia and "the sun-drenched valleys of the Kalamata region of Greece," and wherever the milk came from the mozzarella was made out of. Wherever they were in the world the plants in question drew carbon from the air—it was in the form of CO_2—and used it to synthesize carbohydrates. To do this they drew water up through their roots (or perhaps a parasitic fungi pumped water *into* their roots), drew water up into the Hypersea, that is, up into the bodily fluids of whichever eukaryotic organism it is we're talking about. There, in its chloroplasts, the plant used the energy of sunlight to split the water, and then used the hydrogen it got from the water to reduce the carbon dioxide to carbohydrate. When it did this, it simultaneously created something for me to sink my teeth into *and* completed the global carbon cycle.

That's how H. Robert Horton and his colleagues put it anyway,[41] and I know what they mean, though I'm not sure that you can complete a cycle. You can just move through it, or the *carbon* can move through it. The carbon just goes round and round in this tight little cycle. In respiration—in the process that fuels the lives of my cells (and yours)—the carbohydrates are reoxidized to CO_2 and water. Later on, the CO_2 will be reduced to carbohydrate again in another bout of photosynthesis, and so on, round and round, from photosynthesis to respiration to photosynthesis, from CO_2 to CH_2O to CO_2 again, animals and plants soon seeming no more than bit players in a drama called *Carbon*, hod carriers, hoofing the element back and forth like Laurel and Hardy carrying the pieces of the piano at the end of *The Music Box*. Unless, that is, the carbon gets drawn down out of this bioatmospheric circulation, is reacted with silicate in the weathering of rock, or gets buried in the biomineral reserves of coal and oil, or deposited in calcium carbonate on the ocean floor in a coccolith or shell, lost—in every case—to the swift cycles of the biosphere, to be caught up in the long, slow cycle of the lithospheric circulation, certain—*some day*—to be uplifted as limestone, or, having been consumed in subduction and metamorphosis, off-gassed some eon in the future through a vent in the ocean floor or out the mouth of a volcano.

A crude sketch, this, of the global carbon cycle.[42] It's crude because it ignores so many of the paths carbon takes, and crude because it fails to recognize the way these paths are mingled with those taken by the rest of the elements, each of them splitting from and rejoining another, incessantly, at every conceivable scale. To be able to think about this at all we've isolated the elements—we've tried to—and so come to think of a carbon cycle and a nitrogen cycle and a sulfur cycle and a phosphorus cycle and an oxygen cycle, and all the rest of the cycles.[43] But each of them is an artifact of our effort to understand it all, none of the cycles exists, every one is hopelessly entangled with the others, carbon, oxygen, and hydrogen cycles mutually entrained, here for instance, in this strand of linguine I'm twirling into my mouth.[44]

God that was good!

And soon now these very atoms will split up and before long link into new configurations—who knows where they're headed?—and to imagine for an instant that any of us is an island entire of itself, that any of us is more than a momentary node (not unlike that strand of linguine) in a social network with hundreds, with thousands, of connections to other humans, each in turn linked to organisms beyond counting in the community of Hypersea, and then through the fluxes of matter and energy to everything, *every*thing, on the planet. . . . Listen: this is how Lawrence Durrell put it in *Clea*:

> You tell yourself that it is a woman you hold in your arms, but watching the sleeper you see all her growth in time, the unerring unfolding of cells which group and dispose themselves into the beloved face which remains always and for ever mysterious—repeating to infinity the soft boss of the human nose, an ear borrowed from a seashell's helix, an eye-brow thought-patterned from ferns, or lips invented by bivalves in their dreaming union. All this process is human, bears a name which pierces your heart, and offers the mad dream of an eternity which time disproves in every drawn breath. And if human personality is an illusion? And if, as biology tells us, every single cell in our body is replaced every seven years by another? At the most I hold in my arms something like a fountain of flesh, continuously playing, and in my mind a rainbow of dust.[45]

Yes. *Yes!* This is more like it, though the watched is not alone in play, the watcher too is a fountain, they're entangled fountains playing in a fountain of sheets (in a fountain of cotton) in a fountain of. . . .

Eddies, eddies on eddies. . . .

It *is* in the pattern not in the parts and the piercéd heart too is no more than a pattern, no matter the pain.

But it is also one whole thing, and as the fountain rises from the water so the pattern arises from its parts. It's all a fountain. The crust diving for the mantle is a fountain and all of evolution is just a fountain and the Big Bang fountaining forth, fragments in play, and one day it's trilobites off the shores of Gondwana and the next it's small mammals in the moonlight on Pangea and soon, near what is today Laetoli in what is today Africa, there's a kid walking with a couple of adults through a fine layer of volcanic ash. It was 3.75 million years ago.

I wonder if they were holding hands?

Kingdom: Animalia, Phylum: Chordata, Subphylum: Vertebrata, Class: Mammalia, Order: Primates, Family: Hominidae, Genus: *Homo*, Species: *sapiens*

Humans. Something like them anyway. *Australopithecus afarensis.* Close enough. Where'd they come from?

As though we didn't know. *They came from the Big Bang!* Along with everything else. *Where else?* The stuff of them, their energetic materiality, out of the false vacuum, randomly fluctuating, bubbling, fountaining forth. . . .

C'mon, I don't need to go through this again, do I? You must know it by heart: grandunification, baryon genesis, unification, quarks, hadrons, leptons, photons, protogalaxies, galaxies, the accretion of the Sun and the Earth, continuously, endlessly diversifying, unfolding, evolving. . . .

The mantle settling down and the off-gassing of the atmosphere and the raining out of the oceans, the small steep-sided felsic protocontinents and, out of the atoms that had been created inside the evolving stars (*our atoms*), the chemical evolution of the earliest cells. . . .

The Age of Prokaryotes. . . .

From whence the organelles inside *our* cells. . . .

Episodic cratonization, Ur, the early glaciation of the Pongola Supergroup, photosynthetic bacteria, the formation of Arctica, the Huronian glaciation, and aerobic bacteria. . . .

Ayy, Baltica!

So much is going on. Everything is always changing. Atlantica's agglutinating and anaerobes of every persuasion are being poisoned to death by the oxygen spewed by their cousins. Arctica and Baltica merge to form Nena and there are eukaryotes—*of which we're composed*—and they create meiotic sexual reproduction—*our kind*—and Nena and Atlantica and Ur sign up to create the supercontinent Rodinia. There're the Riphean and Sturtian glaciations and the appearance of megascopic eukaryotes. . . .

Ediacara!

Glaciers move over the land that's beginning to form Gondwana, animals with shells, with skeletons appear, and invertebrates—which have a coelom like we do, and other things—and then come the glaciations of the

Ordovician and Silurian and soon a quarter of all living families dies off in the first of what some people, forgetful of the many extinctions that accompanied the evolution of the earliest prokaryotes and the great dying of the anaerobes, call "the big five," the big five mass extinctions. . . .

But still: a *quarter* of all the families of life-forms that had existed, gone. . . .

Of course they *all* go, species rarely hang around more than a few million years; most species many, many less. We die—it's the price we pay for sex—and species disappear and genera and families vanish and orders and even phyla, nine of them over the eons, of animals, kaput.[46]

Still it went on. Pangea began forming and vascular land plants appeared and amphibians moved between land and sea and then almost a fifth of all living families vanished in the next mass extinction, the Devonian. . . .

There were more glaciers and then seed ferns showed up and Pangea was complete and pow! two-thirds of all living families disappeared in the great Permian extinction, and then dinosaurs materialized and there was the Triassic extinction, another quarter of all living families wiped out, and pretty soon the first little mammals were scurrying around in the night. . . .

We're mammals. And Pangea began coming apart at the seams and flowering plants burst into flower and the rest of the dinosaurs passed from the scene in the Cretaceous extinction, and then there was a Tertiary—okay, a Paleogene—radiation of mammals and then the ice returned. . . .

It started in the Miocene, in the middle of it. Don't forget—

Don't forget! I know. It's ridiculous of me to say that. *So much is going on.* But unless you keep in mind the dynamics of all the subsystems—the migrating Hadley cells and the thermohaline circulation and the cycling of the phosphorus, the colliding plates, the incoming asteroids and the evolution of the marine biota, the changing CO_2 levels and the rising and falling sea levels and the relationship of the eccentricity of the orbit to the eccentricity of the ecliptic and the precession of the equinoxes, the evolving ecosystem of the Hypersea and the sites of active tectonism and fluctuations in the rates of solar insolation—unless you're got some grasp of all this cranking and clanking away in the background, the scene begins to congeal, and then it's like there's a stage, and we step on it and begin to play our part; and if you think about it like that you can't make sense of any of it, not the horrible jobs most people go to or the hot summers we've been having lately or the crazy music kids listen to these days.

It's all like: things used to be okay, the world used to work this way or that way, and it was cool, but now everything is crazy.

The fact is it's been crazy all along, nothing was ever set, we *are* the stage, there aren't any actors. Or the play is one in which the actors build the stage, that's what the play's about, that's all it's about, but it doesn't have any beginning (or it's so long ago nobody remembers) and no end (or at least no end anyone can see). And a change over here means something's got to change over there, everything's connected, everything's *geared*, though there is this sort of clutch you can push in to get your feet moving when you're falling and you don't want to end up kissing the sidewalk.

Everything's . . . geared but there are governors and sinks and devices that charge up batteries and clutches so it doesn't work like a clock but more like a car that's driving itself.

Only it's not going anywhere. It's just driving to be driving.

And what I was going to say was, don't forget that in the middle of the Miocene, Pangea's still breaking up and reassembling into what looks a lot like a freehand sketch of a modern map of the world,[47] except North and South America have yet to hook up, and a lot of fussing with the details needs to be done. Antarctica's already over the South Pole and the landlocked sea we call the Arctic is already in place over the North Pole; and, whether an exclamation point to the gradual decline in global temperatures that can be dated from the Middle Eocene (say 45–50 million years ago)[48]; or more of a comma in the sharper cooling trend inaugurated by the early Oligocene separation of Antarctica from Australia (~35 million years ago)[49]; or . . . just something triggered by the bolides that 14.8 million years ago routed out the Ries and Steinheim Craters (in what today is Germany),[50] the fact is that 14 or so million years ago glaciers began moving in Antarctica and shortly after that forming in Alaska,[51] and frankly the Earth's been quite chilly ever since. So it's not just that humans evolved coupled to a world of complicated change. They evolved coupled to a cold world too.

The Ice Age We Came Up With

I'm talking about the Ice Age though it's not the Ice Age I grew up with. For one thing the Ice Age that's coming into focus these days is a lot older than the one I learned about in school. "Our understanding of environmental change over the recent geologic past is quite different from that which existed a few decades ago," Neil Roberts acknowledges. "The Pleistocene did not, as was once thought, mark the main break-point in recent earth history. Because the change to a colder, more variable climate began well before this, it is more appropriate to refer to a 'late Cenozoic

Ice Age,' "[52] though really it seems to begin in the Middle Eocene with the cooling induced by the Himalayan and Alpine orogenies. Reviewing analyses of fossil foraminifera, diatoms, radiolaria, coccoliths, oxygen isotopes, organic carbon, biosilicates, and ice-rafted rock material, Bjørn Andersen and Harold Borns conclude:

> The general climatic trend observed on land has been verified in the deep-sea records, which show a general drop in temperature from the middle Eocene, 45–50 million years ago, to the late Cenozoic Ice Age period, which "started" about 2.5 million years ago. The early and middle Eocene climate was very warm, and no glaciers existed even in the high-latitude polar regions. The latitudinal temperature gradient, the drop in temperature from the poles toward the equator, was low, as was the vertical temperature gradient of the oceans. The cooling of the climate following the middle Eocene time was accompanied by an increased oceanographic circulation. Both the latitudinal and the vertical temperature gradients gradually increased, and they seem to have increased faster (in steps) during certain intervals. A first prominent step occurred near the Eocene–Oligocene transition, and it corresponds with the first known formation of glaciers in Antarctica. Other steps seem to have occurred about 15, 10, 5, 2.5 and 0.9 million years ago.[53]

Even when we get to *the* Ice Age it's not the Ice Age I grew up with. In that Ice Age there were four massive movements, four great transgressions of ice. Those in North America were called the Nebraskan, Kansan, Illinoian, and Wisconsin glaciations. In between were balmy interregnums, the interglacials. The Aftonian Interglacial came between the Nebraskan and the Kansan glaciations, the Yarmouth Interglacial between the Kansan and the Illinoian glaciations, and the Sangamon Interglacial between the Illinoian and the Wisconsin glaciations. The big problem was correlating these with the Günz, Mindel, Riss, and Wurm glaciations of the Alps, and the Menap, Elster, Saale, and Weichsel glaciations of the rest of Europe. It was a nominal problem. When you wanted to talk about global events, you hooked the local names up into portmanteau like the Weichsel/Wisconsin Glacial or the Eemian/Sangamon Interglacial.

In the Ice Age that's coming into focus these days there were more like *40* glaciations. There was an incessant to-ing and fro-ing of ice, *incessant*. Names no longer cut it and more and more often marine isotope numbers are used. The Eemian/Sangamon Interglacial, for example, increasingly is referred to as . . . *5e*.[54] Here: I'm looking at a complicated figure showing, among other things, fluctuations in deep-sea oxygen isotope ratios over the past 3 million years. During the past 2 million years I count

37 glaciations of smoothly increasing magnitude.[55] I flip to another figure. This is based on cores extracted from the Rhine Delta, which, since it
subsided throughout the Cenozoic, preserves an almost unbroken sedimentary record. It lets me zoom in, as it were, on the first figure. Where
on the first figure there was a single advance and retreat, say of the
Eburonian, I can see that here there were four.[56] I mean: there was ice
coming and going *all* the time, it wasn't four great lumbering transgressions (rapid enough in geological terms) but something more like pulses,
really swift pulses—kicks—with really rapid onsets: zap! zap! zap! zap! only
as the Ice Age progressed, the amplitude of the zaps increased so there
was more ice and it was around longer each time.[57]

The effects were most pronounced near the poles (where there was
nothing to see but ice), and at higher altitudes at lower latitudes (say, in
the Alps where peaks toothpicked above the expanses of ice), but the
pulses would have been felt in every part of the globe. For one thing, each
crystal of ice meant one less drop of water in the ocean, so as the ice
sheets thickened, the sea levels dropped. This made coastlands of continental shelves. It made peninsulas of islands. It all but made it one great
continent again, periodically, once the Americas got close enough, weird,
unwieldy, and transitory though it may have been.

It's a matter of perspective. Afeurasiamerica was whole long enough
for us and a whole lot of other biota to spread from Eurasia to the Americas; for that matter, to spread out over the exposed Sunda Shelf to what
today are Borneo, Sumatra, Java, and Bali; soon enough to swarm over
the Wallacea to Sahulland, to what today, that is, are the individual islands
of New Guinea and Tasmania and the continent of Australia.[58]

To think about these lands as "bridges" cheats our understanding. It
minimizes their character as land. You imagine they're going to be inun-
dated any second, it makes them fragile, fleeting, it turns them into . . .
barrier islands. You get the impression humans had to . . . *slip* across. And,
okay, they weren't cratons, but Beringia, for instance, "was an enormous
continental area extending nearly 1,500 km from its southern extremity,
now the eastern Aleutians, to its northern margin in the Arctic Ocean. It
was an area that could accommodate many permanent residents, human
and animal, and it endured for a longer time [c. 70,000 years] than that
documented for the entire period of human occupancy in America."[59]

But, then, the consequences of the coming and going of the ice have
always been hard to get one's head around. Here: although subtropical
water (*subtropical* now) could reach as far north during a warm interglacial
like 5e as, say, England (forest elephants, hippopotami in the Thames),
during a glacial period like the Later Weichsel, polar water (that's not tem-

perate, that's *polar* water) could tickle the coast of Portugal (tundra along the Seine).[60] With the Oceanic Polar Front way up north, and continental ice nonexistent, the climatic gradient shallows out, the Hadley and Ferrel cells spread themselves. With little moisture locked up in ice, and temperatures high enough to support high rates of evaporation (and so rainfall), things like the rain forest expand and in Africa, for instance, the Sahara all but disappears.[61] But when the ice and front migrate equatorwards, the climatic gradient steepens, Hadley and Ferrel cells get compressed, and the westerlies, the subtropical high, and the easterlies all are shifted: much of what was dry gets wet and much of what was wet dries up. With moisture drawn down into icepack, and with lower temperatures supporting lower rates of evaporation (and so rainfall), things like the rain forest shrivel, and in South America, for instance, it declines to a handful of refugia (repeatedly), to tiny pockets so isolated so long that novel butterflies and toucanets evolved.[62]

It's cold during serious glaciations. Vast swaths of land are sterilized by ice or cold. Snowlines drop as much as 3,000 feet even on tropical peaks like Kilamanjaro.[63] Even in the habitable parts of the world it's comparatively harsh. It's drier and windier. Mean annual temperatures drop $5°-15°F$ in the tropics and as much as $30°F$ nearer the ice. There's a drastic expansion of permafrost.[64] Is it necessary to add that there is much evolutionary activity of plants and animals, of warm- to cold-adapted species (the evolution of the woolly mammoth, the woolly rhinoceros), that there is wholesale migration[65] and a concomitant spate of extinctions, with most of the Villafranchian and Blancan faunas, over the last 2, last 3 million years . . . *gone*.

So, in the grip of a so-far 50-million-year-old episode of global cooling, ice volumes began rising about 2.5 million years ago in geologically brief but biospherically significant kick-ass episodes of continental glaciation, 40 of them, maybe more—zap! pow! bam! slap!

Advancing, the glaciers drew down sea levels (and so exposed and connected landmasses) as well as atmospheric moisture (and so reduced rainfall), at the same time that they sharpened climatic gradients, thus prompting the migration and/or contraction of rain forests, savannas, deserts, temperate forests, taiga, tundra, and permafrost, along with a range of other biospheric responses including extinctions, adaptations, radiations, and further migrations.

Retreating, the glaciers raised sea levels (and so buried and isolated landmasses) as well as atmospheric moisture (and so increased rainfall), at the same time that they smoothed climatic gradients, thus prompting the migration and/or expansion of rain forests, savannas, deserts, temperate forests, taiga, tundra, and permafrost, along with a range of other biospheric responses including extinctions, adaptations, radiations, and further migrations.

Basically this happened—*is happening*—every 50 thousand years. *Whew.*

The Problematic Relationship of Humans and Ice

Now there's a parallel development (*needless to say!*) among other biota (well, all biota), of what we're going to end up calling "humans." Not that there were any humans or even prehumans around in the middle of the Eocene, except to the extent that the earliest microbe and *all* its descendants through which we can trace our history are prehuman. For that matter, there weren't any in the middle of the Miocene either, not unless you're talking about *Dryopithecus* or *Proconsul*.

It seemed for a while we could be descended from either of these apes, both of whom survived the (minor) extinction event of the Middle Miocene.[66] Or maybe not. The record is spotty, though *Dryopithecus* does pass from the fossil record well before the end of the Miocene. The dryopithecines fade from the scene sort of as the ramapithecines come into focus, say between 15 and 10 million years ago, and perhaps—why not?—there was an evolution of ramapithecines from dryopithecines and of hominids from ramapithecines. . . . At least this was a plausible scenario into the 1970s.[67]

Then some hard work in molecular anthropology eased *Dryopithecus* out of the family tree[68] and turned *Ramapithecus* into the female of *Sivapithecus*,[69] and today, mindful of the threatening gap in the African fossil record between 13.5 and 5 million years ago, one "authoritative" story that's being told has it that from probably the Eocene *Adapids* (but maybe the *Omomyids*) say 45 or so million years ago, and via the early Oligocene *Aegyptopithecus* 35 million years ago, *Proconsul* evolved some 20 million years ago to diversify into many species in the Early and Middle Miocene.[70] One branch evolved some 10 or so million years ago into *Sivapithecus* and so to modern orangutans. Another branch evolved maybe 14 million years ago through *Kenyapithecus* into, on the one hand, gorillas and chimpanzees, and, on the other hand, australopithecines (fl. ~4.1 to 1.3 million years ago).[71] There were all kinds of australopithecines.

We have fossils!

Among others, of their footprints at Laetoli. This is the big deal with the australopithecines: they were bipedal. Maeve Leakey, who in 1995 found the 4.1-million-year-old hominid fossil she named *Australopithecus anamensis*, says, "You never get a novel adaptation like bipedalism without a radiation of species, a number of experiments in this adaptation,"[72] and so no surprise to find *Australopithecus anamensis* and *Australopithecus garhi*, and *Australopithecus afarensis*, and. . . .

And these days we're pretty sure we're descended from *Australopithecus afarensis* (4.0 to 3.0 million years ago), who, in one version of this story, evolved into *Australopithecus aethiopicus* (3.0 to 2.2 million years ago), *Australopithecus africanus* (3.0 to 2.5 million years ago), and *Homo habilis* (2.5 to 1.5 million years ago). In this version *Australopithecus aethiopicus* goes on to evolve into *Australopithecus boisei* (2.2 to 1.2 million years ago); *Australopithecus africanus* goes on to evolve into *Australopithecus robustus* (1.9 to 1.6 million years ago); and *Homo habilis* goes on to evolve into *Homo erectus* (1.7 to 0.1 million years ago) and *Homo sapiens*, that is . . . into us (0.5 million years ago to present).[73] Anyhow that's that version of the story. In another, *afarensis* diversifies into *boisei* and *africanus*, and *africanus* diversifies into *robustus* and the *Homo* species. In still another, *afarensis* diversifies into *africanus* and *aethiopicus*, *aethiopicus* diversifies into *robustus* and *boisei*, and *africanus* diversifies into the *Homo* species.[74]

When I say "*Homo* species" I'm referring to the sequence *Homo habilis* → *Homo erectus* → Archaic *Homo sapiens* and *Homo neanderthalensis* (or *Homo sapiens neanderthalensis*) → *Homo sapiens sapiens*. There's as much controversy about this sequence as that descending from *afarensis*—probably more. Probably it's not a sequence anyway, it's something else (a radiation), since probably *Australopithecus robustus*, *Homo habilis*, and maybe even *Homo erectus* for a little while were all around at the same time, as later were Archaic *Homo sapiens*, *Homo sapiens neanderthalensis*, and *Homo sapiens sapiens*. It was something like a radio dial thick with contemporary pop and rap but here and there an oldies station.[75]

Now as we've just seen, nothing was *set* during the evolution of any of these organisms, nothing was standing still. In swooping from the Eocene *Adapids* to Miocene *Proconsul* I swept through 25 million years. That's 25 *million* years . . . of bolide impacts and glaciations, riftings and orogenies (during which the Alps were lifted, and the Himalayas too), extinctions and radiations (*grasses!*). A *lot* happened. Even against *this* standard, the last 2.5 million years—years that saw the adaptive radiation that sifted the larger brain/smaller cheek-teeth *Homos* from the rest of the smaller brain/larger cheek-teeth *Australopithecines*[76]—were unusual. And the last 900,000 years, the years during which *Homo erectus* and progeny spread throughout the world, were especially crazy.[77]

I'm not suggesting this hadn't happened before. It had. I think all glaciations look like this up close and that glaciations are no more than phases in the planetary heartbeat that occur when supercontinent and Milankovitch cycles line up right. The glaciation of the Pongola Super-group took place 3 *billion* years ago, the Huronian 2.4 to 2.3 *billion* years ago, the Riphean 1.0 to 0.9 *billion* years ago . . . *and so on.* The thing is, this time—and see, every time the wheel turns it *re*turns to a situation that's different thanks to its last pass—there were bright ape-like creatures inhabiting the forests of Africa, and every time the pulse of climatic change was felt these apes were presented with the opportunity to preserve their coupling with the environment by drifting to the ground in the. . . .

In the what? I don't know. In the mosaic of woodlands and grass they were living in? In the savanna that pushed into the woods?

The Local Environment in Which Humans Evolved

We really don't know *what* this environment was like. Until very recently the story we told was about how apes became bipedal when they were pushed from the trees by a climatic shift in Central and East Africa that corresponds to the "step" Andersen and Borns find in their data 5.0 million years ago. The step—in fact, a deepening cascade of climatic hiccups—would have shrunk, then expanded, then shrunk, then expanded the Hadley and Ferrel cells—et cetera—which sooner or later would have whapped even equatorial ecosystems. There would have been something about it like tapping a brake.

There were other sources of dynamism too. In East Africa, "the opening of the Rift System during the later Cenozoic led to a diversification of local climates that encouraged adaptation and speciation among the higher apes," says Neil Roberts.

West of the rifts, rainfall was high enough to support tropical moist forest, but to the east, and in the rifts themselves, the climate became drier and more open, savanna vegetation came to dominate. Whether a move down from the trees encouraged bipedalism and other human adaptive traits is hard to know. However, it may be significant that Plio-Pleistocene hominids and modern chimpanzees and gorillas have disjunct distributions in tropical Africa; the apes are found only in moist forests and adjacent woodlands whereas the hominids lived in the savanna lands within and to the east of the rift system.[78]

Peter deMenocal and Jan Bloemendal also note that East Africa experienced significant tectonic uplift during this period, but that "although this would certainly contribute to a gradual cooling and drying during the Plio-Pleistocene, many of the paleoclimatic and fossil faunal records indicate relatively rapid changes that would presumably exclude tectonic mechanisms."[79]

Yeah, like the effects of rapid-onset glaciation farther north.

"We are accustomed to think of tropical rain forests as the stable product of slow evolution over hundreds of millions of years. They are not!" Raup points out.[80] Neither are the semievergreen, deciduous forests and savannas that trail and finger them, each variously expanding and contracting as rainfall slips below or climbs above the magic 1,500-millimeter bar. Rain forest is enormously dynamic, no matter how ageless it might seem to a human dwarfed by the buttress folds of, say, a giant kapok tree, and variations in soils and topography only complicate the dynamism.[81] Every time the rain forest contracted—and again, it *doesn't* really contract: individual trees *die*, it's like miserable, you can't stay living there any more than you could in the South Bronx when it was burning—the predecessors of *Australopithecus* were presented with the opportunity to retain their coupling with the environment by . . . *drifting to the ground*, slipping into the grasses, going bipedal (so they could scope out the scene like very tall prairie dogs—like very tall squirrels—but unlike prairie dogs keep on moving).

Yes, well, maybe.

Except that Craig Stanford says, "The evidence for bipedalism arising in a fairly treeless savanna niche is, for example, considered tenuous."[82] Well, more precisely, is considered tenuous by some at the moment. Stanford says this because Andrew Hill and J. D. Kingston and their colleagues can't find any evidence of a change in vegetation in their analysis of carbon isotope ratios from the Tugen Hills. "Isotopic data do not indicate any abrupt vegetational shifts correlating to global climatic shifts, such as the Messinian salinity crisis at 5.5 to 4.8 Ma or putative cooling at about 2.4 Ma,"[83] which amounts to the claim that East Africa was significantly buffered, at least at this altitude, from the changes induced by higher latitude glaciation. Their data "document a complex vegetational mosaic of C3 and C4 plants over the past 15.3 Ma," that is, that Pliocene hominid habitats in East Africa likely were woodland or a woodland/grassland mosaic rather than a savanna with only scattered trees; and they conclude that "while the course of human evolution was surely effected by environmental change, our data suggest that interpretations of the origins of hominids in East Africa during the late Miocene should be considered within the context of a heterogeneous mosaic of environments rather than an abrupt replacement of rain forest by grassland and woodland biomes."[84]

So the apes descend from the trees without being pushed by drastic changes in the climate. Sure. *Why not?* Margulis and Sagan are pretty circumspect about it:

> At various times and places when nuts and fruit were in short supply our ancestors descended to the ground. Keeping guard over the tall grass required upright posture and fast glances in all directions. Baboons do this today, after which they quickly return to a crouched posture. Such reconnaissance was strongly rewarded: those animals with their heads up ultimately freed their hands to dig for roots, to throw rocks and wield sticks as weapons, to build and explore.[85]

There's not even any need to imagine a short supply of fruit and nuts. If our ancestors were anything like contemporary apes, they spent plenty of time on the ground as it was, even in the rain forests, which, incidentally—what with large tree-fall gaps in states of arrested succession, wild variations due to differences (often minute) in the type of soil and the height above the water table, and the effects of breaks due to streams and topography—are far from uniform stands of trees in any case.[86] Chimps can knuckle walk miles a day in search of fruit and meat and over the years (*over the millennia*) a similarly "arboreal" ancestor became arboreal–terrestrial, then terrestrial–arboreal, and ultimately terrestrial, pushed, pulled, or moved on his own. Ian Tattersall says:

> Though robust, *A. afarensis* was small-bodied and, being bipedal, it wasn't very fast. Presumably, then, this hominid was pretty vulnerable to open-country predators, and as a reasonably accomplished climber it would hardly have refrained from using trees for shelter, particularly at night. . . . On balance a behavior pattern that combined its climbing abilities with its newfound bipedal capacity seems probable.[87]

Something like that.[88] Look: the pulse of glaciation *was* felt down here in the tropics. Annual average temperatures dropped, the snowline dropped, things got rearranged, but it wasn't like having a wall of ice scrape everything off into a moraine, it wasn't like the wall of death surrounding a glacier, it wasn't even like along the coasts where marine environments were alternately desiccated and drowned. There was change but it was . . . *eminently manageable*, the kind your friends try to convince you is an opportunity instead of a loss.

You know: Stand up! Suck in your gut! Get on with it!

Which is what *A. afarensis* did. That volcano was probably just the final straw.

Autonomous Organisms Can Move without Being Pushed

See. I just fell into it myself.

—*What?*

That trap. The idea that there had to be *something* pushing them, like they were rocks, and could only move if gravity pulled them downslope or a glacier picked them up or a kid did or a stream.

You know . . . *inertia.*

Maybe, *maybe* the volcano had always been there, that is, as far as the kid and his parents were concerned, assuming, of course, it *was* a kid and his parents leaving those prints in that fine layer of volcanic ash . . . *always been there.*

It doesn't take long for something to have always been there. If it were there for 4 or 5 years while we were growing up, it was *always* there. When it moves, when it closes, we mourn its loss as hard as the loss of anything: when it comes down in that storm while we were off at basic training, or while we were away at college, or doing those 8–10 months on state. You hear people mourning all the time: the tree in the backyard, the neighborhood grocery, the art moderne office building that used to be downtown.

Change is a matter of perspective. . . .

We think: well, there was all this uplift and it got drier and trees died and there was more grass—something like this lurks at the heart of every story we tell about it—and I guess we conceptualize coming down from the trees like we're working at a table in the library and a couple of a business majors come along and start working through a case study in high whisper, and we can't stand it—finally—any longer and get our stuff together and move.

Okay: two thoughts.

First, it's not like that most of the time. Yes, yes, an earthquake comes and knocks down your house; or ash buries your family and friends and fields. . . . Hey! Maybe you're out of there. But the kind of environmental changes that most imagine force speciation aren't like that, they're incremental, certainly on the scale of anyone's lifetime, maybe *invisible*, certainly invisible as such. I mean, *how many environmental changes to which you're completely oblivious are you living through at this very moment?* Here: I'm looking at a clipping from this morning's paper. It's headlined, "Ecological Changes Spark Rise of Red Maples, Threatening Oaks." The subhead reads, "This Transformation of Eastern Forests Is Related to Suppression of Fire and Disturbance of Earth." It says:

The forests of the Eastern United States are turning increasingly red, and the growing brilliance of color signals a historic change in the ecological character of a vast region stretching from the Great Lakes to the Atlantic and from Canada well into the South.

Eastern deciduous woodlands are famous, of course, for their bright fall yellows, oranges and russets. Now red is coming to the fore.

The soft green springtime hues of hardwoods like oaks and hickories, and the darker green of northern conifers like pines are being replaced by the blazing red buds, flowers and fruits of another, more adaptable and aggressive species: the red maple.

Its fruits rain down profusely in the spring, producing new little trees whose early start gives them a competitive advantage over other hardwood species, which do not drop their seeds until the fall.

If the new-found dominance of red maples continues, it could signal the downfall of the majestic oaks that have been a mainstay of the deciduous forest for most of the last 10,000 years. A wide range of creatures adapted to the oak–hickory habitat could suffer as a result.

The rise of the red maple is part of a larger, continuing transformation of the Eastern forest. The causes, all related to humans' impact on the forest ecosystem include:

First, people have suppressed fire. Before European settlers, forest fires routinely killed the thin-barked red maples while oaks and hickories were protected by their thicker bark. Fire also created open spaces ideal for the oaks' light-loving seedlings. Now these spaces are closing up, and red maples thrive in shade.

Second, red maples spread like weeds onto disturbed ground, and people have disturbed the forest, chopping it up into woodland fragments and cleared fields.[89]

This is a big change. It's like the ones on all those maps they make to suggest the impact of global warming or global cooling. It's a big deal. . . .

Have you felt it?

I mean, you've lived through this. *You're living through it right now.*

This is what living through significant ecosystemic change feels like. It feels like ordinary life. (Because it *is* ordinary life: things didn't stop changing just because we stepped onto the scene.) And what I'm getting around to is that this had to be how vegetative change would have been experienced by our prehominid ancestors, vegetative change and uplift and the opening of the Rift and the cooling, like ordinary life.

But—and here's my second thought about what we do when business majors in high whisper disturb our peace—even when the change to the land is catastrophic, we don't necessarily move. Stay with the library for a second. "Uh, could you please keep it down?" is a common adaptive strategy. Getting the librarian to move the business majors is another. Fighting wouldn't be totally out of the question. And if it came to it, you could put up with the noise. I mean, let's face it, it's not the noise itself that's—finally—driving you nuts. Maybe *you* can't work with Metallica on in the background, but you probably can beside a waterfall, and it's a lot noisier.[90] What's really getting you has more to do with injustice and the intrusiveness of the business majors and your attention and "couldn't they see you were working?" and. . . .

Not to put it too finely, what you do is up to you.

What I'm getting at is that while the invasion of the business majors may *trigger* a response, it doesn't say what it's going to be, it doesn't determine it. It's not like business majors sit down → we move. It's not even that business majors sit down → we get irritated. We could be ripe for distraction. Their case study could be interesting. They could be attractive. We could butt in, ask about the case study or for a date. What their presence triggers in us will depend on what we call "how we feel," that is, on the sum of things like the internal conceptions we have of ourselves and others and so on and . . . our verbal facility et cetera and . . . when we last ate ad nauseum and . . . the levels of our various hormones and so forth . . . and how sore we are from sitting, gravity, that is, and. . . .

Taken together, this "how we feel," this "condition our condition is in," is the way we have of referring to the state . . . *our structure is in*. It's our structure that conditions the course of our interactions (and the structural changes that interactions may trigger in it [i.e., in us]); and because it is, this structure functions as a buffer, as a clutch (what we engage when coming to land we run to keep from falling [or when business majors invade our space]). This structure is fundamentally metabolic (it's oriented toward maintaining its metabolic integrity [toward *keeping us going*]), but in multicellular organisms with elaborate nervous systems like ours (among others), this metabolic imperative expresses itself at each level of integration in ways specific to the level at which it's integrated. That is, whatever the structural coupling proper to each of its cells (busily maintaining the ongoingness of *their* individual lives in the various niches comprising the larger body), each multicellular organism also maintains a structural coupling and ontogeny adequate to its structure as a composite unity.

We've been here before. When we floated up from those ions of sodium and calcium streaming across the membranes of your trillions of cells to your pushing the cart up and down the aisles of the grocery store shopping for All-Bran, we switched from talking about a first-order autopoietic system (a cell) to talking about a second-order autopoietic system (a multicellular organism—me, you). What I want to stress here is the way your plucking All-Bran from the shelf—or later, having drifted, Barbara's Shredded Oats—is determined by *your structure*, not by shelf space, packaging, advertising, et cetera, which may trigger but can never instruct your behavior.

That is, at every level, at that of the cell, at that of the organism, at that of any higher order autopoietic unities into which the organism might enter, it is *its structure* that determines the interactions into which it will enter, not the environmental trigger. "Ontogeny," say Maturana and Varela,

> is the history of structural changes in a particular living being. In this history each living being begins with an initial structure. This structure conditions the course of its interactions and restricts the structural changes that the interactions may trigger in it. At the same time, it is born into a particular place, in a medium that constitutes the ambience in which it emerges and in which it interacts. This ambience appears to have a structural dynamics of its own, *operationally distinct* from the living being. This is a crucial point. As observers, we have distinguished the living system as a unity from its background and have characterized it as a definite organization.[91] We have thus distinguished two structures that are going to be considered operationally independent of each other: living being and environment. Between them there is a necessary structural congruence (or the unity disappears). In the interactions between the living being and the environment within this structural congruence, the perturbations of the environment do not determine what happens to the living being; rather it is the structure of the living being that determines what changes occur in it. This interaction is not instructive, for it does not determine what its effects are going to be. Therefore, we have used the expression "to trigger" an effect. In this way we refer to the fact that the changes that result from the interaction between the living being and its environment are brought about by the disturbing agent but *determined by the structure of the disturbed system.*[92]

Nothing about this is even unique to living beings. It's true for all structured systems—which is to say it's true . . . for everything, everything with which we can interact.[93] It's true for computers, it's true for cars. The illusion that we're instructing them when we turn the ignition key or press the gas pedal can be sustained only as long as—as we choose to put it—they're working. When they're not, when the battery's dead or the gas tank's empty or a linkage is broken—that is, when the structure of the car is incongruent with our "instructions"—then it becomes all too clear that it really *is* the structure of the disturbed system, *and the structure of the disturbed system alone*, that *determines* what changes in state it will make.

By turning the key we *perturb* the state of the car, and this *can* trigger the car into running, but as the jerks who won't stop cranking the starter prove, running is really *up to the car*.[94] This is true for the environment too. We can perturb it, but we can't instruct it, not tell it what to do. It's true for every structured system, that is, every entity, including the apes in the trees and us at work in the library. *It's our structure that determines what course of action we will follow when a trigger has been pulled.*

There are constraints. When the guard comes through at closing time shooing everybody out, sure, it's still up to you, *but the bar is higher.* When the air conditioning goes out, *the bar's in a different place.* And if someone blows the building up or an earthquake knocks it down. . . .

The Turnover Pulse Hypothesis

So, thinking again about our australopithecine ancestors coming down from the trees, in the first place they wouldn't have been any more aware of the kinds of changes—high-latitude glaciation, the opening of the Rift, regional uplift, creeping savanna—to which we typically attribute changes in their behavior than *we're* aware of the changes triggering changes in our behavior. (*Oh, yes!* You think *we're* immune?) Whatever it was would have felt a lot like the "rise of red maples" feels to us today.[95] And if that's the way these changes were experienced, as we experience maple seedlings sprouting in the lawn in the spring, why not imagine that our australopithecine ancestors made the changes they made—for example, toward terrestrialism—in response to triggers continuously pulled by no more than the fluctuating mosaic of riverine gallery forest and more open savanna they'd been living in all along (as subject to annual as longer term flux) *without* invoking the forcing effects of climatic change. Since any changes triggered in our australopithecine ancestors could only have been those permitted by their structure, the changes bespeak plenty of *pre*adaptation. Meaning that these ancestors had to have things like hands, adapted to their life in trees, that once they stood up could be used, could become useful, in other ways. Obviously they had the brain circuitry to work the hands already, and. . . .

One of the things at stake here is something called the "turnover pulse hypothesis." Hill traces its origins to the early work of Croll and others relating orbital irregularities to climatic change,[96] work Alfred Russell Wallace picked up on to relate to rates of faunal change.[97] Darwin himself suggested that bipedalism might have arisen from "some change in the surrounding condition,"[98] so we're not talking new or marginal ideas here. More recently it was C. K. Brain who raised the possibility that hominids arose as a response to a Pliocene drop in temperatures.[99] Today the degree to which extrinsic factors influence the pattern of change in vertebrate and hominid evolution remains a major issue. Hill says:

These ideas, which incorporate a strongly environmental approach to faunal change, with respect not only to hominids but to terrestrial fauna generally, have been made explicit and refined by Vbra in her turnover pulse hypothesis and its variants. In its simplest and most original form it suggests that "speciations and extinctions across diverse lineages should occur as concerted pulses in predictable synchrony with changes in the physical environment, chiefly in global temperature." Although hers is a general theory, one important and prominent example focused on the apparent radiation of hominids. . . . [100]

It makes a great story, doesn't it? There's a high-latitude glacial pulse, the rain forest contracts, the apes drop from the trees, they stand up: pulse, contract, drop, stand; pulse, contract, drop, stand . . . until one of these transitions takes and . . . *humans!!*

Except as we've seen, (1) there's no real evidence for it and maybe even some against it[101]; and (2) it ignores the . . . yes, the *autonomy* that autopoiesis grants what in the end, call them apes though we will, are really no more than gigantic networks of self-referencing, autocatalyzing, macromolecular reactions producing the very molecules needed to make the reactions run so they can run again. It *is* the whole point of the membrane isn't it—without which there would be no life as we know it—that it isolates and insulates the reactions taking place inside from the flux of matter and energy in the field outside? And in big metacellulars like our australopithecine ancestors this original membrane had been wrapped by membrane outside of membrane outside of membrane, the outermost of which was adaptive primate culture. They had many ways of responding to the trigger—if there was one—of a contracting rain forest.

One of which *might have been* standing up.

Making a Niche

One of which *was* standing up because they *did* stand up, some of them anyway. What we don't have (maybe can't have) is the data to attribute this to a change in climate. At the end of a review of suid and hominid origins, Tim White says:

The hominid fossil record is not complete enough to test any hypothesis linking hominid origins to global climate change (even if anthropologists could decide what they meant by the phrase origins of the family). Biochemically based divergence estimates between living forms are poorly time-constrained. In short, we have no data of reliable precision to indicate when hominids arose relative to global climatic change.[102]

Which is cool, no data either way, no data capable of saying hominids arose *thanks* to a change in the climate or *independent* of it, though I'll tell you what it really is, it's ... I don't even like the *hypothesis*. It betrays ... I don't know ... a significant failure to appreciate what it is that makes life-forms different from rocks. It's not, dude, like ... their materiality. *Their materiality goes without saying.* It's the structural dynamics of the matter in question to produce—down elaborate pathways of self-referencing, auto-catalyzing, macromolecular reactions—the very molecules needed to make the reactions run so they can run again.

This is it.

Now it's not that this gives them free will, neither the simplest chains of reactions nor early life-forms nor our australopithecine ancestors nor us. That's not it. That's not it at all. Free will is a red herring. I'm not even sure what it's supposed to be.

And it's certainly not that this makes them independent of the environment, any of them.

No, it's the fact that the circular organization of the reactions obligates them to assume an orientation toward the world that is neither unfocused nor inert nor reactionary. Because the chains of reactions require the participation of only certain ions (like calcium and sodium) and have to be isolated from others (like cesium and lithium), they end up shaping their immediate environment into a niche—that is, into a phase of the environment that is structurally coupled with them—or they lose their coupling and cease to exist. What this does is *corrode* the distinction between the organism and the environment, and it does this at each level at which the life-forms are integrated: individual cells in their individual environments, the multicellular organism as a whole in its. The structural coupling of organism and environment *in the niche* makes hypotheses like the turnover pulse hypothesis hard to frame.

It's not that I doubt it's all run by gravity and solar insolation and the decay of the radioactive elements inside the Earth. No. It is. *No doubt.* It's just that once gravity and solar and radioactive energy operating on the Earth's substance kicked self-reproducing enzymatic chains into action—as they did ~3.8 billion years ago—they brought into being (as they had previously brought into being Hadley cells and the oceanic circulation), a novel phenomenology, one we recognize as biological. Maturana and Varela put it like this:

> The emergence of autopoietic unities on the face of the Earth is a land-mark in the history of our solar system. We have to understand this well. The formation of a unity always determines a number of phenomena associated with the features that define it; we may thus say that each class

of unities specifies a particular phenomenology. Thus, autopoietic unities specify biological phenomenology as the phenomenology proper of those unities with features distinct from physical phenomenology. This is so, *not because autopoietic unities go against any aspect of physical phenomenology*—since their molecular components must fulfill all physical laws—but because the phenomena they generate in functioning as autopoietic unities *depend on their organization and the way this organization comes about*, and not on the physical nature of their components (which only determine their space of existence).[103]

Among the phenomena this autopoietic organization brings into being is the niche; and with respect to the construction of this niche, that is, with respect to the organismic shaping of the world, there is no difference between the simplest protometabolic chain of reactions and the great specialized elaborations of integrated interacting chains of chains like us. Or rather there is a difference and it's quantitative but not simple, for though we may be active, we're not numerous (not compared to bacteria), and so we have yet to attain the ability to affect the atmosphere that, say, early oxygen-generating bacteria had (though we're working on it).

See. *They* weren't rocks. A rock may oxidize but it doesn't go actively looking for oxygen. Organisms do. Organisms select their environments. Organisms modify their environments. Organisms determine which aspects of their environment are relevant and which environmental variations can be lumped or ignored. *They* do it. They do it themselves. Organisms don't just react, they respond to their environments actively. Organisms shape the very form of environmental signals.[104] This is why the oaks are disappearing, because in shaping *our* niche we clear forest for subdivisions (okay, okay: and we clear forest for fields too and all the rest of it) and we're careful about fire; and because in shaping *their* niche red maples seed lots of trees in the spring. Though the oaks could surprise us. . . .

And if you say, "But they don't know what they're doing," I need to ask, "*We* do?"

Baby, we don't have a clue. We're just doing our thing.

Choosing to Come Down from the Trees

I know you're going to think I'm nuts but I gotta go through this one more time. *There's so much at stake.* Beyond the mutation, adaptation, selection, and speciation that led from our australopithecine ancestors to *Australopithecus afarensis* (and so to us) lie deeper issues of causation and change. Beyond them lies . . . what?

I'm not sure.

Maybe it's that I'm trying to understand the dimensions of my own autonomy. I've never been convinced by those who justify their actions—say, evicting tenants or failing students or firing employees (or going to war)—by saying "I was forced to," "I had no option," "My hands were tied." On the face of it, it's not so simple. It's not that I'm unaware of my own continuous adaptation to changes in the world around me, but I resist the idea that behavior, and so change, is extrinsically driven. *Shaped, prompted, suggested, triggered?* Sure. *Driven?* Not on your life.

In saying this I'm not thinking of denying extrinsic reality, its changes, and the necessity I have to remain structurally coupled to it. What it's to think is of dancing instead of being bullied.

Yeah, this is good. Look, what the case usually is, I think, is that we see the organism as a little kid cowering at the feet of a looming environmental bully. The kid has his hands up to shield his face. The bully has his fist cocked. I think this image stands behind most models of organism–environment relations. *The environment pushes the organism around.*

Okay, here's another: Fred Astaire and Ginger Rogers dancing to "Let's Face the Music and Dance." It could be "Cheek to Cheek" or "Isn't This a Lovely Day," but the words "There may be trouble ahead, but while there's moonlight and music and love and romance, let's face the music and dance" have a resonance in this context the others don't.[105] Anyhow, Fred's leading and Ginger's following, he's the environment, she's the organism, and the thing is . . . *she doesn't have to follow his lead!*

I mean all he can do, with the frame he's making with his hands and arms and the momentum of his body (and his choreography and the music) is trigger her response. And most of the time, because she wants to retain her coupling, she goes with him. But there is a moment in "Let's Face the Music and Dance" when, having circled the stage, turning first one shoulder and then the other toward each other, that Ginger continues in a series of *chaîné* turns, her hands uplifted, and Fred follows *her* lead with his arms encircling her waist.[106] There are actually a lot of moments like that, and because there are, the dancing of Fred and Ginger is a really good way to think about the relationship of organisms and their environments.

That said, and the objections of Hill to the shrinking of the forest being noted, listen to John and Mary Gribbin tell the story of how our australopithecine ancestors came down from the trees:

> How does a tree-dwelling primate adapt when the forests in which he lives begin to shrink? There are two obvious solutions. Either you retreat into the heart of the forest, and carry on your business of finding

food to eat and a safe place to sleep as before, or you step out into the new surroundings and try to make a living in a new way that suits the changing world. Without the individuals concerned making any conscious choice along these lines, this is the way that evolutionary pressures work to select new varieties from a common ancestral stock at times of environmental change.

There is no need, we are sure, to labor the point—you see what we're driving at. When the forests shrank, some apes became more "ape-like," sticking with the trees and, if anything, becoming more efficient at finding resources there in the face of increasing competition for dwindling resources. Their descendants are the gorillas. Other apes, originally members of the same species, scrambled a living in the more open woodlands nearby, still among trees (although not thick forest), but forced to find food on the ground as well. As evolution fitted them better for this lifestyle, they became chimps. And a third branch of the family, perhaps descended from the individuals who were *least* adept at the old lifestyle and were pushed out on to the plains by competition from their cousins, had to find a completely new way to live, or die. They developed upright walking, were forced to eat almost anything that came to hand, and learned the value of sharing their food amongst a family or larger group. Eventually, they developed large brains, and became humans.

All this must have been a gradual process. Stating it so baldly makes it sound like an overnight transformation—with one bound the primitive ape leaped out of the forest, stood upright and became human. In fact, the adaptation is gradual—unnoticeable from one generation to the next, but building up over hundreds of thousands of generations, and millions of years of time. The apes that stayed "apes" need not actually have retreated into the shrinking jungle—they were probably the ones that happened to live in the bits of jungle that were still there after the environment changed (and remember, this time the climate changed *slowly*). The apes that "became human" did not physically move out of the jungle, but over a long period of time the climate in the valleys where particular groups of apes lived dried out and the trees slowly disappeared. In each generation, individuals that learned to cope with the changing conditions a little better than their contemporaries would be the ones that found the most food and had the best chance of rearing off-spring. If upright walking, for example, made it easier to find food and carry it back to the family, then natural selection would favor upright walkers, and the characteristic bipedalism of the human ape would evolve.

One interesting piece of corroborative evidence, that it was the changing climate that made humans out of apes, comes from the monkey line. . . . [107]

Though actually what putting everything this baldly makes patent is that the climate didn't make anything out of anything. Look, the environment, here reduced to the climate, is Fred, is leading. The apes are Ginger, are following. They dance. Even though Fred only makes one set of moves, the dances turn out entirely differently, this one staying up in the trees; this one up and down, up and down; this one on the ground.

What's up with this?

What's up is that the climate could tickle the apes, but could not tell the apes what do; and what this means is that "evolutionary pressures," "select," "in the face of increasing competition," "forced to find food on the ground," "pushed out to the plains," "competition," "forced to eat almost anything," "learned to cope," and "natural selection would favor" are just metaphors too, and not very good ones.

Look, pressure's real, but unless we're talking about "the application of continuous force by one body upon another that it is touching, compression"—and remember the apes are experiencing this climatic change (if there was one) as we are experiencing the rise of red maples—"pressure" is a metaphor masking our ignorance, our complete inability to conceptualize the relationship of ontogenic to phylogenic change.[108]

That's a sort of time-framed way of thinking about it. A more space-framed way would be to say that we really don't have a handle on the relationship of the individual to the whole.

And we don't.[109]

The Seedpod of a Dandelion, the Immense Indifference of the Milky Way, the Bedroom Curtains Blowing in the Wind

Is this some way of insisting that we're free, free to choose, the metaphor of the dance, the talk about the trigger, the deal with the business majors in the library?

Not actually. Not really. I don't know what freedom even is, not the way we usually talk about it. The metaphors are more, well, an acknowledgment of my ignorance, or of our ignorance. They're one of the forms this ignorance takes.

For example, I don't think the apes were pushed from the trees, but talking about them *choosing* to come down from the trees is just as stupid. A big part of our problem is that while we're talking about the *apes* we're thinking about *us*.

It's so easy to say the apes were pushed from the trees, especially the way the Gribbins put it, with natural selection doing all the heavy lifting in the background, as long as you forget about us. But I can't help seeing that we're identically situated, living lives today not much different from those our ancestors led—10 years ago? 100 years ago? 1,000 years ago?—and I can't help wondering what's pushing us?

If something *is* pushing us, how can we imagine we're masters of our fate? What happens to the notion, to the sense we have, that we're in control, that we make choices, that we're autonomous, that we have anything at all to say about the way the world works, all those drivers in all those cars—

Pushed to drive, probably by their genes. I mean, if Steven Pinker is right, there must be a driving-to-work-by-yourself-in-a-Lincoln Navigator gene.

Which is nonsense, because genes are molecular descriptions of proteins inscribed in the course of *organism–environment interactions*—

"Which confer some kind of benefit," some would add—some kind of reproductive advantage—but I don't know about that. Benefit to whom? Benefit to what? The cell? The individual organism? An ecosystem? Gaia? Benefits can only be calculated from some point of view, there have to be goals, some kind of teleology, and how do you decide where to stand? Besides, there are all kinds of genes that don't seem to confer a whole lot of what we usually think about as benefits on the individuals carrying them, recessive—

And this is where people often introduce the term "defective," but again, *defective*? From whose point of view? The gene's? I've never understood how we could talk about "defective genes" when at the same time we talk about "random mutation" as the source of evolutionary variation. Except for "defective genes" we'd all be anaerobic prokaryotes, we'd all be chemical heterotrophs. We're mutants, freaks, we're eukaryotic freaks. I mean, from the perspective of a chemical heterotroph, "normal" is a single-celled scavenger of environmental sources of ATP, which would make us, even at the cellular level, incredibly freakish. Do you see how this language traps us? Defective! Benefits! It's like hurricanes *helping* and *hurting* the environment, it's garbage talk.

Which is what *pushing* and *choosing* are: garbage talk.

Push, choose: things just don't work that way. It comes to saying either that it's something *outside* of things that acts on them, pushing or pulling or forcing them, gravity or the mob or inflation; or that the force—and here we are again with that idea, no motion without a force—or that the force comes from somewhere within.

And perhaps it seems there's no other option, that this inside/outside talk follows from the development of the membrane long since cited as the mother of all alienations. It's easy to stir the bitterness in the cup I mixed for myself, but I just, I just don't know—

It's too simple. The opposition denies the reality of the coconstruction of the inside and the outside, of their cooperation . . . their structural congruence, in Maturana and Varela's terms.

What if we were to rethink the membrane? Obviously the membrane was thrown up by the outside, that is, by what comes to self-define itself as the outside in the act of throwing up the membrane. That is, the membrane comes into being in what *only after the membrane comes into being* we'll think about as the outside. Before that it's all one. The membrane, and so also the inside it brings into being, is a piece of the outside.

The inside comes from the outside. In fact, *where else?* And what does the membrane wrap up? It wraps up stuff that used also to be outside—

Which, exploiting the membrane, then increasingly differentiates itself from the outside—

But still: it's the outside wrapping up the outside, the outside wrapping the outside up . . . so that the outside can differentiate itself. Though only in terms of its organization. Structurally it's the same stuff. And furthermore the differentiation of the inside from the outside that then ensues occurs only through the connivance of the outside which continues to support it, that is, that tolerates it in the first place, and then nurtures it, and now I'm thinking about the niche that is also a kind of membrane.

We tend to think about the niche as something organized by the organism—turf management, nest building—but it's also a supportive environmental concentration of organismically relevant stuff. It might be useful to think about these concentrations as arising from the interaction of, at one level, extrinsic chemical gradients with respect to which the organisms are opportunists. This would make the niche an environmental given, a gift—

Though "gift" is just as bad—

Though it may be closer to what's going on than thinking about the organism and the environment as opposed in any way.

The niche as a gift *is* garbage talk, but it makes me feel better. Where I'm trying to get to is a way of thinking about the environment that . . . wraps itself up inside itself (and *that* would constitute an organism); but at the same time a way of thinking about the organism as something that . . . turns itself inside out (and *that* would constitute the environment); so that the interpenetration that is so emphatically a reality. . . .

What, if in everything I've said above, I switched every use of the term "outside" to the term "environment"? So we'd have "the inside comes from the environment," but also "it's the environment wrapping the environment up," and the—wait, wait! Wow! We could also replace "inside" with "organism." That would give us . . . "an organism is a piece of the environment that the environment's wrapped up"—

And also, "The membrane—and so also the organism it brings into being—is a piece of the environment," and "it's the environment wrapping the environment up so that the environment can differentiate itself," and "the differentiation of the organism from the environment that ensues occurs only through the connivance of the environment which continues to support the organism."

This is something like it! It's like what Maurice Sendak had Mickey say at the end of *In the Night Kitchen*, "I'm in the milk and the milk's in me. God Bless Milk and God Bless Me."[110] Because we can just as easily replace "organisms" with "humans," and "environment" with "land," to get "the human comes from the land," "humans are pieces of the land the land's wrapped up," "the membrane, and so also the human it brings into being, is a piece of the land," "it's the land wrapping the land up so that the land can differentiate itself," and "the differentiation from the land of the human that ensues occurs only through the connivance of the land that continues to support the human."

We are the land and the land is us.

Think about it. I don't know what else you're doing now besides reading this, but if there's not a cup of coffee beside you or a soft drink, there was or there soon will be. That sugar, it had to get *into* your coffee somehow, into your soda, the land somehow had to rise up and pour itself—thanks to container freight and supermarkets—*down your throat.* Isn't that *in effect* what happens? Thanks to container freight and supermarkets via the Big Bang, gravity and the electromagnetic force, the accretion of the terrestrial sphere, cratonization, the spread of cyanobacteria, its differentiation into . . . bushes of coffee, *Saccharum officinarum* . . . you, me, us.

But then the apes too are environment, pushing themselves down from the environment. The apes too are wrapped up pieces of the land, including climate and all the rest of it, and so it's, like, the land, I don't know, *breathing?*

The apes coming down from the trees was the land breathing? Is this too . . . *insane?*

I'm struggling with this. We *too* are the land breathing, and our agonies and indecisions and our sense that we're making a decision, and the decisions we make, it's all the breath of the land, slipping through the land's mouth, the land's teeth and tongue and palate shaping. . . .

We are the breath.

The breath of the land. We're already an eddy in the great currents, a lake in the Hypersea, a link in the carbon chain—now we're the breath of the land.

It's not too much. We *are* the breath of the land, and all our aspirations and the rest of it are the form the breath takes. Maybe it *is* more garbage talk, but it's more interesting anyway, which is the point, really, of all this history.

Since each turn of the screw changes the conditions on which the next turn is based, you can't *learn* from history. The most you can hope for is a loosening of the rigidities. You know: it wasn't like this once, so it doesn't have to be this way now. Open some room for maneuver.

And that too must be a form of the breath.

It may be the *essence* of the breath.

8

The Land Covers Itself with Humans

"You never get a novel adaptation like bipedalism without a radiation of species, a number of experiments in the adaptation," Maeve Leakey says, and so, as we've seen: *Australopithecus anamensis, Australopithecus afarensis, Australopithecus garhi, Australopithecus* . . . .

These days we're pretty sure we're descended from *Australopithecus afarensis* (lived 4.0 to 3.0 million years ago), who, in the version of the story I'm following, evolved, maybe into *Australopithecus garhi* (lived 2.5 million years ago)—though these fossils have just been found and exactly where they fit is unclear—but almost certainly into *Australopithecus aethiopicus* (lived 3.0 to 2.2 million years ago), *Australopithecus africanus* (lived 3.0 to 2.5 million years ago), and either *Homo habilis* (lived 2.5 to 1.5 million years ago) or *Homo rudolfensis* (lived 2.4 to 1.6 million years ago). *Homo habilis* or *Homo rudolfensis* evolved into *Homo erectus* (lived 1.9 to 0.1 million years ago), archaic *Homo sapiens* (lived 0.6 to 0.05 million years ago), *Homo sapiens Neandertalensis* (lived 0.3 to 0.03 million years ago), and *Homo sapiens sapiens* (from 0.5 million years ago to the present). These dates are subject to revision with the turn of a trowel, and the exact, even the general, relationship among the various *Homos* is almost entirely up in the air, though that there was always plenty of overlap is indisputable.[1]

Edging Over, Mile after Mile, Generation after Generation

Most of this story unfolded in Africa (though this too could change with the turn of a trowel), the early part of it most likely east of the Rift Valley on the high East African plain, where the snow lay bright on the peaks of Meru and Kilimanjaro and Kirinyaga, the game trails to the waterholes were smooth, and the ancestors of zebras and wildebeests grazed in unfathomable abundance,[2] though, as we get into it, elsewhere in Africa too, down around the Cape, and in the Ahmar Mountains of Ethiopia; but better than 1 million years ago, maybe as long as 2 million years ago (since we've uncovered a 1.8-million-year-old *erectus* fossil in . . . *Java*),[3] *Homo erectus* had begun to edge out of Africa.

Why "edge"? Because "migrate" and "move" carry too much baggage, too strongly suggest the idea of people packing up and moving—*moving out*— whereas what happened here was more a matter of . . . *edging over* as their numbers increased and ecological opportunities, responsive to more global—yes (*sigh*)—climatic changes, incited the advance (and retreat) of ice and grass and forest and associated faunas. At 20 miles a generation—which is barely moving, is really just . . . over the hill—humans could have made it from the Rift Valley to the Dordogne (in southwest France) in 10,000 years.

There was no rush. They weren't going anywhere.

And this is why I kept chewing on the apes coming down from the trees, because every time there's any change—like now, this edging out—all the usual suspects are going to line up: climate → vegetative change, over-crowding, competition. When I trot them out—and I don't want to deny their agency, I just want to complicate the chain, throw in a clutch—I want you to remember the apes coming down from the trees, want you to think about how and why you respond to stimuli, about what makes you move, about what it's like to move . . . *on your own*. Mindful of Hills's caution that isotopic data fail to record abrupt vegetational shifts in this part of the world over the past 15.3 million years,[4] I am mindful nonetheless that when the ice sheets spread themselves in the north 1.9 and 1.8 million years ago, temperatures and snowlines here *did* drop; and that there had also been a change toward larger volumes of ice, and steeper climatic fluc-tuations, ever since *Homo habilis* had shown up 600,000 years before.

As complicated as I feel about it, I still admit the salience of Joseph Campbell's observation that "noteworthy is the coincidence of periods of accelerated evolution with glacial advances," *afarensis* into *habilis* during the Donau glaciation, *habilis* into *erectus* during the Günz–Nebraskan, *erec-tus* into Archaic *sapiens* during the Riss–Illinoian, and archaic to modern *sapiens* during the Würm–Wisconsin.[5] And, okay, the timing's *not* well constrained, and we have a lot more fossils today than Campbell had (and a lot more glaciations), but still it's hard to miss the pattern not just of the anatomical changes, but of the dispersal. It's almost as if glacial pulses had *pumped* people out of Africa, *habilis* up and down the Rift Valley, not necessarily by the Donau glaciation, but by those accelerating pulses characteristic of the climate of the latter Pliocene; then *erectus* up the Nile and on around the Mediterranean and the Fertile Crescent and into India and China and Sundaland by the deepening pulses and more extensive glaciations of what we used to think about as the Günz–Nebraskan and the Mindel–Kansan glaciations; then archaic *sapiens* and Neandertals pumped into most of Africa, Spain and southern England, the Caspian region, the South China coast and Borneo by the *slap! slap! slap!* of the Riss–Illinoian cluster; and finally modern *sapiens* propelled around the world by the zingers corresponding to marine isotope stage nos. 3–6, the old Würm–Wisconsin glaciation.[6] Though we can only *guess* why *erectus* began edging out of Africa (climate → vegetative change? overcrowding? competition?), *erectus* bones began being fossilized in Southeast Asia and China a million years ago, and in Europe 300,000 years after that.

And maybe it wasn't the glaciers, though it's hard to imagine they didn't have something to do with it.

There's an ugly battle in paleontology—unusual in science only by virtue of its virulence (or maybe even the virulence has always been part of science)—about what happened next. The multiregional hypothesis, most vocally championed by Milford Wolpoff,[7] holds that modern *Homo sapiens* evolved from *Homo erectus* in situ, that is, *in* Africa and *in* Southeast Asia and *in* China and *in* Europe; that is, *African Homo erectus* evolved into modern Africans; *European Homo erectus* through (or beside) Neandertals into modern Europeans; *Asian Homo erectus* into modern Asians; and *Indonesian Homo erectus* through the Ngandong people into modern Aboriginal Australians.

In this story *humans* originated in Africa; *modern humans* all over the place.

The objection to this otherwise "unimprovably simple" story is the genetic improbability of the parallel evolution it requires.[8] Actually, the multiregionalists don't claim thoroughgoing isolation of the gene pools. They imagine that weak flows kept the developments in synch.[9]

Against this multiregional hypothesis, the out-of-Africa (or rapid-replacement) hypothesis—hypothe*ses*, actually, there are more than one of them—agrees that *Homo erectus* edged out into the rest of the world, but argues that *Homo sapiens* evolved in Africa from those *who stayed behind*.[10] It's these *sapiens*, supposed to have edged out of Africa no more than 200,000 years ago,[11] from whom all *living* humans are supposed to be descended, these having (rapidly) replaced the descendants of the original *erectus* pioneers thanks to possession of the right stuff (better tools, more language): in southwest Asia as long as 100,000 years ago; in southern, eastern, and southeastern Asia 60,000 years ago; and in Europe 60,000 to 35,000 years ago.[12]

An elaboration of this hypothesis, a *multi*migration variant, posits a fork in the African *erectus* tree, with the migrating *Homo erectus* on one tyne and a stay-at-home *Homo ergaster* on the other tyne. In this version, stay-at-home *ergaster* evolves into the archaic *Homo sapiens* who moves out of Africa after *erectus*, leaving *ergaster's* archaic *sapiens* descendants in situ to continue the evolution into true *Homo sapiens*. These true *sapiens* subsequently follow their archaic cousins in a third wave around the world, again replacing everything before them. Colin Tudge pushes even this idea and, thinking through the evolutionary and migratory history of horses, proposes a reciprocal cascade model, *waves* of migration, but with continued evolution of the migrants, who ultimately may even spread back home, and with introgression.[13] This is my favorite (*entirely unsupported*) model. Unfortunately, while "the replacement hypothesis has the advantage of not demanding a parallel evolution in many remote areas," Cavalli-Sforza says, it "does not explain claims of regional continuity outside Africa."[14]

Sigh. . . .

So, okay, we don't have the details all worked out, but if we stand back the picture is clear enough. Five or so million years ago, for whatever reason—or just for the hell of it—our australopithecine ancestors in East Africa began to come down from the trees.[15] For a long time, maybe as long as a couple million years, australopithicines retained a profound arboreal orientation (perhaps to escape predation[16]), but somewhere around 2.5 million years ago (perhaps as a consequence of a cascade of events set in motion by the onset of extensive continental glaciation[17]), they gave up on the trees altogether, to give rise to a new genus, *Homo*, whose slower maturation permitted the growth of a brain large enough to resolve the problem of predation *without* recourse to the trees.[18]

Something like that.

Tools! We have tools!

The Humans Articulated by the Land Articulate the Land into Tools

Homo was unlikely the first to use tools. Well, *Homo wasn't* the first to use tools. Gulls use them, raccoons use them, chimps certainly use them. It's hard—it's impossible—to imagine *Australopithecus* didn't use them too, and indeed there's little doubt that *Australopithecus garhi* butchered ante-lope and wild horses.[19] Still, if the big deal with *Australopithecus* was bipedalism, the big deal with *Homo* is the use of the hands *Australopithecus* freed to make (not merely find) and use tools, perhaps precisely what was needed to fend off the wily carnivores the slowly maturing—because big-brained—*Homo* infants would have attracted. It's not as fashionable as it once was to talk about australopithecine bipedalism freeing *Homo* hands whose exercise with tools elaborated existent brain circuitry which helped evolve hands better adapted to making and using tools (and so on), but undoubtedly this was part of story too.

Of course, because tools made of things like the sticks chimps use to dip for termites aren't well represented in the fossil record, we're talking about *stone* tools, and because *stone* tools, *stone* ages. Somewhere in here, precisely as geological time took over from astronomical (from astrophysi-cal) time, after the accretion of the terrestrial sphere, anthropological time takes over from geological time after the appearance of humans—why not? it's our story—though for the first 2.5 million give-or-take 10,000 years anthropology reduces to . . . *stones and bones.*

And let's admit that the Old Stone Age, that is, the Paleolithic, is all but coterminous with the Pleistocene. Running from 2.5 million to 10,000 years ago, the Paleolithic *does* go on and on (and on), but the subsequent ages we distinguish get briefer and briefer (and briefer). In northwestern Europe, probably for no good reason, we distinguish a Middle Stone Age, that is, the Mesolithic. It's allowed to run from, say, 10,000 to maybe 5,000 years ago, and so lasts no more than 5,000 years (against the 2.5 million years of the Paleolithic). It eases more or less smoothly out of the Paleolithic, though during the Mesolithic *Homo* does push its Paleolithic accomplishments into a markedly wider range of environments. Outside Europe we don't distinguish a Mesolithic. Outside Europe the Paleolithic debouches smack into a New Stone Age, the Neolithic, that runs from, say, 10,000 years ago until whenever metal replaces stone tools. This var-ies widely with location, from 8,500 years ago (in, say, Anatolia, where the Neolithic lasts no more than 2,500 years) to the present (in, say, the part of the Philippines occupied by the Tasaday where the Neolithic may be said to have lasted *at least* 10,000 years [more on this below]). In general, it's during the Neolithic that people really begin to manipulate the niches they live in, *really* begin to manipulate them; that is, that they begin the se-rious domestication of plants and animals, that is, heavy-handed niche construction.[20]

As I say, the subsequent ages get briefer and briefer. When people first smelted copper, from 8,500 years ago on (again depending on loca-tion), a Copper–Stone Age, the Chalcolithic, opens. The Chalcolithic, in many but by no means all places, fades into and inaugurates a Bronze Age. The Bronze Age is variously observed to open from 5,500 to 3,800 years ago, rolling from the eastern Mediterranean to China, and lasting until iron took over.[21] The Bronze Age then begat an Iron Age, *yes*, and the Iron Age begat an Age of Steel, and the Age of Steel begat an Atomic Age, the one we live in.

Everything rushes up when you're coming in for a landing.

The names are actually kind of silly, and they're ripe for revision. Because the Paleolithic does go on (and on [and *on*]), and because our brains are blowing up and our tools evolving so rapidly, the age does get chopped. In fact, it gets minced.[22] The Paleolithic's Early or Lower (2.5 to ± 0.2 million years ago), Middle (± 200,000 to ± 40,000 years ago), and Late or Upper (± 40,000 to ± 10,000 years ago) periods are further distinguished into phases or cultures, the Early (or Lower) Paleolithic into the Oldawan and Acheulean; the Middle into the Mousterian, Abbevillian, Clactonian, Tayacian, Micoquian, and Châtelperronian; the Late (or Upper) into the Châtelperronian, Aurignacian, Gravettian, Périgordian, Solutrean, Magdalenian, and Azilian. Yes, there's some overlap. This isn't rocket science. (Rocket science isn't even rocket science.) These are *cultural* phases now. Hell, they're *cultural* ages.

Sad, but not surprising, to say, anatomical and cultural markers don't always match up. That is, the stones don't always line up with the bones. We'd like to be able to say something like *Homo* = tool use, *H. habilis* (and/or *H. rudolfensis*) = Oldawan tool culture, *H. erectus* (and/or *H. egaster* et al.) = Acheulean tool culture, and so on; but, demonstrating how little we understand about what was going on (as well as how few bones and tools we've uncovered), it's not like that. Or it is sort of, but the edges are rough.

For example, early Oldawan technology turns out to be hard to distinguish from that of the later australopithecines, messing up the neat *Homo* = tool equation. The recent discovery of antelope and wild horse carcasses disarticulated and defleshed by *A. garhi* in Ethiopia's Awash Valley with stone tools 2.5 million years ago puts them "across the river and through the trees" from the 2.4 million-year-old Oldawan tools known from Gona, as well as from the 2.3 million-year-old chert and basalt flakes associated with a *Homo* maxilla from Makaamitalu, the oldest known association of *Homo* with stone tools. That's "across the river and through the trees" in spatial *and* temporal terms, making it more than a little likely that the behavioral changes associated with the emergence of the *Homo* clade from *Australopithecus* had a lot to do with lithic technology and enhanced carnivory.[23]

As well as ice. . . .

In East Africa, *H. habilis* (and/or *rudolfensis*) made the crude Olda-wan hammerstones, the unifacial and bifacial Oldawan choppers, and the savage Oldawan flakes for nearly a million years. That's a long time. Around 1.4 million years ago they began to make the wider range of more specialized tools we call "Acheulean" (in other parts of the world up to 500,000 years later), adding the neat bifacial Acheulean cleavers and Acheulean hand axes to the Oldawan repertoire, and marking everything they touched with a degree of standardization in form and technique lack-ing among the Oldawan.[24] The 1.3 million-year-old assemblage found along with a mandible of *Homo ergaster* in Konso-Gardula, Ethiopia, marks the onset of a technological uniformity that lasted for another million years in a region that ultimately stretched from Africa through the Near East and India and most of Europe, though it didn't reach Europe until 500,000 years ago.

It's hard to tell what we're seeing here: a global fluctuation washing through the biosphere (*humans* in motion)? or one washing through some kind of culture sphere (*flaking technology* in motion)? Of course we don't see either . . . in motion, just a few bones here and a few tools there. It's a lot like trying to reconstruct *Pulp Fiction* from a handful of stills, and most likely the Oldawan was kind of . . . *cloud of culture* . . . thrown up by *Homo habilis* (and/or *Homo rudolfensis*), one trailed by an Acheulean cloud thrown up by *Homo erectus* (and/or *egaster*), with its variants: the Abbe-villian with its deep flake scars and jagged edges (a local European vari-ant), a later Clactonian cloud (in England), a Tayacian cloud (associated with remains of *Homo heidelbergensis* in France), and finally a Micoquian cloud with its thin cross sections and its very narrow points. Though Acheulean tools have showed up elsewhere too, they never reached the Far East (where a tool tradition called the "chopper chopping-tool indus-try" dominated), either because East Asians could make better tools out of bamboo (for instance) or because what we used to call the Riss glaciation pretty effectively isolated them from developments in the rest of Eurasia. Or both.[25] But it doesn't do to forget that ice is advancing and retreating (and advancing and retreating) from poles and mountains throughout this period (*zap! pow! bam! slap!*), nor that a culture that functions in one locale with one set of resources (etc.) might not in another.[26]

Furthermore, they coexisted for 300,000–400,000 years, *Homo habilis* and *Homo erectus*, that is. That's 15,000 to 20,000 generations. Can you imagine that? *Homo habilis* with its Oldawan tools, and *erectus* with its Acheulean tools, *erectus* already in Europe, in Asia, with the cave bears and *Pelorovis* and *Megaloceros* and the mammoths, the birch and pine giving way to elm and oak, the elm and oak to alder, hazel, yew, and hornbeam, and back again, at least as the Ice Age deepened; far away—in any case—

from the land in East Africa from which *erectus* sprang, the whole thing
and all its parts in tumbling motion. . . .

So much is going on!

No organism is inert in its medium (or it's not an organism), but the
evolution of culture gave *habilis* (and/or *rudolfensis*) and *erectus* (and/or
ergaster) an ability to shape its niche that was different from that of other
animals, an ability to expand its niche until the niche became indistinguishable from . . . *the land*. Bacteria could change the environment. Bacteria *did* change the environment, radically so. They made it green in the
first place. What happened later with the oxygen was the least of it.

But somehow this human *making* of things to affect *other* things (like
weapons to kill animals), this organismic construction of a *relay* independent of the organism, this seems to be really different. Or maybe I'm just
too close to it. A big part of the story was fire, which was another part of
the whole thing that happened with this Acheulean stone industry, that
happened with *erectus*. You could beat back the cold with fire, you could
beat back predators, and break down plant and animal fibers, and burn
off toxins, and bake—*okay,* smoke cure—meat, and . . .

There's evidence of the use of fire in East Africa at 1.4 million
years,[27] and around the same time in South Africa.[28] This would be
around the time Acheulean tools (i.e., *erectus*) start replacing those of the
Oldawan (i.e., *habilis*), but the evidence is spotty and there's not another
hit until inside 500,000 years ago when signs of fire start showing up everywhere. At Zhoukoudian in China the relatives of Peking Man used fire
between 460,000 and 230,000 years ago all the time; and there are nearly
coterminous sites in Brittany and along the French Riviera. This means
erectus leaves the Early Paleolithic with stone tools *and* fire *and* cooking
meat *and* sharing it. . . . [29]

And after the fire is out, spreading sand or dirt over the embers and
sleeping on it. . . . [30]

Although maybe Neandertals were the first to do this—they seem to
have made the first hearths too—and a bare 60,000 or so years ago,[31] deep
in the Middle Paleolithic (100,000 to 40,000 years ago), with the Mousterian tools they'd been making for 100,000 years by then, wherever *they*
came from. These smaller, more specialized Mousterian tools (i.e., these
Neandertals, perhaps even these *sapiens* [for remember that the tools are
probably only clouds of culture moving with people]) began to supplement, and then to supplant their Acheulean antecedents (i.e., *erectus*)
nearly 200,000 years ago throughout most of their range in Europe, in the
Near East, in Africa; though, as *erectus* overlapped *habilis* for thousands of
years (*for thousands of years*) there was extensive overlap here too. (The
"rapid" in "rapid replacement" is relative.) With their new Levallois flak-

ing techniques Neandertals in Europe made fine new scrapers and points. Outside Europe *sapiens* did the same. Between 100,000 and 75,000 years ago, North African Aterian knappers squared off the tangs of their points and hafted them to wooden shafts as spear points and arrowheads. They may even have taken their tools to the grave with them, unless the ones we've found among their bones were no more than litter in the dirt . . . assuming that Neandertals buried their dead. There are enough sites—La Chapelle-aux-Saints, La Ferraisse, Roc de Marsal, Amud, Kebara—where even skeptics have to admit the dead were buried,[32] but to what end is harder to say, and whether as a sign of intelligence or of stupidity even harder. I mean, let's keep it real: if human intelligence has an origin, human stupidity does too.

So *stupidity* then, and *fine stone tools* and *hafted weapons* and *fire* and *cooking meat* and *sharing it* and. . . .

. . . and *speech*, speech is the big deal.

Articulating Breathing into Ontogenic Communicative Behavior

It's like *Australopithecus* stands, *Homo* makes tools, *sapiens* talks.

And with speech . . . *or:* with fine stone tools and hafted weapons and fire and cooking meat and sharing it and burying the dead *and* speech . . . the (*gasp! faint!*) human brain. *The brain!* Oh, I know it's amazing and wondrous and intricate beyond its own fathoming but . . . you know . . . so are the globèd peonies and the tiger-moth's deep-damask'd wings; and I have a hard time convincing myself the brain's more intricate or even more intelligent than an equivalent volume of soil. Just think about *its* billions of interacting bacteria and probing miles of root hairs—with their overlapping rhizospheres—and the hyphae and mites and beetles and seeds—in their unfolding spermospheres—and fungi and earthworms and protozoa and roundworms and centipedes and springtails and slugs and snails and sowbugs and millipedes and larvae of flies and flatworms, all so linked through chains of reciprocal consumption, all so hooked up, all so tied together by the mycorrhizae and the filamentous actinomycetes branching everywhere that the mass develops an operational integrity, cycling elements tightly, simultaneously linked dynamically to the rest of the biosphere, to the atmosphere, to the hydrosphere, to the lithosphere. And you may say they're all, like, individual animals, but so are the neurons living in your brain (and neither can survive on their own), and if you insist the soil can't paint *The Last Judgment* or write Beethoven's *Fifth Symphony* I'll have to note that neither Michelangelo nor Beethoven could support a tree, or for that matter a tomato plant, and it's six of one, a half dozen of the other, you know, especially when you're hungry.

Not to knock the brain, but it's no pinnacle of anything[33]; and of course we don't have to choose, we can eat tomatoes *and* look at paintings. We can have it all, *but only because we have it all*. That's how it comes, whole. I get tired of this singling out of things. It's not like we could have brains without soil. . . .

It *has* been growing, though, larger and larger. It occupied 450 cm³ in *afarensis*, 500 cm³ in *habilis*, 825 cm³ in *erectus*, 1,200 cm³ in Archaic *sapiens*, and 1,500 cm³, more or less, in *Neandertalensis* and *sapiens sapiens . . .* and of course (curse of the academic): *why?* What would select for this?[34] And Harry Jerison says it's *not* the rise of technology that drove it, but probably the evolution of language,[35] and Stanford says that "the intellect required to be a clever, strategic, and mindful sharer of meat is the essential recipe that led to the expansion of the human brain,"[36] and Leiberman says that "we can't claim that hunting was the key for the evolution of human language and thinking," it has to be speech.[37]

But what if we stopped all this either/or business and said *language* = *sharing* (which we sort of do in "Let me share this with you . . . "), or even *language* = *sharing* = *cooking* = *hunting* = *making tools* (where the relationships are transitive, the *sharing* depending on the *tools* with which the meat was butchered, for example, and the *cooking* involving *tools*—and fire *is* a tool—and the *hunting* using *tools* as weapons, etc.), what if we tried to see it as a *complex*, as a *whole*, like we see an *ecosystem*; and tried to see the entire ecosystemic complex as a kind of *thinking*, the thinking arising from and informing the behavior, the behavior arising from and shaping the niche, the niche arising from and shaping the land . . . and then reciprocally, the land in its atmospheric–oceanic–tectonic–biospheric wholeness making the niche possible, which in turn informs the behavior (which can only arise in the organismic wholeness of the niche), which is to say, informs . . . the talking–cooking–sharing–hunting–toolmakingness of human life.

Godamnit it's hard!

To say it all *whole*,[38] that is, so that it's not the land making this animal with this brain and these behaviors, but the land whole at this point in its evolution *in* this animal *with* this brain and these behaviors and *vice versa* . . . [39]

Though whatever the claims of hunting and sharing and shaping tools and cooking, certainly it was speech at the end. Lieberman says:

> The reorganization that makes voluntary control of speech possible is one of the defining characteristics of the modern human brain. It un-

doubtedly had occurred 100,000 years ago in anatomically modern fossil hominids such as Jebel Qafzeh and Skhul V, who had modern vocal tracts. It probably had already occurred in fossil hominids such as Broken Hill who may have lived 125,000 years ago, may have been ancestral to these fossils, and had a vocal tract that was almost completely modern. It may have occurred in the African fossil hominids (who have yet to be unearthed) ancestral to the early specimens of anatomically modern *Homo sapiens*. The dating of adaptations for syntax and abstract cognitive processes (which appear to derive from the human prefrontal cortex) is more difficult to determine. The archaeological evidence associated with Jebel Qafzeh and Skhul V is consistent with their possessing a fully modern human brain—one adapted for complex syntax and logic—but an earlier origin cannot be ruled out. Although much of the enlargement of the prefrontal cortex may derive from the specific contributions of speech and language to biological fitness, it enters into virtually all aspects of behavior. Therefore any cognitive activity that enhanced biological fitness could have contributed to its development. . . . [40]

Such as, one would have to imagine, shaping tools and hafting them to shafts for spears and organizing the hunt and cooking the meat and sharing it out to kinfolk defined ever more elaborately and burying them with yarrow and cornflower and St. Barnaby's thistle.[41]

Singing, you know they were singing . . .

By 30,000 years ago they were all gone.

What happened to them?

Another ferocious battle: Neandertals evolved in place, or were rapidly replaced, or gradually died out, unable to compete with more modern peoples. Geoffrey Clark and John Lindly think the record shows "clear evidence of cultural and behavioral continuity between Middle and Upper Paleolithic," and Richard Klein says that "the relatively abrupt and radical nature of the change provides far better support for the 'Out of Africa' hypothesis than for its multiregional alternative," and Allison Brooks says that "there is no archeological evidence of a wave of people flowing out of Africa into Europe," and I could go on but there's no point.[42]

What we know is that in the Late (or Upper) Paleolithic, from 40,000 or 35,000 (and maybe much earlier in Africa[43]) to 10,000 years ago, this hunting–cooking–sharing–speaking people began diversifying and specializing its toolkit (*blades! bones! hafted microliths! burins!*) and making clothing (and *beads!*) and building shelters and . . . well, *painting*, all over the walls of Lascaux and Altamira and Périgord (*figurative painting!*), spreading out—humans all *over* the Americas—and, at the end, making pots and planting wheat.

We're talking *sapiens sapiens* now, us, *sans* sneakers.

There's this rabid radiation of culture in the Upper Paleolithic: Châtelperronian, Aurignacian, Gravettian, Périgordian, Solutrean, Magdalenian, Azilian. More than anything else this sets the age apart from that—from those—preceding it, this *stirring*, changing, innovating, restlessly exploring, pushing-the-boundaries character of it. Some of this is an illusion created by proximity. From a distance little of this would be visible. I'm not being cynical. This is *the* lesson of history. A hundred million years from now it'll be good work to distinguish the Stone Age from the Metal Age from the Information Age. The whole couple million years may well be denominated the Age of the Artifact (if anyone, if anything, cares). Observers may wonder why it took us so long to build . . . whatever it is we end up building . . . on the obvious accomplishments of . . . *the amphibians*. I don't know. I do know I'm leery of claims about "slow" and "fast" when we're talking about the past.

That said, Oldawan tools don't seem to change much over the course of a million years, while the Châtelperronian—which doesn't last 10,000 years—added blade and bone to an already evolved Mousterian technology. Some locate the Châtelperronian in the Middle Paleolithic (it's neater that way because it's distinctly Neandertal),[44] but most front-load it in the Upper Paleolithic, say from 40,000 to 34,000 years ago, and assume the Neandertals evolved it out of contact with inmigrating *sapiens*, Cro-Magnons (if that's what happened), with their Aurignacian scrapers and ivory beads, their tooth necklaces and figurines, their far-flung cousins settling out across Beringia and the Sunda shelf into the Americas and over the Wallacea into Sahulland. They had boats. They'd had boats a long time by now.

Waves of people on the move, waves of them.

Dripping tools everywhere they went.

In Europe the Aurignacian ran from 40,000 to 28,000 years ago (in the Near East from 32,000 to 17,000), replaced, augmented, supplanted by—*folks: these are names we've given to variations in assemblages of tools*—the Gravettian, with its backed-blades and bevel-based bone points and voluptuous Venuses, which, lasting from 28,000 to 22,000 years ago, scarcely hung around for six millennia.[45] The Solutrean didn't hang around for two, with its freakishly fine flaking, 21,000 to 19,000 years ago (though compare for brevity the Rock Era, which, running from Elvis to *Sgt. Pepper*? Elton John? The Bee-Gees?, lasted less than 30 years).

During the Magdalenian—six millennia, 18,000 to 12,000 years ago—there were paintings all over the walls of Lascaux and Altamira. You've seen the pictures. They made microliths that could be hafted to all kinds of shafts, and multibarbed harpoon heads of bone and spear-throwers too. They wiped out the mammoths and the woolly rhinoceroses, and the Azilian folks who succeeded them ate mollusks and birds and small mammals exclusively.

We're right at 10,000 years ago.

The ice was long gone, not even a memory.

We're Still Living in the Ice Age, in the Stone Age Too (and in the Ages of Atoms, Molecules, and Bacteria)

The ice is gone but the Ice Age isn't. We may want to pretend the Pleistocene is over, but it's not. [46]

Holocene!

Like calling it new makes it new. It doesn't. we're still in the Ice Age, we're in an interglacial, and the ice is coming back, and when it does it's going to come back fast. The ice sheet's going to be thicker than Toronto's CN Tower is tall, and it's just going to scrape all that stuff—well, like Toronto—into a garbagy, icy heap and deposit it somewhere south of Buffalo.

I'm not saying you need to start packing. We'll have generations of warning, even as unwilling as we are to pay attention to the signs. The "fast" in "come back fast" is relative too.

So, no need to panic, but to keep it real we need to stop thinking as though the world started a few months ago and everything that happened in the past only happened . . . *back in the days*. It didn't. It's still happening. we're in media res.

And it's not just the Ice Age that's not over. The Stone Age isn't over either. Or, if it is, it ended, like . . . just yesterday.

Again, I know you're not going to remember the dates—and they're all going to be pushed back anyway—but I hope you noticed the overlap, the way *Homo erectus* only *gradually* disappeared from the record, the way Neandertals only slowly slipped into it, the two coexisting for thousands of years, *erectus*'s Acheulean assemblages only turning into Neandertal's Mousterian over time; and at some point, at a time when *erectus* fossils no longer appear in the strata, and *all* the tool assemblages possess Mousterian traits, we can say, there, yes, we've left the Lower Paleolithic and firmly entered the Middle Paleolithic, but exactly when this took place is impossible to say. It didn't take place *exactly*, no matter how many lines we draw on diagrams.

It was only in 1971 that *we*—that is, readers of *National Geographic*, university folk—had our ignorance of the Tasaday corrected, and, *whatever their history*, when we started photographing them they were using the very same Stone Age tools as those who 20,000 years earlier had left the frontal bone and toothless jaw at Tabon Cave on the adjacent Philippine island of Pelawan, tools indistinguishable from those associated with the Niah skull of 40,000 years ago, that is . . . with *Old Melanesian Culture*, conventionally dated from circa 50,000 to between 8,000 and 5,000 years ago, conventionally, but evidently inaccurately.[47] It's just that we resist a dating like circa 50,000 years ago to 1971.

Which may be when the Stone Age finally petered out,[48] if it has, if there are not still other peoples fiercely holding onto their Stone Age ways.

Who could blame them?

We have this notion that now is everywhere but it's not. *"But today the world's tied together with satellites and telephones."* Right. The fact is *half* the world's population still has never made a single phone call.[49] And when everyone has (which, god forbid), half the world's population will be onto something else. Yes, electromagnetic energy zips along at 186,000 miles a second, but people who don't have enough to eat rarely invest the energies they have in cell phones. And often those who have plenty to eat don't bother either.

What time *is* it?

It's not the time you think. It's baryonic time. It's atomic, it's molecular time. It's galactic, it's stellar time. It's earth time. It's bacterial time. It's eukaryotic time. It's animal time. It's mammal time. It's primate time. It's *Homo* time. It's Stone Age time, or it just ceased being Stone Age time, and we can't say, speaking of the planet, that it's any other time yet. Other ages may have opened, ages we can't imagine in their unfolding, but they have yet to become planetary phenomenon, not one of them, for not yet all of us eat domesticated plants and animals, or read and write, or are caught up in the urban scene (even in its most generalized gravitational field), or are participants in the Information Age—whatever *it* is. It's an almost vanishingly small number of us who are wired. If you think otherwise, you're too locked into your parochial little perspective, you need to get out and take a deep breath.

And it's not just that not everyone is as caught up in the whole Internet thing as some would like to convince us everyone is, but that we are not defined in any important way by the latest fad we've jumped on; that, living as we are in all the ages that have yet to close, we *are* baryons, we *are* atoms, we *are* molecules, we *are* star dust, we *are* earthly, we *are* cells, we *are* animals, we *are* mammals, we *are* primates, we *are* humans, we *are* Stone Age peoples, or if not, *just not*, though frankly it beggars belief to imagine that having just left the Stone Age, we aren't still essentially caught up in its rhythms.

The Stone Age: *now.*

We talk about geological time. What about evolutionary time? we've been what we now think we are so briefly it scarcely figures. I came of age during the war in Vietnam. My grandfather Wray was in World War I. My great-grandfather Poppa was a kid during the Civil War. Our birthdays fell on the same day and once when I was old enough to remember we even celebrated them together. His great-grandfather Thomas was kicking around North Carolina during the Revolutionary War. That's so far back in the boonies of time most kids today can't keep it clear of the Age of the Dinosaurs, those who've even heard of the Revolutionary War, and if you think most kids have, you're nuts. Yet taking a generation at 20 years, the whole national history of the United States has taken up only . . . 12 generations.

All of it, Washington and the Delaware, and Dan'l Boone, and cowboys and Indians, and the Great Depression, and sockhops.

Columbus sailed the ocean blue 25 generations ago. Julius Caesar was murdered 100 generations ago (we're back with Cleopatra). Two hundred and fifty generations takes me back to the Chalcolithic, to the Copper Stone Age, but the Stone Age lasted 250 *thousand* generations. All of what history books want to call "history" amounts to a *thousandth* of no more than our history *in the Stone Age*, for 999 parts of which we were hunters and gatherers, or gatherers and hunters, mostly gatherers for most of that time, whatever the effect of the hunting on development.

Coming Down from the Trees, We Spread Out and Started Hunting

And though I insist on the salience in all of us of this hunting and gathering history, I have no interest in pretending it was static. We know our brains were growing, but the fact is our bodies had been growing, too.[50] Simply as a function of surface–volume relationships, larger bodies are relatively cheaper to maintain.[51] This is to say that as absolute food requirements rise, relative food requirements decline; and because of this relationship, a selection pressure favoring larger bodies arises in every taxon.[52] Now, life in the trees imposes a constraint on body size, and, as Robert Foley says, "For primates the price of release from the high energy costs of a small body size is terrestriality."[53] What this means is that we came down from the trees so we could get bigger and reduce our relative energy costs.

How many reasons does this make now?

Whether it's *the* explanation or not (but there *is* a strong trend among primates toward terrestriality with increased size[54]), our increased body size—which is indisputable—had all sorts of consequences. First of all, larger bodies need more food. This is simple and straightforward. The need for more food puts pressure on home ranges—making them larger— and population density—driving it down. However, at the same time, larger bodies need *less food relative to body mass*. This means they can live lower on the food quality chain. They can, for example, switch from fruit to leaves, which their increasingly long intestines lets them digest. we're standing way back here and painting with a broad brush.

As the need for more food, even of potentially lower quality, in- creases home ranges, it increases mobility, with everything that increased mobility implies for adapting to a wider range of resources and, ulti- mately, cooler climates. Enhancing this adaptability to cooler climates is the greater heat retention and activity that come with larger bodies.[55] Greater strength and speed also come with larger bodies, and both im- prove predator resistance and/or avoidance. All of this drives the growth of the brain, since any improvement in its ability to support the greater flexibility in behavior required for the exploitation of increasingly com- plex resource patches will be powerfully selected for. Longevity increases with size too, and this increases reproductive span and developmental pe- riod. Longer prenatal periods, in turn, raise reproductive costs, which lower birth rates and produce greater birth intervals. This places greater emphasis on the survival of each young. Partly as a result of increased care of offspring and partly as a result of increased longevity, increasingly complex social behavior arises. This in turn supports increasingly effec- tive foraging and, as the metabolic costs of larger brains rise (and so a need for higher quality food), hunting, with its high-quality protein (and we know what hunting drives).[56]

Hunting arose in the context of seasonal stress in the increasingly savanna-like environment in which our australopithecine and later *Homo* ancestors were living. In the early stages of hominid differentiation, groups pursuing a variety of ways for getting through the dry season— which would have varied with local conditions (often widely)—must have been scattered throughout the tropical wet–dry regions of Africa.[57] In or- der to make it through the dry season, and especially the dry season in dry years, early hominids would have had to opportunistically clamber up and clatter down the food-quality chain, taking advantage of whatever was offered—termite eruptions, fruits, eggs, fledglings, flowers, young leaves, fish stranded in drying ponds, roots, grass seeds, herbivore young, or whatever—as the seasons revolved from wet through dry and back again. From the get-go this seasonal variability would have promoted high mobil- ity with complex patterns of concentration and dispersion.[58] From the very beginning our hominid ancestors had been . . . *edging on over*, keeping their population density low, smearing australopithecines up and down the Rift Valley, extending the *habilis* range, moving *erectus* up the Nile and . . . out into the world.

What I'm trying to say is that it wasn't a matter of these hominids just hunting and gathering for 250,000 generations like we imagine Pa and Ma and Laura and Mary might have in the Big Woods (if it hadn't been for the coming of that thresher),[59] very gradually figuring out how to make better and better use of stone for tools, and finally fire and pottery and civilization. Instead it was a continuous ongoing evolution of behavior—and let's say . . . of character—triggered no doubt by echoes of distant glaciations, but *no less* by selection pressures driven by things as fundamental as the energetics of surface–volume relationships. Each subsystem unfolds according to the beat of its own drummer, and so, as the biodynamic selecting for larger body mass began pushing australopithecines out of the trees (I can't believe I'm saying this), it did so in a seasonal wet–dry environment enmeshed in a climate system geared to the Milankovitch and supercontinent cycles. It's the intersection of these—and other systems (among them the biodynamics of other local life-forms)—that constitutes the historical situation in which hunting arose as a way to smooth out variations in food availability by resource switching in a highly seasonal environment.

But meat isn't berries with higher quality protein. It doesn't just sit there waiting to be picked and it doesn't come berry-sized either. Your typical carcass is a lot of food, and so meat offered an opportunity for sharing. There's a raging debate (*of course!*) about whether this promoted the development of reciprocal altruism,[60] even in the form of what Glynn Isaac liked to call "tolerated scrounging," but Isaac's argument that the division of labor evolving from the inclusion of meat into the hominid diet *obligated* a mechanism for food sharing is cogent despite its critics.[61] Reduction of risk is the fitness advantage offered by sharing (which was the point of switching resources in the first place). A characteristic "vigilant sharing" increasingly distinguishes hunting and gathering populations,[62] whether or not it's "the result, at the level of the social system, of the behavior of sociable and self-interested individuals whose motivations include a strong desire to get enough for themselves coupled with a strong desire to make sure that no one else gets more than they do."[63]

Hunting = Tool Cache? Home Base? Settlements?

Isaac saw that such sharing depended on a place to do the sharing. A home base, therefore, would be central to the development of early hominid foraging, or if not a home base, perhaps the tool caches Richard

Potts sees early hominids dragging carcasses to for butchering.[64] Foley says:

> What we can see here is the convergence of several processes. On the one hand is the need for shelter on the ground, on the other a change in foraging behavior brought about by the development of tool-making as an aspect of meat-eating. The conjunction of the two were the antecedents of the home base—in other words, a central focus for several activities and behaviors. The principal consequences of this would have been a gradual shift away from multiple use of stone caches as part of an orthodox pattern of primate foraging, towards one in which foraging involves starting from and returning to a central place, as is the case with modern hunter–gatherers.[65]

Until recently many felt neither butchery nor the use of home bases much predated the appearance of *Homo sapiens*,[66] but the evidence of butchery on the part of *Australopithecus garhi* at 2.5 million years suggests the possibility that *all these developments*—the expansion of the diet, hunting, labor specialization, the use of stone tools, some kind of sharing, meat eating, shelter—were set in motion when australopithecines got too big to continue living in trees.[67] I like the way Tudge envisions those he calls *Australopithecus hooliganensis*:

> I find it easy to envisage the first australopithecines in their semi-tropical Mio-Pliocene Sherwood Forest employing very chimpish techniques: hurling themselves en masse from the trees in an orchestrated frenzy, armed with sticks, and beating a monkey, a pangolin, a small antelope, or a pig insensible before it had time to gather its wits. And such creatures, noisy, cooperative and frenetic, would in practice have had little to fear from the megapredators like leopards and hyenas. Such predators tend to be specialists. They do what they do. These hyperactive hordes would have been outside their ken: too unpredictable, too weird. The early australopithecines, one might suppose, would have been creatures to leave alone.[68]

Considering that he's describing creatures between 2.5 and 5 million years old, I find the description wonderfully familiar, less the droogish quality, than the loosey-goosey opportunism. Leaping from the trees onto clueless prey, they also pick up anything they find, stick it in their mouths—how else does a frugivore become an omnivore?—and sooner or later in the mud.

Agriculture!

Okay, so I'm crashing through 5 million years of history. Yes, but not really. The thing is, their larger size and the seasonal habitat *did* conspire to make mobile opportunistic omnivores out of these australopithecines, setting in motion the cascade of adaptations we recognize in the record as a melange of anatomical and behavioral—all right, *cultural*—changes: *habilis* and *rudolfensis* butchering meat with their Oldawan flakes and choppers and slowly turning into *erectus* and *ergaster*, who added their Acheulean cleavers and hand axes to the mix and cooked their meat and gradually turned into Neandertal and *sapiens*, who used their Levallois techniques to craft their Mousterian scrapers and points and sat around the hearth, using their modern vocal tracts and language to gossip and lay down the law and gradually turn into *sapiens sapiens*, with their Aurignacian, their Périgordian, their Magdalenian burials and cave paintings and pottery and wheat and books and TV and rocket ships, for we *still* are *sapiens sapiens*. I don't see the stasis some paleontologists do in the million-year sway of the Oldawan, of the Acheulean. Those folks were busy evolving politics and poetry, 99% of it anyway, that is, sharing, and the modern vocal tract, and a brain to use it. Scientists struggling with humility like to quote Newton's "standing on the shoulders of giants." Well, whose shoulders do you think *we're* standing on?

So: opportunism. Leaping on prey and beating it senseless. Stooping by a pond as the dry season sets in and grabbing that flashing thing flopping around and chomping on it. Many must have died this way, but the cupboards they opened must have sustained many more. The value of sticking things in your mouth. The value of poking sticks into things, into termite mounds, into beehives. Trying out this and that. Throwing rocks into water. Skipping stones. Chewing on your pencil.

Five million years of this.

You bend over, pluck a stem, and stick it in your mouth.

Insouciant.

You bend over, and stick it in the ground. . . . [69]

Drawing a Line in the Ground and Calling One Side of It a Garden

So, okay, if the Ice Age didn't end, the ice did, it was gone, or going, at least for a while. *Sapiens sapiens* was busy pushing species after species over the brink of extinction. About the very time sown grain was first harvested in the Middle East, the great cave bear, the mammoth, the woolly rhinoceros, the musk ox, the steppe bison, the giant Irish elk, a number of kangaroo-like species, and the diprodotons were hunted—by mankind—off the face of the earth. Probably this is no more than coincidence, or maybe it was the climate.[70] The great hunters were an unlikely group to invent

gardening, and certainly the hunters of the North European plain, the
Clovis crowd in North America, and early Australians didn't invent it in
the Fertile Crescent. But the loss of this megafauna could not but have
helped create circumstances favorable to the subsequent reception of the
culture of grain, and in any case it's a reality it would be disingenuous to
overlook. Whatever it was, it wasn't "progress" that led to agriculture. "As
far as I'm concerned," Robin Fox says,

> We reached our peak as a species in the Upper Paleolithic, when we
> were the top carnivore, knocking off whole herds of mammoths. Now
> we're like a car with a cruising speed of eighty, doing a hundred and
> forty, and we can't keep that up for very long. We're still paying for the
> consequences of the Neolithic Revolution, when we went from being
> hunters to agriculturalists—the Great Leap Backwards, as I call it, which
> sent us to seven days of backbreaking work. Since then what we've
> mainly perfected are ways of exterminating each other, most of the
> world is in a debased peasant condition, and you and I belong to an elite
> that lives on the back of billions condemned to dawn-to-dusk food gath-
> ering. In what sense is this progress?[71]

Put this way, it is hard to disagree, and indeed from Genesis to Fox this *is*
the consensus: farming was the end of Eden.

In Arnhemland the women of Gidjingali can feed themselves by gathering and processing food for an hour and a half a day.[72] When a visitor
suggests he might plant crops, a hunter–gatherer in the Kalahari asks,
"Why should we plant when there are so many mongongo nuts in the
world?"[73] And these are peoples working the refuge margins of the Earth.
Better watered places, especially along interphases or abrupt changes in
elevation (coasts, estuaries, streams debouching onto plains from hills,
streams entering lakes or leaving lakes, or just with springs and high water
tables), and especially across transitions in biomes (where, for instance, a
tropical broadleaf forest gives way to savanna) would have been paradise
indeed. Early man flourished—*evidently!*—without planting a thing for millions of years; nor is there much question that the move to herding and
cropping was initially disastrous to nutrition, health, and longevity, that
the biblical story of the loss of Eden was grounded in life.[74]

So what used to be obvious—I mean, when I was coming up it was *so*
obvious that settled farming had it *all over* unsettled hunting around (how
else to explain the rightness of "our" farming the Indians' former "hunting" grounds?)—now requires an explanation: *why* the move to so inferior
a way of life? Again, there are way too many assumptions lurking behind
these apparently guileless questions. Personally I'm attracted to the idea
that early peoples were just used to sticking things into things, and that

this time something came up that had been seeded in a mudbank and pretty soon: you know, you invest energy in sowing, it's satisfying to see the plants come up, and pretty soon you're sitting out there beating off the birds and the beetles, and investment begets investment . . . you're trapped.[75] Hell, it's as good an explanation as any. It may be that Calder's got it right, that "we were seduced into cultivation because at first it was easy":

> As the mud flanking the wadi dried after the floods, it cracked in the sunshine, and feminine laughter no doubt disturbed the water birds when the women came with a bag of seeds to invent crops. Perhaps it was a waste of good food and nothing to tell the men about—yet it only took moments to poke the seeds into the ready-made cracks in the mud. A crazy experiment, but exactly what was needed, not just to make use of the self-cultivating riverbank, but to divorce the grasses from their wild relatives so that they would evolve under human hands to suit human purposes. The women knew little of plant genetics, but the grain grew and ripened in their riverside garden before the sun parched the ground entirely, and when they came back with stone sickles, they may have felt a certain goddess-like pride.[76]

The pride is the easier to understand as large parts of the picture are left in shadow, though something like this pride must have played its part. We're not talking about strangers here. Whoever made this move was as like us as can be without wearing designer clothes. But one has to imagine that Calder himself does not garden so schematic is his picture of these women: they put seed in the ground, go off, and later return to harvest what they've sown. Contrast such a fantasy with Carl Sauer's view in this passage where he attempts to explain why the first gardeners had to be sedentary:

> Above all, the founders of agriculture were sedentary folk. I have already said that groups move as little as their needs of food, water, fuel and shelter require. Mobility as a dominant character goes with specialized hunting economies or with life in meager environments. Growing crops require constant attention. I have never seen primitive plantings that are not closely watched over until the crop is secured. A planted clearing anywhere is a feast set for all manner of wild creatures that fly, walk and crawl to come in and raid fruits, leaves and roots. What is food for man is feast for beast. And, therefore, by day and night someone must drive off the unbidden wild guests. Planting a field and then leaving it until the harvest would mean loss of harvest.[77]

This is something more like it, especially if we admit that Calder's image of wheat on the riverbank is probably anachronistic in any case, admit that the earliest gardens were as likely to be what Edgar Anderson has called "dump heap gardens," circumscribed places, but gloriously open habitats, perhaps refuse heaps or garbage dumps originally[78]; or no more than especially fruitful places, regularly harvested (and ultimately enclosed), as among the oak and pistachio forests in the foothills of the Zagros Mountains in what is today Iran, or among the gazelles of the Levantine corridor and the Jordan Valley.[79]

Better and better dating techniques help make it pretty clear that this is where domestication first took, in the Fertile Crescent, maybe in the Jordan Valley, about 10,000 years ago. Today most agree that it would have happened rapidly, within fewer than 15 generations, inside 300 years, say, from maybe 10,000 to 9,700 years ago. It was a pretty cool place to live. Pines and cedar, oak, fir, and pistachios cloaked the hills. There were edible fruits and nuts. There were almonds and there were rich wild grasses. There were wild oats and wheat and spear grass and wild barley. There were wild legumes and wild pigs and goats and cattle and sheep. There were ducks and eels in the shallow lakes. Mediterranean climate. Figs with brioche and coffee on the terrace—

Okay, okay, maybe not *yet*, but I want you to feel the tremendous continuity running through us to these early agriculturalists—*it was only yesterday*—and through them to the rest of the Stone Age and through it to our australopithecine ancestors in the savanna and through them to our entire primate history, our history as mammals, as land animals, as multicellulars, as eukaryotes, as prokaryotes, as chains of self-referencing autocatalytic, reactions, as molecules, as atoms, as baryons. . . .

Can you do that, feel it all, in the breeze and the bread?

Emmer wheat. People had harvested it with stone sickle blades for years where it grew wild on the hillsides. Einkorn wheat. Even today the wild ancestor of einkorn grows in massive stands from Anatolia down through the Zagros. Among the more than 150 seed-bearing species hunter–gatherers ate along the edge of the Euphrates Valley, wild einkorn was among the most common. Why not? Thriving on the open slopes where the scrub oak woodlands fade into the grasslands, einkorn was so easy to harvest a small family could have collected enough in 3 weeks to live on for a year.[80] Imagine the nice weather and the country tawny and green and the grasses rippling in the sunlight and everybody moving together on the hillsides. It's hard to think about it as work.

Ditto barley.

The simple step of harvesting these grains with stone sickles once they were ripe would have been enough to select for plants with semi-tough rachises, rachises that would hang on to their seed instead of, as required by life in the wild, scattering it to the wind when touched. In a few centuries (or less) this nonshattering grain would have been dominant in the fields people worked.

That's it.

That's the domestication of grain.

As for the next step, deliberately planting the seeds, that's . . . another story.

It could have been the climate. The Younger Dryas had just come to its abrupt end[81] and the drier conditions during the Dryas might well have reduced available food resources and so inaugurated thinking about a way to reduce seasonal variability (don't forget, as thinkers these people were indistinguishable from you and me).[82] Or it might have been the pressure of a growing population, and that could have had the same effect. At least as long as 12,500 years ago some of the region's inhabitants were living in permanent year-round settlements, and tossing away the worn sickles and grinding bowls and slabs and cupstones they would have used to collect and process grass seed and—who knows?—this might have set in motion cascades of changes that could have pumped up the population. Mark Cohen has influentially argued that population pressure was an ever-present and ever-growing problem for hunter–gatherer societies,[83] but you know what? None of these ideas fares well in single-armed combat with the data,[84] and they suffer from pushing around—more or less as though they were an ideal gas—what were rapidly turning into complex and wildly adaptable social structures.

Third-Order Autopoietic Unities

That is, what were turning into elaborate third-order autopoietic unities.

To call a social system a *third-order autopoietic unity* is to draw attention to the chains of continuity linking it to *second-order autopoietic unities*, that is, it's to pay attention to the bounded quality of its self-producing autonomy. Recall that *first-order autopoietic unities*—cells, prokaryotes, eukaryotes (i.e., self-bounding networks of self-referencing, autocatalyzing, macromolecular reactions producing the very molecules needed to make the reactions run so they can run again)—are essentially distinguished from . . . *rocks* . . . by the fact that the phenomena they generate in functioning as autopoietic unities depend on their organization, specifying these ions and not others[85]; and that the *circularity* of this organization, requiring the specified ions again and again, orients the unity toward the world in a way that is neither unfocused nor inert nor reactionary. Recall as well that *second-order autopoeitic unities*—metacellular organisms (like us) arising from the structural coupling of first-order autopoeitic unities (i.e., cells)—are essentially differentiated from unicellulars by having a structural coupling and ontogeny *adequate to their structure as composite unities*. That is, multicellular organisms have a macroscopic ontogeny, like ours, and not a microscopic one like our cells.[86] Now, precisely as cells couple with each other to form multicellular organisms, so multicellular organisms couple with each other; and this coupling can acquire, over the course of their ontogenies, a recurrent nature. "When this happens, the co-drifting organisms give rise to a *new phenomenological domain,* which may become particularly complex when there is a nervous system."[87] These new phenomena are *social,* that is, those "in which the individual ontogenies of all the participating organisms occur fundamentally as part of the network of co-ontogenies that they bring about in constituting third-order unities."[88]

What?

Listen, you can't understand a bee without knowing about the hive, that's all.

That is, the life of a bee (*individual ontogeny*) takes place within the context of (*occurs fundamentally as part of*) the hive (*network of co-ontogenies it helps to constitute*).

Just as. . . .

The life of one of the cells making up a bee (*individual ontogeny*) takes place within the context of (*occurs fundamentally as part of*) the bee (*network of co-ontogenies it helps to constitute*).

So there is this apparent nesting, the cell arising within and coconstituting the bee, the bee arising within and coconstituting the hive, the hive arising within and coconstituting the niche, the niche arising within and coconstituting the environment, *except*—as we've seen—the bee is part of the cell's environment, as the hive is part of the bee's, and the niche part of the hive's, that is . . . *they're all environment from some perspective,* which as we all know amounts to saying *they're none of them environment from some perspective,* not even the environment's. This . . . *corrosion* . . . of the environment as an independent variable amounts to insisting that the organism and its environment are mutually defined, is tantamount to acknowledging that the whole thing is *clutchy,* is *all* a matter of clutches—of relays—and that the more clutches there are involved (the deeper the nesting), the less likely it is that any change in one part of the whole—like the dropping of the atmospheric temperature—is going to produce some *necessary* change in another part of the system. The nested structure of autopoietic orders ensures alternatives *at every level.* Just because it's freezing out doesn't mean my blood is going to turn to ice. If I start to shiver, I'll get up and put on a sweater. Pretty soon the thermostat'll kick the furnace on.

Not that I can't freeze to death, but if you thought I had trouble with changes in the climate pushing our australopithecine ancestors out of the trees, I've got tons more with the idea that 5 million years later their descendants—*with* their stone tools and their houses and their baskets and their language and their big brain and more acorns than they could possibly eat—could be *made* to start farming because it got drier, or it got wetter (yes, there are adherents of both positions), or because there were more of them. Not that that's *not* what happened (we don't *know* what happened), but I just have trouble with the idea. And not that, all clutchy—*as we inhale all those atoms formerly part of Leonardo da Vinci's body*—everything doesn't impinge on everything else. *It does.* It's just that 10 to 12 millennia ago the Natufians had as many ways of responding to the trigger—if there was one—of a desiccating environment as we have.

There is Brian Hayden's early five-step model[89]; and the four-factor model of McCorriston and Hole[90]; and the subtle two-step transition model of Donald O. Henry with its climatic *triggers* but complicated clutchy response trains[91]; and Hodder's domestication *of society* idea—symbolic domestication preceding social domestication preceding economic domestication—with the domestication of grains and animals an easy extension of the domestication of . . . women[92]; and Hayden's later model with intragroup economic rivalry forcing the move to agriculture in support of bouts of emulous feasting (with grain domesticated to produce beer for the feast!)[93]; and . . . *like that!* . . . the appearance in the archeological literature of societies that resemble mine (and yours), with structures of signification linked to complicated power struggles between haves and have-nots and across genders and generations,[94] the adoption of more intensive methods of production (climaxing in agriculture), "match[ing] the desires of dominant groups in society, in that the new relationships trapped people within relatively fixed economic and social structures on which they came to depend"[95] and. . . .

9

Humans Cover Themselves with the Land

And it's, like, I *know* these people who are beginning to appear in the archeological record, and trying to understand the slide into agriculture becomes no different from trying to understand our descent into automobility. . . . [1]

And *maybe*—thinking about how things seem to happen around *me*, everything operating in the patchy way it does—*yes*, okay, sure. . . . There *were* shifts in habitat across a couple millennia—okay. And doubtless population growth too. But withal there were abundant resources, and an evolving society constituted of people body-, brain-, and hormone-wise pretty indistinguishable from us: artifact-rich, signifying, cooperative (above all!), competitive too, and knowledgeable, and curious, used to sticking things in their mouths and other places. . . . [2]

They were different. The past *is* a different place. But they weren't *that* different. Compared to what we've seen back in the days . . . they really weren't different at all.

Hunting and Gathering, Fishing, Growing Grains

Why did they start planting wheat?

Well, they did: in the Fertile Crescent about 10,000 years ago, emmer wheat and einkorn wheat and barley; and then they planted wheat's early companions, peas and lentils and chickpeas and bitter vetch and flax (all around the same time but probably one here and one there[3]); and gradually they domesticated sheep and goats and pigs and cattle too. Around this assemblage they established the village farming way of life, or *a* village farming way of life; though we don't need to imagine they stopped hunting and gathering for a *long* time,[4] or that all the folk in the region joined them, many continuing—perhaps preferring—to hunt and to gather (and not all the apes came down from the trees either). So there was a range, a multiplicity, a patchiness of lifeways: hunter–gatherers, and doubtless here and there Sauerian fisherfolk,[5] and village agriculturalists; and by the time these last were both growing grain and raising animals— 8,000 years ago at the latest—they had in place an economy that could support, well, nearly anything.

And soon enough it slips down the Tigris and Euphrates: *plows! irrigation! city states!*

Meanwhile . . . a thousand or two years later (though a turn of the trowel could push this back), just about the time this Middle Eastern assemblage was slipping into Europe along the Mediterranean and beginning to spread across the plains to the north, all the way across the continent in China, in the vast woodlands that then stretched from the future location of the Great Wall all the way to the sea (broken only by the rolling grasslands of the north), where the Hwang Ho descends from the western highlands not far from its union with the Wei, the P'ei-li-kang were already living in large villages and growing foxtail and broomcorn millets and raising pigs and dogs and chickens; and *at the same time*, in *at least two distinct reaches* of the vast freshwater landscape of shallow lakes and marshes that characterize the Yangtze after it leaves the mountains, the Ta-hsi and the Peng-tou-shan in the Hupei basin (as early as 8,500 years ago around Lake Tung-t'ing[6]), and the Ho-mu-tu along the coastal plain of Hang-chou Bay in the Yangtze delta region south of what is today Shanghai, were slashing and burning forest for paddy rice and water caltrop and fox nut and raising pigs and dogs and maybe even buffalo.[7]

In the Americas not even a couple thousand years later, near 5,000 years ago in central Mexico, and closer to 4,000 years ago in the eastern United States and the south central Andes (though given the state of archeology in these regions, who are we kidding?[8]), there was a whole corn and beans and squash world from Ontario to Argentina, with potato and quinua (and llamas and alpacas and guinea pigs) up in the Andes, and manioc and sweet potato and who knows what all else in the Amazon lowlands.[9] Smith guesses that when the methodological dust settles corn will prove to have been domesticated between 6,000 and 5,000 years ago along the Rio Balsas, reaching Tehuacán 4,700 years ago, and the southwestern United States and South America 3,200 years ago.[10]

This is a case where better dating has pulled the dates in, even as they steepen the transition from hunting and gathering as practiced, say, at Zohapilco in the Valley of Mexico. This would be in keeping with what happened in the Middle East and China. Everywhere it seems to have been a rapid process, happening over handfuls of generations, and here not just with corn, but with the common bean and the lima bean and the runner bean (but shortly later, maybe along with corn, in the heart of modern Jalisco); and with the endless varieties of *Cucurbita,* though this is harder to say since almost everything about the early cultivation of this plant is open to question.[11]

As it is about *Chenopodium quinoa* and the potato, and the domestication of the llama, alpaca, and guinea pig, though as I said, between 5,000 and 4,000 years ago, give or take. . . . [12]

At numerous sites in the Ohio and Mississippi Valleys in the east of what is today the United States, long before corn made its appearance, folk were growing another *Chenopodium,*[13] and a squash and goosefoot and sunflower and marsh elder and. . . .

Simultaneously in Africa pearl millet was being raised near Dhar Tichitt along the southern margin of the Sahara in modern Mauritania, sorghum around Adrar-Bous in the Chad–Sudan savanna where its wild ancestors still run rampant, and African rice at Jenne-Jeno near the great bend of the Niger.[14] Around Lake Turkana in what today is Kenya they were herding cattle,[15] and north and west of there raising guinea fowl and, listen, early Africans domesticated over 40 genera and species of plants, finger millet and *noog, enset* and oil palm, pigeon pea and sesame, black fonio and yams. . . .

It Happened Everywhere

Another great fluctuation, all this domestication, all this . . . niche construction, because that's what it is, just a way of niche-ing up the world, turning all that . . . environment into . . . land, into *Lebensraum.* Nothing new in this (for what is stone shaping but a shaping of a niche?[16]), no great wondrous miracle—*The Invention of Agriculture (gasp! faint)*—but just another step, and not necessarily forward (as we've seen), but then . . . half of the steps in any dance are backward.

Look, everyone did it. It's not like when I was coming up, when supposedly it was invented in the Fertile Crescent, which was therefore the *fons et origo* of civilization, and therefore also the locus of . . . what? Human greatness? Even in the Fertile Crescent it was invented all over the place, emmer domesticated in the Zagros, einkorn in the Taurus, barley in the Levant, and, if we had the resolution, probably time and again (it's a matter of *harvesting*, after all, with stone sickles once the grain is ripe [that's *all* there is to it]), each grain domesticated many places in its range (as barley was likely domesticated in the Zagros too, at Ganj Dareh, as well as at Gilgal in the Levant), and then sheep in the Taurus at Çayönü and Abu Hureyra and Cafer, and goats at Ganj Dareh (again, in the Zagros), and pigs in the north and the coastal portion of the Crescent, from Çayönü south to Labweh (but also in China!), and cattle maybe in the Jordan Valley, and this still leaves us with bitter vetch and lentils and peas and flax and chickpeas to account for. . . .

So maybe as many as *12 independent*—though doubtless overlapping and communicating—*origins of agriculture* in this *single one* of what are now universally regarded as *no less than seven* areas[17] where independent domestication of plants and animals arose: (1) in the Fertile Crescent; (2) along the Yangtze, where we already know rice (alone) was domesticated in two different places, up around Lake Tung-t'ing and down around the delta (not to get into the domestication of water caltrop and fox nut and water buffalo); (3) along the Hwang Ho, where the P'ei-li-kang domesticated foxtail and broomcorn millets and pigs and dogs and chickens (and again, probably not in a single place); (4) in central Mexico—*all over the place* in central Mexico—corn along the Rio Balsas and beans in Jalisco (and how many varieties of beans in how many individual domestications!) and all kinds of *Cucurbita*, et cetera; (5) up in the Andes (and what about the desert cultures and those in the jungles down the slopes?), quinua and potatoes, alpacas, guinea pigs, and llamas; (6) in the Mississippi and Ohio Valleys, another *Chenopodium* and goosefoot and another squash and sunflower and marsh elder; and (7) in *numerous* sites in Africa scattered across the southern margin of the Sahara from the Senegambia to Ethiopia, sorghum in the Chad–Sudan savanna and pearl millet around Dhar Tichitt and, as I said, African rice and finger millet and oil palm and pigeon peas and yams . . . over 40 genera and species of plants.

In 40 separate acts of domestication?

You know: *who knows?* But it must be clear we haven't touched them all, manioc and sweet potato in Amazonia; the camel along the Red Sea coast of Arabia[18]; banana and taro and arrowroot and breadfruit and orange and lime and lemon and tangerine and grapefruit and another yam and coconut and the sago palm and elephant ear and cloves and pepper and nutmeg and *maybe* even rice in Southeast Asia, in the Mekong, Red, and Chao Phraya valleys, in Indonesian, in New Guinea; the horse in Inner Asia; the—

What am I missing? How about: cotton, sugar, cacao, coca, hemp, the opium poppy, the avocado, peanuts, soursop, guava. . . .

It doesn't stop you know—it's still going on—but these *all* from . . . *back in the days*, every one of them potentially a solitary independent domesticate (as unlikely as this might be), over time, with a little bump and grind, mucking up into the assemblages that come to typify this part of the world or that: beans and tortillas and chile, for example, in Mexico; in Lebanon, pita and bulgar and lentils and olives and—

Olives! I forgot olives!

And India! And Assam, the Brahmaputra Valley, Burma—[19]

Anyhow another planetary fluctuation, the independent domestication of plants around the world, *everywhere*, between 10,000 and 5,000 years ago, not necessarily . . . *farming*, not right off, more or less horticultural for a while—small in scale—but not necessarily only gradually easing into larger scale swidden systems either; depending, you know, on circumstances, the farming systems carried by Linearbandkeramik cultures (the so-called LBK cultures), for instance, stomping farming populations across central Europe so fast radiocarbon dating can't resolve their movement (it's like one day it's *back in the days* and the next it's full-scale farming, *colonization*, like what would later happen in Kansas after the land rush); while in southern Scandinavia there was this gradual transition through a number of stages (there's a full expression of a kind of Mesolithic, it was a question of adoption)[20]; and so for the rest of the world, this way and that, but in the beginning, you know, mostly the plants were domesticated . . . *incidentally*.

Incidentally by basically affluent hunter–gatherers living in affluent societies in large permanent communities beside lakes or springs or marshes or rivers—with all *sorts* of stuff to eat—and plenty of water for the crops to come.

They were living well, the folk who first domesticated plants. There are theorists who can't imagine people in comfort changing their ways (so much for R&D!), who draw attention to (putative) hard times just past, to the effect the Younger Dryas *must have had* on the Levant, for instance, or to the Middle Holocene cooling that pushed the Sahara into the Sahel (etc.). It's like they imagine memories of these events haunted their "survivors" as memories of the Great Depression haunted the grandparents and parents of my generation, and so . . . what? Memories of hunger prompted domestication? How was that supposed to work? Hoarding → harvesting? *What*? Besides, it's only three generations later today and for the kids in the malls these days (hell, for their parents), the Great Depression might as well have never happened.[21] And the Great Depression was a sharply defined event easily connected to material well-being, not substantially looser changes in ecosystemic webs responding to climatic forcing. I don't buy it, any more than I buy the idea that the Pei-li-kang and the Dhar Tichitt folk were inspired by the example of their less-well-off neighbors to, again . . . *what?* Domesticate rice?

"Domesticate fast, Johnny, or you'll end up a starving like our neighbors."

The ploy may work now and then, but not often.

But these are fantasies in any case, *of insecurity bondage,* and what the archeological data reveal, in site after site after site, are broad spectra of wild plants and animals in the diet—healthy people—thanks to their access to rich aquatic habitats, healthy and . . . *sedentary,* enough food (enough protein!) in the flux to drive all kinds of elaborations, serious clothing and permanent shelters and public spaces (and trade) and the ritual life to go with it (the societies Hodder and Hayden fantasize), and . . . well, the leisure and self-confidence and—hey!—*ambition* to notice, hmmm, "Check out this grain, Shem. What's up with this nonshattering rachis?"

Okay, okay, this is a fantasy too. But necessity is *not* the mother of invention (or rarely) and hunters and gatherers driven into a hunger corner aren't suddenly going to imagine what rich harvests they could have if only they'd thought of planting crops, they're going to be thinking about beating Shem to that elderberry bush down by Buzzard Creek if with it all dried up like it is there're any berries on it at all.

And whether they need to share them—that's what they're going to be thinking.

Agriculture Is Just Another Way of Coupling with the Medium

So here's the story I'm telling: 4 billion years ago these self-bounding networks of self-referencing, autocatalyzing macromolecules pulled themselves together, one thing led to another (we've been through all this), and now these sedentary, more or less well-off *Homo sapiens* living in these well-watered places with their stone tools and their myths and songs and knowledge and curiosity and increasingly complicated social lives—so that there were *all kinds* of vectors in play—*fooled around* (say, like Thomas Edison fooled around) to end up harvesting so much stuff, and sooner or later planting it, that whether from pride in competence, or due to the symbolic value of taming the nonhuman, or because it only takes a generation or two for humans to get used to things ("but that's the way we've always done it"), or because it was something that made sense to and enhanced the power of "opinion leaders," or all of the these and then some, they settled into what would become . . . our agricultural way of life.

It changed everything.

Let's say . . . *again*. Let's say it changed everything . . . *again*. I mean, it's not like things hadn't been changing ever since we came down from the trees. We'd been getting bigger for one thing, and we've seen how that put all kinds of pressure on ranges, making them larger, and on population density, driving it down. The easiest way to keep population density down—so everyone can eat—is to edge on over; and as we've seen that's what the australopithecines and the early *Homos* were up to, spreading up and down the Rift Valley, almost to the Cape in the south and almost to the Red Sea in the north, and then *erectus* with those Acheulean hand axes and cleavers on up the Nile Valley and . . . well, to cut the story short, by the time *sapiens* started planting wheat and rice and corn there were people just about everywhere, with their densities up already thanks to the fancy stone work they'd learned how to do over the previous couple million years, and the spears and bows and arrows they'd figured out how to make, the nets and the snares, *language!* for god's sake, and the container revolution—pots and baskets—and clothing and permanent settlements in lots of places, all of which made it possible for us to be bigger *and* to live closer together (we could eat *all over* the food quality chain), so that, what? 10,000 years ago there were already 5 or so million of us, billions and billions already having left their bones on the planet.

So when agricultural foodstuffs made even higher densities possible, it was like there was this huge global base in place and the whole population rose together, increasing fifteen-, twentyfold in just a couple thousand years.

It changed everything.

Let's say . . . *again*. Let's say it changed everything . . . *again*. One thing worth considering about this history is how language links brains together so that even as our individual brains were getting larger and larger, our collective brain was increasing in size exponentially. We're so hooked on individuality we often miss the fact that when I talk to you—or when you read these pages—what's happening is that my brain is plugged into your brain. There's a lot of science fiction that develops this idea with actual plugs (check out *The Matrix*) or with minuscule virus-like nanodevices (like in Neal Stephenson's *The Diamond Age*), but we have language so we probably don't need these things. Of course we'd been communicating via actual viruses and bacteria all along (and with gestures and no doubt some trophallaxis), but language permitted a new high-speed level of co-ordinated behavior across an ever-widening range of behaviors by hooking brains together via mouths and ears and later hands and eyes into a—call it a society and/or a culture—novel ability to shape its niche. From this perspective, agriculture was just the first *really big* thing these linked brains got up and running.

It's important to be clear about exactly what agriculture is. In fact it's just another way of shaping the immediate environment into a niche, that is, *into a phase of the environment with which we're structurally coupled.* In terms of organism–environment relations, agriculture is continuous with the transformation of the local environment that an individual cell establishes in chemical gradients by its transmembrane transportation of only specified ions. It's not at all nuts to think about that as a kind of *cellular domestication of the environment.* At a more macrolevel there's the spermosphere, the region of soil under the influence of—*organized by*—a germinating seed (the spermosphere is its niche); or at a still more aggregate level, the ectorhizosphere, the biologically tumultuous region surrounding—*and structured by*—the root, or, if it has one, its mycorrhizosphere, which can extend the influence of a root substantial distances (as trade can extend the influence of individual human societies). Ultimately there's the rhizosphere, all the soil shaped by roots. Nests and burrows are a face of the same phenomenon at still another level. So are home territories, the beating of paths through grasslands and forests, the game management of carnivores, the domestication of the environment by hunters and gatherers (the transformation of the local environment they effect in biological gradients by their consumption of only specified plants and animals), their use of fire. . . .

There's a continuum in the way organisms from single cells to societies make their beds, which since they're all biological phenomena is only to be expected. What agriculture produces is a niche *that's able to absorb as much energy as humans with big connected brains can throw at it,* which given that the niche throws more back (though at a steadily decreasing rate of return), humans seem willing to do: whether they put it in *over* the plow, or *making* the plow, or making the *steel* for the plow, or *mining* the ore for the steel for the plow, or *laying the tracks* for getting the ore to the mill to make the steel for the plow, or *driving the engine* of the train that runs on the tracks that takes the ore to the mill where they make the steel for the plow, or—as my friend Dread does—*driving the van* that picks up and carries to their motel the engineers who drive the train that runs on the tracks that takes the ore to the mill where they make the steel for the plow. . . .

Something *like* this has been going on for 10,000 years now (longer if you include all the years of increasingly intensive niche shaping we engaged in as hunters and gatherers), and though it obviously has to come to an end some time (if only when the rate of return hits zero), and though personally I can't stand the ceaseless and rapacious alienation of the land around me,[22] in fact there's nothing *in the record* to say it has to stop anytime soon.

More's the pity.

More's the Pity

Why?

Because when I'm out cycling, you know, riding along a road I cycled the week before, and I come around a bend where last week there were trees, and now there's a wasteland of red clay and Caterpillar tractors and a sign telling me another Food Lion is coming, there's a dull pain in my stomach. I can't stand it. I hurt. Obviously other people don't feel the same way. Or maybe they just deal with the hurt differently. Count their money. Whatever. I don't know.

Yet I'm hardly mindless of my place in this huge history at the end, or any rate, at *this point* in the 15-billion-year-long transformation of stuff. It's not like things haven't been changing from the first 10^{-43} second—

And the thought that the changes then were ... *natural* ... doesn't cut it. *We're natural!* And so the changes *we* make are natural. There's got to be a different rhetoric. That's the whole point of this history. We're descended in one unbroken line from those self-bounding networks of self-referencing autocatalyzing macromolecules—

Do I need to go through it again? *We're natural.* Bottom line. Take a look at the literature referred to in the notes. That we're natural is the heart of Vernadsky's point about life being *the* transforming geological force on the planet. It's the point of Westbroek's *Life as a Geologic Force.* Life pushes stuff around. It cycles minerals. It pollutes the atmosphere with oxygen. It covers it with lichen and fungi and grass and trees. It makes soil out of rock and water and air.

And we're just another life-form.

Which was the point of *Man's Role in Changing the Face of the Earth*— one of the classics—our long-standing, historical, land-changing agency: burning forests, planting grain, hunting animals to extinction. Damming rivers, cutting roads, building cities. We are a geological agent and this is the form of our agency. What's the difference to the trees whether it's a rock slide or a bulldozer that takes them out?

Except that if it had been a rock slide I wouldn't have had anything to say about it. But because I delude myself into imagining that with the bulldozer I did—or *could*—have had something to say, to stop it, to stop the bulldozers, I feel—I sense—*my own agency*; and the hurt isn't just, or maybe it's not at all, the trees coming down (they're a second-growth scrub on former cotton fields), so much as my own guilt, my own complicity in the transformation.

Which is not always easy to see, but: (1) I shop. So I contribute to the culture of shopping that drives the supermarkets in the strip malls. (2) I moved here. So I can hardly complain when other people do the same and want to shop. And (3) a lot of the work is done by former students, planned by them anyway, zoned by them, or designed by them, the working drawings done in their offices, the landscape architects, the architects, the planners *who took my courses*, hundreds of them—

And while I may not have encouraged my students to design parking lots, the point is they graduate, they get licensed, they hang out their shingle, they starve. Or they accept jobs they may be ashamed to let me know they do to pay the rent, the grocery bill. Some of them. Some of them get pleasure out of practicing a competence. Some of them *like* parking lots. But most of them, it's, they take the jobs they can get. And I can see that. It's no different from the rest of the pathways we've been looking at. It's like the chemical pathways reaction chains worked out. It's like the way the cells mutated to deal with the ultraviolet. It's like the pathways followed by life coming onto the land, or the pathways followed by the apes coming down from the trees, or the pathways humans followed into agriculture.

Look, let's say you're reading, and suddenly the noise level rises. People come into the room talking loudly, or a jackhammer starts up outside—sort of like the business majors in the library—and assuming you continue reading, *which path do you take?* Do you expend some psychic energy to block out the sounds, concentrate harder—

Or do you move? Find a quiet place? Ask the noisemakers to shut up? Turn on your portable white-noise generator, drown them with "silence"?

Each of these is a distinct pathway, and each pathway couples with the environment in different ways. At the very least, the concentrate-harder one, you *stay where you are*. The move one, you *go somewhere else*. Because of this the path you choose ends up coupling you with different environments. And say, for example, because you moved to the environment you did, you met someone who changed your life. There's a hint of the butterfly effect there, but the fact is you put yourself in the environment—or the environment embraced you (to turn it around)—in a different way, with different potentials, et cetera. And which path you follow, or take, or which path draws you in (to turn it around again), depends on the potentials and the constraints. Let's add a constraint: you *can't* move (you're supposed to meet someone there, or you're serving a detention, or in 15 minutes you have to take something out of the oven). Then, the pathway most likely to draw you in will be the concentrate-harder one or asking the noisemakers to turn it down. But you can't concentrate. You have too much on your mind and the reading's insufficiently compelling. So you have to ask them to shut up. But let's add the constraint that you're afraid. What do you do? New pathways open. You blow up. You buy earplugs, give up reading. The point is, molecules built themselves up ±3.8 billion years ago in precisely the same way, ions bonding in a flux. And wrapping themselves up in the membrane was like moving into a private office where the likelihood of being disturbed while you're reading is reduced. It's the same with the rise in oxygen, with coming up onto the land, with coming down from the trees, with the domestication of the environment (which is a lot like building a membrane), with commercial development, inventions, the articulation of scientific hypotheses, the writing of a book, a conversation. It's *exactly* how conversations flow. And it's how students with no real interest in designing parking lots end up doing so.

Put this way it can seem so simple, but it gets harder when you start getting lots of pathways tangled up together, at numbers of levels. It gets nuts almost immediately.

For instance, the *reading* in the example above embeds the whole issue in, to pick just one vector, the larger history of reading, the privatization of experience (the expectation I have as a reader that I should be able to do something by myself), the commerce of writing, the technology of printing, and the ways all this was transmitted to or reproduced in you, learning to read, in school. To say nothing of the history of taste and desire that propelled you to choose whatever it was you were reading. And that's just the reading side, or part of the reading side—

Then the noisemakers have their whole history and their expectations which is part of how they see the place where they meet you, so that it's part of the history of public spaces—

All these things are intersecting and we haven't even alluded to the metabolic level—everyone has to eat—or the chemical level—

Which may be the hardest to see, but reading is connected to paper which is tied to trees which are major players in the carbon dioxide budget which is tied into global climate. Reading—the dissemination of technology, of descriptions of machines, of methods for making, for doing things, on paper—has been a major, a huge, an unfathomably huge, vector in the human organization of its way of being. We shouldn't kid ourselves. If humans do contribute to the early onset of the next episode of glaciation, reading will have played an indispensable role. It's not some innocent bystander. Reading is us!

And we're the breath of the land—

So reading is the breath of the land. No, let me get this right: reading is like the teeth or the tongue. It helps shape the breath of the land. And with respect to the Food Lions, even though I personally will never shop in one, I'm definitely complicit here, and I feel that—

And maybe it's liberal guilt, but whatever it is, it's real. And just as natural as driving a bulldozer. Look: there are people who don't even like trees. And that's natural too: trees come down in high winds, smash in their roofs or squash their cars. Or the trees drop things on their cars that stain them or leaves that have to be swept up or seedlings that have to be pulled up. Trees are a pain in the neck. And if you don't need shade because you've got air conditioning, what good are they?

Some people think trees are beautiful, but it's not hard to dismiss them as "tree-huggers." In any case there's no absolute place to stand to discriminate between the tree-huggers and the car-lovers, each is making a judgment, and neither is more nor less . . . *natural.* You grow up in this culture it's unnatural *not* to love cars. These judgments—trees are good, cars are more important—are really just names we give the orientations toward phenomena taken by organisms. Like, in protoctista, they're like . . . "Okay, let's move in and engulf this thing," digest it, whatever. Which would be " . . . Umm, good!" Or, let's not. Which would be "Yech!"

This is pretty much what judgments amount to in us, though they may not all be about eating. But "She's fine" or "Or that was a great movie"—and their analogues—are pretty obviously related. In us I think we surprise orientations in ourselves—"Wow! I'm feeling terrific!"—and name them—"This is good!"—which now, as judgments, we can reflect on and talk about, share with others, in language that seems to be about the world instead of about us. Clearly judgments about land change are of this type. They acquire an agency all their own, and it's, like, the sum of judgments that carries the day: "To bulldoze or not to bulldoze—"

What I'm suggesting is that the decision is rarely some kind of spontaneous, individual conclusion. It takes a shitload of capital to buy the land, rent the dozers, and pay the labor to support this whole system: lawyers, brokers, government, the whole schmear. There's a huge permitting process. And the whole thing moves somewhat coherently according to the sum—or some averaging over, or taking into account of—a whole range of judgments on the appropriateness, on the rightness, on the adaptiveness of these courses of action at many levels. I guess we call this "politics."

I'm not saying this makes it good or anything. I'm certainly not saying it makes it fair. Or even adaptive in the long run. It may kill us. Or help to kill us: a failed experiment, a pathway leading to a dead end. But on the other hand, it's not some rogue gnawing at a tree with his teeth.

My hurt—whether for the trees themselves or for my guilt in not speaking up more loudly for them or whatever—is the system. . . .

No: let me get this right. I'm saying that my hurt *is the land,* expressing *through me* or *in me* its disquiet with this course of events, even as it's expressing other things through the bulldozing, through its construction of these things, these strip malls, whatever. Because they too have to be made by the land. Though maybe "expressing" is the wrong word. Or maybe it's the right word. I mean, we are a piece of the land wrapped up so that the land can differentiate itself, we are the land, at least the breath of the land, and when I hurt, it's like the land's taking a deep breath for a change—

The thing is, since we *are* the land—we and everything else—in the only form in which the land exists, it's going where . . . we—and everything else—take it. But this is why we're fighting. Because we don't know where we want it to go (the *land* doesn't know where it wants to go [it doesn't *want* to go anywhere]). And because we're all yanking it in different directions, it ends up following the vector of our indecision.

Not because we're not *going* anywhere, but because *this is how* the land goes where it's going to go. You could think of them, all the judgments, all the positions, as mutations. There's the strip development mutation. There's the anti-strip development mutation. And some are going to maintain their coupling with the environment, and some are going to lose it. Or, more likely, both are going to maintain their coupling, and they'll be like different species, coexisting, complicating, enriching the niche. I hate strip malls, but what if I'm just an elitist jerk?

What I'm saying is that this must be what it felt like to have lived through one of those flexures when things changed, like, the apes coming down from the trees; and change probably *always* feels like this. Look: life, the land, the universe, is always changing, so ordinary daily life *has to* be one of change. So therefore everyday life feels like change and change feels like everyday life. We just don't think about it that way. We feel it's different when the kids go off to college but not when they switch from middle school to high school. Or we know it's different, but they're still home, so it's not that different. *But it is.* And let's face it, universal education as we know it is nowhere near 100 years old. So it's been like five generations have taken us from assuming few would go to school—when it wouldn't have been an ordinary, daily, defining part of everyone's life—to being something "everyone" does. So, at the very time our lives are changing with the kids going off to college, the assumption that kids will go off to college is changing too. And it's changing as part of the continuing changes we associate with the words "industrial revolution" which remain embedded in the continuing field of changes we associate with the words "agricultural revolution"—

Conventionally, the "industrial revolution" is supposed to have been over long ago, but actually it's far from finished washing its way around the globe. Or even through a culture like ours. Even the factory phase hasn't ended. Look at Research Triangle Park in the Raleigh–Durham area. It fronts as some sort of 21st-century think tank and research center, but among other things what really goes on out there is IBM assembles all its PCs. It may not be a steel mill, but it's a factory, with folks on assembly lines earning just above minimum wage. . . .

Anyhow, the little changes like your kids changing school are geared into these bigger more obviously landscape changes. It's all about the land, always, but it's hard to notice the change, especially at the life-cycle level; and therefore, the way change is an inbuilt, ongoing, daily part of our lives; the changes, at every level, shaped and driven by the judgments we make, and the opinions we express, and the positions we take. And vice versa.

What I'm saying here is that I don't know whether it's the judgments that drive the other behaviors, or the other behaviors that drive the judgments. I don't really know how those things are related. But I have to imagine it's always *felt* like this—even as folks were imagining they were living unchanging lives—at least since *sapiens* began to burn over vegetation to select for this or that plant, or to drive game. It's certainly felt like this ever since we started to protectively tend these plants at the expense of those—which was way back in the days—and since then it's been just one increasingly laborious change after another: replacement planting (initially in the wild), transplanting, sowing, weeding, harvesting, fertilization, drainage, irrigation, land clearance, systematic soil tillage, breeding, plowing, harnessing animal labor, harnessing water energy, harnessing fossil fuels, making artificial fertilizers, manipulating genes. . . .

I just recited David Harris's evolutionary continuum of people–plant interactions. There's nothing necessary about it. Or unidirectional. Or deterministic. It just relates level of energy input to ecological effects. Robin Fox would say, I'm sure, that it's an index of our increasing slavery.

And the cropping's gotten more and more intensive: from forest fallow to bush fallow to short fallow to annual cropping to multiple and continuous cropping to greenhouses. . . .

And more and more dependent on fewer and fewer plants. . . .

And that's where I was heading when I said more's the pity. I was thinking about the rapacious, ongoing, unceasing alienation of the land on the part of humans that has attended the agricultural revolution, the *ongoing* agricultural revolution. And I was saying more's the pity that there's nothing in the record to say the rapacious alienation has to stop anytime soon. Yesterday the UN announced that the world population—they meant of humans—had passed 6 billion, and that we were adding 78 million to the planet every year. This is the unbearable pressure that in the end is driving all the rest of it, and what I was saying is that there's no reason to believe that we won't keep on increasing our numbers.

And probably it can't go on forever, but there's nothing in the record to say when it has to stop. I'm beginning to fear our only hope for a smooth landing is some incredible viral attack on corn and wheat and rice simultaneously—pow! starvation all over the place, infectious disease, not enough workers, collapse of the industrial system—

That would be a smooth landing. Other scenarios are much rougher (we *all* die off).

People keep trying to guess where the edge is. Malthus, for example, almost 200 years ago, promised all sorts of horrible things in his essay on the principles of population growth, and, yeah, maybe, maybe not—

How can you tell from inside? How do you measure misery? I measure it in my gut when I see another Food Lion going up. It makes me sick. And it makes me sicker to think there's no reason for me to imagine any end to this in my lifetime. It could be Food Lions and Wal-Marts and Burger Kings without woods or fields from here to Durham and here to Henderson and here to Fayetteville, and subdivisions named Whispering Oaks everywhere else.

But it's nothing *new*, this change. It's been going on for 15 billion years.

Resource Switching, Sharing: Storing, Trading

When I was coming up one of the facts they seemed to think we couldn't get enough of was that it took only one farmer to support 25 of us. They always thought this was, like, a . . . miracle, but living in Cleveland as I did, down along the Cuyahoga when they still made steel there, even as a kid I could see it was only thanks to the other 24 supporting him (and each other *in* supporting him). Later I began to realize it was also thanks to our, like . . . *total dependence* on the few crops the farmer could grow that way, and all the coal and oil it took (to feed the fertilizer stocks, fuel his tractor, ad infinitum), and . . . later still . . . that we hadn't gotten the bill yet (we still haven't: wait and see) for the soil lost and killed, for the herbicides and pesticides and genetic fragility and. . . .

I mean, the incredible diversity of life-forms flourishing in nonagricultural ecosystems derives directly from the returns *each* gains from their *mutual* association. As the human niche is increasingly focused on the support of . . . *the human*, this diversity has declined as a matter of course. It's hard to hear other voices when one person's hogging the floor. That's what it boils down to. The flexibility and resilience of the system drops directly as its diversity (the problem with hearing only one point of view), precisely as its susceptibility to destabilization rises.

Due to rust or blight, for instance. Or old-fashioned climatic forcing. . . .

So, okay, if the point to the *resource switching*—way back when the first hominids started hunting—was to smooth out variations in food availability across a highly seasonal environment (which was made smoother still by *sharing*), if that was any part of the point, then dumping all this energy into fewer and fewer crops was nuts. A single bad year could wipe out everything (as almost happened to Ma and Pa Ingalls and their kids in *The Little House on Plum Creek*).

And okay, agriculture was probably never intended—

Hell, there wasn't any *agriculture*. *Agriculture* is solely a retrospective point of view. From up close it was "Hey! Let's try this!" and meanwhile there were plenty of fish behind the weir. It might not have even been about *food*. It might have been fish poisons they were growing, or dyes (though if so it hasn't shown up in the archeological record yet), or flowers to put on the altar, and, *okay*, let's say it *was* all about magically getting a handle on the world, running a little piece of it as a metaphor for running all of it, or at least being able to count on all of it *as if* you ran it. Or maybe it was like when you've finished unpacking and you set up . . . whatever it is . . . that turns the house into a home. Something like that, and it started back in the Paleolithic with the figurines and cave painting, or before that with the first stone shaping, *getting a handle on it*, and as it got more and more involved, the handle had to get more and more able to . . . *handle it all*, and domestication was about that, really was magical-symbolic, and the . . . cornucopia of food that resulted was . . . an added dividend.[23]

You know sometimes, when you're contemplating stuff you don't absolutely need at the moment, how you decide to stick it on a shelf in the closet, saying, "You know we might need this some day." And maybe these *sapiens* didn't have a closet but they had the containers—they'd had the containers for centuries by now—and that's how they ended up *storing* the food.

Anyhow, the logic of storing food is that it spreads the environment across time: the grain ripe *then* is at hand *now*. Well, if that's the logic of storing food, the logic of trading it is that it spreads it across space: the stuff that comes from *here* is at hand *there*.

It makes the niche much, much larger.

Which by bridging biomes and climate zones makes it much, much harder to have a bad year, unless the whole planet's going under.

That's the logic anyway. (It's also the logic of hedge funds.) In practice it's hard to pull off.[24]

Lately every year's been a bad year—*in a lot of places*. Which is really no surprise since we're running the whole enterprise these days on basically four grains plus soybeans.[25] I mean, *this is beyond monoculture*. It's a system massively susceptible to a blight or a virus or an infestation or a drop in rainfall, which, hey! global warming. . . .

But you know the drill.

Drought isn't something that hasn't happened before. *It's happened all the time*. It was precisely to smooth out the irregularities in food supply caused by highly seasonal environments (read: tropical wet–dry savannas; read: dry seasons) caught up in longer term fluxes of greater and lesser extremes (read: climatic cycles; read: droughts) that the hominids started food switching in the first place. So, no, there's nothing *new* about drought. But what's new—and more today than 10,000 years ago—is, well, there are so many of us now that small game hunting is no longer an option (we've usurped the game lands to grow grain), and the waters are overfished or polluted, so forget the Sauerian fallback position, and. . . .

What it comes down to is food switching's no longer an option.

Stitching the Niches Together

No, if it weren't for trade it would be death unimaginable.

Not that there isn't death aplenty (*has been from the beginning*), or that we could have grown such an agricultural system without trade. I mean it's not like trade comes along to save humans from their plowing addictions, it's that the two grew up together, the one prompting and pushing and pulling and making possible the other.

Listen, I want you to grow stuff so I can ship it.

What?

Well, I've built this railroad and. . . .

But of course it was only the agricultural surplus that fed the guys who laid the tracks. . . .

There was nothing *new* about trade, or at least about the long-distance movement of . . . well, *good stone*. And maybe "trade" is the wrong word for periods about whose economic organization we know . . . nothing. But if not trade, movement for certain. The stone for the tools Louis Leakey found at Olduvai when he first visited the gorge had been transported there from several kilometers away,[26] and throughout the Paleolithic fine-grained rocks like chert and obsidian were moved increasingly long distances all over the world.[27] So there's this ancient history of moving *things* increasingly long distances. Trade had little to do with food at first, directly anyway, though, c'mon, what was the stone for? In all these early agricultural communities—

And still that's *our* vision: *agricultural communities*. Maybe they were planting some things, but they were still hunting too, and there were dooryard gardens and small mammals to snare and fishing holes with fish in them and crustaceans. They weren't *agriculture* communities. They weren't sitting on a one-legged stool like we are but on a bench with dozens of legs even if some of them were pretty spindly. So in the beginning and for a long time, it wasn't food per se moving along the trade routes, or if it was food it wasn't the kind of food you stoke your belly with (though soon enough, and before it, salt). Within a few thousand years of the domestication of grain a thriving trade in the Fertile Crescent already was shipping copper from the Wadi Faynan in the mountains of Jordan across the wastes of Wadi Arabah into the Beersheba Valley.[28] They were smelting the ore in towns up and down the coast and making adzes and awls and axes and chisels. This was another change that like the planting of crops took place—from an aeonic perspective—all over the globe at once. *Okay*, maybe a *smidge* later in the Aegean, but at nearly the same time in the Indus Valley and, in general, the metal working shadows domestication (*is probably the mineral face of domestication*), not everywhere (just as by no means everyone's jumped on the agricultural bandwagon), but people in many places slipping and sliding out of the Stone Age into the Chalcolithic, into the Metal Ages generally.

So first stone moved, and then metals. . . .

And with the metals the logic began to move. Slowly it ceased being simply about resource shifting. Slowly it became about ensuring that those involved in the moving had something to move. In a way it was like they had shifted from working the land directly to working those who worked the land, though in fact this is an illusion, and all they'd really done was shift up yet one more trophic level in the niche. But for them, then, working the niche meant making sure they had something to shift, some mineral, some foodstuff, something it would pay them to carry on down the line or out a long distance. That is, yes, the domestication of minerals, but almost coterminously . . . *the domestication of the movement itself*.

Domesticate: to train it to the use of the people who use it. . . .

Most trade goods didn't move too far in the beginning—they stayed for a while inside a radius of say 60 or so miles.[29] But the arsenic- and antimony-rich copper ores cast into a crown 5,500 years ago at Nahal Mishmar had traveled 800 miles at least,[30] and as long as 5,000 years ago they were using cloves in the Euphrates Valley, *cloves* now, from the Moluccas in the East Indies, thousands of miles away.[31] The speed with which the radius of trade goods grew is an index of the rapidity with which the logic shifted. By the time the Bronze Age opened—which it or its analogues did at different times in different places—Baltic amber was all over the place, in Hungary, in the Balkans, and in Greece; and Greek pottery was in Italy, in Sardinia, and in Spain; jade was all over China; gold was moving in Africa; turquoise from Santa Fe was moving throughout Mesoamerica; remains of tropical parrots and macaws from Mexico have been found in Anasazi graves; salt was everywhere. . . .

Trade, raid, tribute: people hooking up in any case, smearing out the niche.[32]

Trailing domestication . . . an aspect *of* domestication really, making ever more of the globe our home *wherever we lived*.

Taming the Farmer

Which is to say that trade is a form of domestication.

Our image of domestication is a kind of taming. That's the root. The wild horse is tamed by the brave young kid who just hangs on to the bridle until the horse is willing to do his bidding. And depending on the story we want to tell, we make this more or less brutal, the horse beaten into submission, or gentled into a loving willingness to work with the kid (but with the horse's spirit left intact so that later in the movie he can save the kid from some horrible disaster). But in either case something not under our hand is brought under our hand.

Though there's a covert reciprocity we don't always show in the film, providing the horse a place to live, making sure it's fed.

The argument I'm making is that this is something *life-forms* do. *Life-forms* domesticate their environment. It's what I said not too long ago about agriculture. I said, in terms of organism–environment relations, agriculture is continuous with the transformation of the local environment that an individual cell establishes in chemical gradients by its transmembrane transportation of only specified ions. This gives the niche—the cell's immediate environment—a tensioning, a character, that's tuned to the metabolic organization of the cell.

I said, it's not at all nuts to think about this as a kind of cellular domestication of the environment and went on about spermospheres, the ectorhizosphere, et cetera. It's a kind of taming of the environment, a restructuring that's supportive of the organism's continuity in it, of its structural integrity, its structural coupling.

You could see it the other way too: the environment supports the organization of this nexus of activity—call it an organism—that helps the environment maintain itself in a sustaining fashion. That's Gaia. Another way of thinking about it is that if the environment can't support the organism, it kills it. Though another way of putting that is that the system evolves. I talked about "pruning" a long ways back: organisms that can't shape a supportive niche lose their coupling, they go extinct. Which is to say, the environment fails to sustain itself in *that particular fashion*. We miss this mutuality by focusing on the organism.

Where I was going with this was to say that the niche construction the cell achieves is exactly the transformation of the local environment that metazoan heterotrophs effect in biological gradients by their consumption of only specified plants and animals. They work to maintain these gradients because, when they change, their own numbers can drop, and the numbers of other organisms can rise. This can lead to a seesawing that becomes a stable characteristic of the environment, stable because it happens over and over again. It becomes an environmental heartbeat. (Stability doesn't mean quiescence; it means ongoingness.) And though in this hypothetical example the numbers of specified organisms rise and fall, they *keep on* rising and falling. Over time, of course, this whole system shifts in *its* niche, and what we see is extinctions, or better yet, continuity with change, because it *is* one unbroken chain, now to then.

The cell domesticates its environment by tuning it. The seed tunes the soil around it into the spermosphere. Plants and animals tune the environment around them into niches by the selective way they have of relating to it, exuding this toxic substance, building a nest of these substances in that place, eating those other organisms, and so making space for something else. We're an animal and this is what we did, selectively harvesting—just gathering—this wheat in those stands as we did. Agriculture as we practice it is a simple extension of gathering, which is basically taking—like other animals—plus storage and sharing, which isn't unique to us either. This taking or gathering or collecting shapes the niche, and it's not all that different from the way birds spread the seeds they eat by scattering them, or carrying them and spreading them around in their feces. They promote the spread of species they like. *That's* domestication.

What that does, what domestication does, is change the character of the niche. This is what's really important. It makes the niche more congruous with the organization of the bird, or with us, or with whatever. Storage smears this more congruous niche over time, so that if the niche offers wheat now, at the end of the summer, then by means of storage the niche can offer wheat year-round. That's another aspect of domestication. It changes the temporal structure of the niche. And trade, I was saying, opens the niche up, pushes its boundaries out. Trade achieves this by dedicating—as in, say, a dedicated phone line—people, which is to say, labor (which is to say, focused, niche-transforming energy), to move outside the home range by provisioning them to be able to devote all their energy to moving.

The advantage here is that the niche acquires resources it didn't have. I compared the effect of trade to the mycorrhizosphere before. It has that effect. It spreads the niche out, though it's dilute at its farther reaches, because it's extremely selective out there. This is another way of saying that traders trade in only certain things; the full niche doesn't flow through the line, only things that, from the perspective of the ongoingness of the organisms supporting the trade, will be supportive of their continued structural coupling (i.e., that will sell).

What happens once traders emerge is that there are two different organisms that couple with their environments differently: the farmer exploits grasses and herbs and animals; the trader exploits farmers. So the traders have to tame the farmers as the farmers tamed the plants and animals.

This is because the traders live off the skim; and they have to be able to assure themselves that the farmers are going to produce stuff to flow. Because with no flow, there's no skim. And this process was more or less brutal, the farmers beaten into submission, or gentled into a loving willingness to work with the traders (but with the farmers's spirits left intact so that later in history they'll go to war for the traders).

Even in the best cases it was pretty grim; and it's this logic that has driven, *that drives*, road construction, the encouragement of exotic tastes, telephone systems, stock markets, railroad construction, FedEx, derivatives markets, television, the Internet—if not life itself, then certainly this endlessly filiating network. And its logic is such that only when the *whole world* is connected up will it be connected enough, and then only if things can flow instantaneously! Which is why Mosammat Anowara Begun has a Nokia cell phone to her ear in the Bangladeshi village of Chamurkhan.

It also means there can be no end to our alienation of the land. None. New Food Lions . . . everywhere.

Globalization

Food moving . . . *soon enough* staples, like the barley and oil, for example, sesame and linseed, shipped by Mesopotamia down to the Gulf ports between four and five . . . *millennia* ago.[33]

We always go, like, "Dude! They were doing all this way back *then?*" but it's more a problem of our perspective and our parochialism than of their achievement. It's no big deal for us: why should we imagine it was for them?

"But, dude, we've got cars and jets—"

And *they* had canoes—bark and dugout—and skin floats and skin boats and boats made out of reeds and outriggers and mtepes and dhows and junks and galleys and round ships and feluccas[34]; and, without getting into it, many, many reasonable people think folk had been using these to cross the oceans (and surely the way they moved through Beringia was by coast-wise boating), across the Pacific, for example, from Indonesia to South America,[35] voyaging . . . *to and fro repeatedly* . . . as they did a similar distance (without raising the same number of eyebrows) across the Indian Ocean to Madagascar; and from Europe to North America (probably well before the Vikings), and from Africa to South America—

Anyhow, whatever the merits of these arguments (which are not insubstantial), we've got documentary evidence of relatively large ships plying the Nile and eastern Mediterranean 8,000 years ago, and not too much after that in China.

Wondering about why people started trading, about what they thought they were achieving, is like wondering why *Hynerpeton*'s descendants started walking on the land. "Avoiding predators was just one possible motive," opines Kerri Westenberg. "A short walk might lead them to meals. Or perhaps landlocked pools offered safe places to lay their eggs."[36] It's not quite "Your guess is as good as mine"—because there're all kinds of constraints we do know about—but it's every bit "For the same reason the apes came down from the trees" or "Like they started planting crops. . . . " The facts are, the goods were en route, and early on they were able to move great distances.[37]

Globalization is nothing new.[38]

Those cloves, for example. Five thousand years ago when the trade took off (if that's the way to put it), cloves grew only on five of the Moluccas, on Ternate, Tidore, Motir, Makian, and Batjan—little dots really, off the coast of Halmahera,[39] itself no big deal, between Sulawesi and New Guinea. Trade early on would have been from one island to another, cloves from Tidore (and nutmeg and coconut) say, for sago from Halmahera, with the Halmaherans turning around and trading some of the cloves with the Javanese for rice. It's silly at these dates to talk about China (because as such it doesn't yet exist), but it's hard not to imagine that the Ta-p'en-k'eng along China's southeast coast and on Taiwan weren't trading with the Javanese and the Sumatrans (or maybe the route was through the Philippines); and that the Javanese and the Sumatrans weren't trading with folk along the Coromandel and Malabar coasts—well, they were, we know they were—cloves and Javanese pepper and Sumatran camphor for what? copper? gold? lapis lazuli?[40] I see them coasting around the Bay of Bengal at first, but given the nature of the monsoon, soon enough across it.[41] Here the cloves would have been caught up in the growing trade between the Harappan culture evolving in the Indus Valley and those al-

ready flourishing in the Euphrates Valley, a heavy trade conceivably canti-
levered off the long-established trade in grains and oils that long since
connected Mesopotamia with the Arab ports along the Persian Gulf.[42]

Ports! Another retrospective construction! Let me say: with the fishing and pearl diving *sites* adjacent to springs and marshes along the west
side of the gulf, trade with them, and with the ports on the other side
along the Makran coast with their easy connections up into and over the
Zagros.

Ports!

It's easy for us to imagine there were, you know, face-to-face exchanges that might have developed through some unscheduled routine
into, what? an intermittent but scheduled . . . *market?* Maybe the evolution
of reciprocal ceremonial exchanges? Finally the development of trade for
. . . *gain?* Administered soon enough, markets established and overseen,
ports (*ports!*) soon enough . . . regulation consignment-based trade, entre-
peneurship, all the rest of it; but in any case, 5,000 years ago, *incidental
stuff,* "trickle trade,"[43] moving things on down the line.[44] And the story
I've told about the cloves moving from Tidore to Halmahera to Java to the
Coromandel and Malabar coasts and up through the Persian Gulf (on
reed rafts probably) into and up the Euphrates Valley might make you
think that's what this was, moving things on down the line. But I'm not so
sure about it. Trying to imagine the beginning—how, in other words, this
trade *self-organizes*—is like trying to imagine which atom first hooked up
with which in the self-organization that led to the earliest protocells; and
it misses the point, really, that there's not some first gesture from which
the rest follow, *but a simultaneous reaching out of hands that grasp each other
to form . . . a knot? a ring?*[45]

And, so, *okay,* originally the cloves *did* go to Halmahera for sago, and
the Zagros–Persian Gulf links *were* made as part of an early metal trade.
The thing is, embedded in any trade, no matter how local, *is the logic of
trade,* which once it's established is autocatalyzing, because, as we've seen,
traders don't *make* things, but live off the gradients that exist between
those who do. Goods flow down the gradients—well, *traders move goods*
down the gradients—and they live off the flow. They dip. (It's like the ATP
our cells scrape off as oxygen tumbles down that glycolytic chain.) Given
their dependence, soon enough traders find themselves *maintaining* gradients (you know, "artificially")—if only to ensure their own survival—*exag-
gerating* gradients (e.g., by diverting stuff, holding it back), even *creating*
gradients (e.g., by monopolizing production at one end, and then killing
it off to force trade).

Sound familiar?

It should. The point is, it wouldn't take much to precipitate the larger
trading structure through which the Mesopotamians would be able (or en-

couraged) to press their demands on Tidore (however elaborately buffered), and, in this way, structure the entire sequence of links. *Suddenly the
chain–the ring, the knot–gets taut.* Can you see this? These incipient local
links just beginning to form but even before they get too far, this larger
structure crystallizing out of them down through which all the "I'd-be-willing-to-give-you-these-for-those" can flow. And once there's a flow, well, "a
person can go far by just skimming off a tiny bit of it."[46]

Mercantilism!

What I'm trying to get at is that we don't have to imagine this *slow
gradual* development of trade, but acknowledge that as soon as the food
surpluses generated by agriculture had released energies formerly focused on field and stream, trade followed (just like the domestication of
minerals did), linking together all these high-energy, surplus-generating,
experimenting-like-mad cultures—as conservative as they also undoubtedly
were[47]—which in this way were exchanging not just tin and mace, cinnabar
and cinnamon, but solutions to common problems achieved by the locally
linked networks of brains and thereby—c'mon, say it—expanding the
phase of the environment with which we were all structurally coupled,
and so affecting life at every level.

Soon enough trade was steering the juggernaut.

So, no surprise to find that a straight-up consignment-type, long-
distance clove trade soon succeeded the earlier, more *articulated* exchange, the dhows (or whatever they were) that followed "the cinnamon
route" slipping through the Sunda Strait to hitch a ride during the winter
monsoon on the Equatorial Current, straight across the Indian Ocean to
Madagascar, perhaps as long as 3,000 years ago, afterward trading in the
Rufiji Delta (probable site of ancient Rhapta), before coasting north to
the Cushite polities of Punt (and later Aksum), where the cloves (and cinnamon and aromatic Indonesian gums) would have been transshipped for
Egypt (the Mediterranean and beyond), hitching themselves here to the
well-established trade of Egypt and Punt in ebony and myrrh that sailed
the Red Sea or crawled along the Nile.[48]

Eddies on top of eddies. . . .

It's easy to make these trade connections sound more substantial and
regular than they were, especially by yanking the temporal dimension and
letting the routes from every era pile up. (They were regularly broken or
they subsided and years passed before they were revived.) But it's even
easier to underestimate the degree of connectedness. The whole thing's
been linked for a long time, not just in this Asian realm I've been describing but almost everywhere. I could as easily be talking about the trade
across the China Sea, or up and down the Usumacinta River, or along the
Sahel.[49] There's nothing special about cloves either. The list of things that
moved is endless.

And the cross-linkages! There were ties, for example, across the Red Sea between the Cushites on the one side and the Arabs on the other. And sure, these waxed and waned, and political dominance occasionally slipped from one side to the other, even as it slipped north and south—for example, Aksumite influence at times reached all the way to Tanzania, to its shell and turtle trade. At the same time both Cushites and Arabs maintained ties with others, the Cushites, on the one hand, inland to Kush, to the Kingdom (that is) of Nubia (which for a while there, 2,700, 2,600 years ago, controlled the whole Nile Valley[50]), as well as to the East Indies. On the other hand, the Arabs were linked to Mesopotamia and its heirs and assigns (to Assyria for one, which dominated the Fertile Crescent at the time the Nubians were controlling the Nile), and to the East Indies independently, trade that not only tied them to the Malabar and Coromandel coasts, but also to the Indus Valley.

So all these places were linked and cross-linked, what today we call China to the East Indies and so to Madagascar and the Indus Valley and the Malabar and Coromandel coasts, and all these through the Nile, the Red Sea, and the Persian Gulf to the Euphrates Valley and so to the Fertile Crescent, and—via the Silk Roads—*back to China*[51]; and all linked *out* too, laterally into Inner Asia, Europe, the Sahel; and everywhere, cities, all the cities we've never wanted to admit existed, Napata and Meroe and Rhapta and Aksum and Adulis and, you know, there is no end to the naming of ancient cities. . . . [52]

And so they weren't New York, but then there weren't 6 billion people on the planet either.

The dominant idea of the world as a bunch of backward spots isolated until touched and united by Alexander, or Rome, or European colonists is *so* much baloney. There were strong local links everywhere, like those between Java and Halmahera, Mesopotamia and Harappa, Punt and Egypt, and they'd been invigorated for millennia by a kind of long-distance tensioning that gave the world a "global" character way, *way* back in the days, as if what with the winds aloft and alow and the ocean currents and the continents doing their bump and grind around the planet, microbes and seeds in the air, birds flying hither and yon and *oh! those atoms flowing in and out of Leonardo's lungs!* it hadn't had one all along.

Modern Times (Nothing New)

Salt moved south and gold moved north across the Sahara from the minute the camel penetrated the region some 1,600 years ago,[53] not a whole lot of either at first (and the caravans carried other things besides), but the Muslim world—which was a major consumer of gold—behaved in relation to this region's fringes as an enormous zone of demand; so that from the seventh century on (like 1,300 years ago) the gold was being caravanned from West Africa. It came from the Asante fields of the Gold Coast, from the deltas, that is, of the Komoe, Tano, Pra, and Volta Rivers. It came from the Senegambian mines of Galam and Bambuk. It came from the mines of Bure in the Niger inland delta. It all moved through the ancient kingdom of Ghana (later Mali) which waxed fat on the trade . . . *tons of gold* . . . every year from the ninth century on and—I have no intention of even glancing at the convoluted history of Islamic coinage but—the "flood" of gold in Christian Europe in the 11th century was of substantially West African origin. *Tibr* they called it, the purest in the world.[54] What did Europeans do with it?[55] They coined it, and traded it with the Muslim south, sole supplier of . . . cloves . . . cloves and silk and porcelain and pepper and you name it, all wonders of the East.[56]

So the salt flows south across the desert and the pepper floats west across the sea, and gold from the Akan fields of the West African Coast ends up in Hsing-tsai in the coffers of Kao-tsung—*what's new?* And it was pretty obvious you could go far by just skimming off a tiny bit of the action, as throughout and after the European crusades the Venetians made clear; and the Portuguese stuck there on the Atlantic fringe—basically out of it, marginal to it all—worked it out that if they could somehow figure out how to get into the Indian Ocean by sailing south around Africa, they could cut the Venetians out of it, the Venetians! hell, the Turks too, and the Arabs, cut 'em all out—*move everything by water*—which may be part of what the Chinese had in mind when the 15th-century Yung-lo emperor of the Ming sent his admiral Cheng Ho on *massive* tribute-collecting voyages into Southeast Asia, the Indian Ocean, the Persian Gulf, as far as East Africa which he reached 50 years before Vasco da Gama.

And what would have happened had Cheng sailed around Africa and "discovered" Europe? Nothing would have been much different, not from a global, not in an aeonic, perspective. Maybe the Chinese would have "discovered" the Americas. There probably wouldn't have been an African slave trade then, and it would have been the Chinese "miracle" chauvinists wrote about instead of the European one, and we'd all learn Du Fu and Li Bai instead of Dante and Shakespeare in the 10th grade, but I can't imagine much else being different.[57]

We still would have made the move to steam and trains and planes and automobiles. Squinting and standing back aways, the picture would still look the same. All those Europeans would still have spilled out of the continent, gone to homestead those new lands the Chinese were opening up. But it would have been Chinese gangbosses driving the Anglos on the track-laying crews.

It doesn't stop.

I mean, this *is* the story of the Neolithic, of the Metal Ages, of modern times, the thread that makes a whole cloth out of what up close want to pass for a pile of patches, the busyness marking the growth and withering of communities and kingdoms and nations and empires, the triumphal arches and manacled slaves, the massacres and massed brass braying.

What stitches them together is the seamless and unceasing chopping of trees and burning of brush and turning of soil as—well fed—our numbers have not ceased increasing. Five million, did I say, when 10,000 years ago we began this experiment with planting? and before the first flush had faded, when the village farming thing was just beginning to wash around the world and suds up into a faint premonition of urbanity, say 6,000 years ago, *arrggghhhh!* the numbers are *insane*, you want to, like, duck and shield your face: *86 million*, like . . . (snap of the fingers, twist of the wrist) *that*.

And now, of course, all *those* people could become farmers. Or worse, traders and protoindustrialists dedicated to supporting (while generally ripping off) the farmers as—ultimately—we all do, me and you included, eating what the farmers produce, or slaving down at the mill to make the steel for their tractors, or working in the capital markets that can't be separated from making steel these days, or writing books for the traders' leisure reading,[58] or, in the early morning before the traders come to work, emptying their wastebaskets and sweeping their floors—*oh*, we're *all* still part of it, *still!* even after all these years! just as much a part of it as when each of us ran with the rest of the troop after the gibbering monkeys in the trees aloft . . . for they too serve who stay at home . . . a little *distanced* perhaps . . . slightly out of it . . . stretched . . . thin feeling (like Gollum too long with the Ring), alienated (in that vocabulary), and why not? Six thousand years later there are *6 billion* of us, the land 10,000 years ago first sown with seed no longer exists, and according to an item in a "Harper's Index," the average distance anything *you* eat has traveled to reach your mouth is 1,500 miles.

No wonder you feel a little stretched.

You are.

But was there some stopping point we missed? No, I don't think so. The logic of domestication is simple, but it contains no codons for turning it off, and so as we've seen it moves from plants to animals to minerals to farmers to traders, for the merchant bankers domesticated the traders, as the capital market boys have domesticated the merchant bankers, and . . . and . . . and. . . .

And we can run out these sequences: hunter–gatherers → fisherfolk → planters and/or seed agriculturalists → pastoralists → animal husbandry and/or hydraulic agricultural systems → market agriculture → industrial agriculture; or, camps → permanent settlements → towns →"Oriental" cities → mercantile cities → industrial cities → service cities (or "edge cities" or whatever they are); or, Paleolithic → Mesolithic → Neolithic → Chalcolithic → Bronze Age → Iron Age → Middle Ages → Modern Times; or, social/linguistic revolution → agricultural revolution → industrial/medical revolution . . . but call them what you will, each of them is just a another way to spell: "D*O*M*E*S*T*I*C*A*T*I*O*N," which, as we know, is no more than a way to construct a niche that can take anything we can throw at it.

So far.

En route, the trading required to link all the little domestic niches into the planet-scale multiniche niche we live in spawned . . . writing . . . as a way to keep track of what went where, to rationalize it,[59] as if we could get the energetics of at least this part of the process self-consciously in our hands; and as an added dividend, like the cornucopia of food that resulted from domestication, we ended up with books,[60] which not only link our brains together but . . . *freeze-dry them* . . . and so we can see as never before a little more of our world, back (that is) a ways (all the way) and maybe a little bit forward, and, seeing where we are, maybe make a little something different out of our lives than's been true for a while. My having written this and your reading it are, after all, the way the universe is expressing itself—through our lives—in the here-and-now.

What's it saying?

I don't know. It may all be hopeless. Then again, it *is* hopeless. If you've read this—and bought *any* of it—I can't imagine you still entertain the notion that there's any point or direction or meaning to any of this. And once you've grasped that, what's to look forward to? The next meteor? The return of the ice? The day the Sun runs out of hydrogen and starts consuming its helium and all of a sudden . . . the Earth's *in* the Sun?

Not that there's any serious fear of our being around by then. No species has survived anywhere near that long. *N*owhere near that long. *So what's the point?* Well, there isn't any *point.* Except what you make out of it through your struggle. *It just is.* I like Jack Harlan:

> Agriculture did evolve in several parts of the world and formed the bases of civilizations in both the Old and New Worlds. Towns and cities arose. Societies became stratified with ruling families, merchants, artisans, standing armies, navies, priestly castes, entertainers, healers, laborers, the poor and the slaves. History tells us of conflicts, slaughters, conquests and defeats, oppression and unrest, plagues and pestilence, crop failures and starvation. Instability of agricultural systems has led to untold misery as populations rise in a series of good years to exceed all possible support in a series of bad years. Many of the conflicts can be

traced to the instability of agriculture and most to overpopulation. Now there are more than five billion of us living on the produce of unstable systems.[61]

Six billion of us, actually, and counting. It would help, I suppose, if we shared what we've got. Trade and its analogues does move stuff around, but never equitably. It's insane today what some people have to eat and others don't. It's not, though, you know, that we're not living in harmony with nature. *We are nature.* Bottom line. But nature is a profligancy without reason, nosing every pathway open, and harmony is *our* idea. This means harmony's an idea that arises in nature (since it arose in us and we're natural), and though not all the ideas that arise have their match in the world (wouldn't that be something), maybe through us, maybe through us. . . .

[*Sounds of celestial laughter.*]

Yet all the outcomes in the record don't exhaust the possibilities, and the record is rich with extinctions.

Hey: this instability is nothing new. Maybe the form is, but the instability isn't. Life's unstable. So's the rest of the universe. That's the deal, and so our history is one of unfolding instabilities. No rest for the weary. Ever.

I don't know: *what do we want?*

What do you *want?*

The Land Lives in a Network of Unholy Complication[2]

Is this to say that trade is like the wind, a naturally occurring flow induced by differences in, let's say, the density of cloves, as the wind is set in motion by differences in temperature? Is it because there're a lot of cloves in Tidore that the clove pressure drives them to Halmahera?

One problem with the metaphor is that the wind doesn't operate this way. It's more complicated than I originally made it out to be. I happen to feel it's really hard to think about the land unless you know something about the circulation of air and water and rock. It's essential to understand that what you throw into the stew goes somewhere else. Whether the *you* we're talking about is you, or some bacterium, whatever's tossed off doesn't vanish. It slips downstream.

Look at the tail Chernobyl wagged. Look at the plumes of pollution you can see in rivers. And these are, like, the really obvious ones people have published all over the place. Thanks to its great plume of dust you can see a car on a dirt road miles away. It's trailing an even larger plume of invisible gases and heat.

Everything's like this. Unless you have some understanding of how air flows, and water—at least—you can't begin to think about *where you are*, because where you are is always upstream of somewhere else.

What this means is that you're not where you seem to be, you're in a larger where that includes *at least* what's downstream of you, downstream in the streaming of anything. Of course you're in someone's else's downstream. And your collective upstreams and downstreams are upstream and downstream of others still. There's no stopping this. The whole planet is downstream of all that solar radiation (without which nothing), and the Sun and the Earth together formed downstream of a sequence of supernovas that had previously enriched the interstellar medium with the heavy elements, and we're continuously bombarded by space stuff—

A question that arises is how knowing this, how knowing that everything came from the Big Bang, helps us deal with our real pragmatic problems now? To a lot of people it seems . . . irrelevant. Immanuel Wallerstein, for instance, wrote an article about this called "Hold the Tiller Firm" in a collection of essays Stephen Sanderson edited.

Wallerstein said something like: "And in this sense everything is determined by the Big Bang if there was a Big Bang. But while it's salutary to remember this, it is not very useful to build our analysis on this quicksand which will very rapidly engulf us. Once again the question is pragmatic." He was really attacking Andre Gunder Frank who doesn't take the modern world system back to the Big Bang as I've done, but does take it a lot farther back than Wallerstein. For Frank, as for me, the modern world system is not something that began 500 years ago but thousands of years ago . . . well, for me, billions. . . .

Wallerstein's right about the pragmatics. You *do* have to choose the unit of analysis appropriate to your problem. On the other hand if, analytically, you're not open to the openness of the system you've bounded—and Wallerstein admits all systems are open—you may not only "miss" the "source" of the problem (or the "cause" of the behavior), but in attributing it to some *inner* dynamic, distort your understanding of that dynamic to make it generate the consequences you think you observe.

We all do this when we think we're doing something wrong, and then it turns out everybody has identical headaches, and they're caused by the vapors from the carpet in the workroom. The problem wasn't inside as we'd imagined, it was outside; and so the solution had to be sought outside as well. The headaches turned out to be something talk therapy couldn't cure, but that interior design could.

This is not always the case—I have no brief against talk therapy at all—but these *are* grounds for criticisms of Wallerstein's contention that the world system begins around 1500. That he finds the world system dawning then is only because he's closed off his European system too much from the rest of the world, and so misconstrues, among other things, the role of American silver. He looks inside Europe because he can't see Europe *in* the world.

And how do *you* keep your vision open?

All this Big Bang history is a way of helping to build an *open* framework for thinking about where we are, for thinking about what it means to be *here*, on this porch, for example, breathing this air, looking across at that apartment, watching the sunlight flash from the windshields of those cars, embedded deep within these nested systems. Where *are* we? It's to keep thinking moving along these lines that I told the simple tale I did, among other simple tales, about atmospheric circulation and ocean currents and continental drift.

And of course there *is* convection and there *are* Hadley cells, and so there *are* trade winds; and cloves *can* move downwind; but it's none of it simple and we don't really know how any of it works. In one of the notes—

Look: all the doubts and second guesses and equivocations are in the notes. The goal was to keep the text clean, to keep it a *simple* story—like a kid's book, *that* simple—and with the repeated recitation of the eons and epochs making it that self-assured too. There's no point trying to have conversations about this stuff with people who think the *world* is 5,000 years old and we're all descended from Adam and Eve. But you know how some of them can recite the books of the Bible? Joshua, Judges, Samuel, Kings? Like Robert Duvall has his kids do at the beginning of *The Apostle*?

If we could teach our kids a history of the land that simple, they'd have a spine for time that would give them some landmarks, a basis for a conversation. It can be complicated or eroded as necessary, but we need a *foundation* for talk—for talk that's not just emotional—about this planet that is our only home. I was thinking, maybe if we could get people chanting: "Supergrandunification, Grandunification, Baryon Genesis—" you know, like a rhyme, like a rap—

Supergrandunification, Grandunification, Baryon Genesis—
Unification, Quarks, Hadrons, Leptons—
Photons, Protogalaxies, Galaxies, Accretion of the Sun and Earth—
Mantle, atmosphere, oceans—
The steep-sided felsic protocontinents (I love that phrase)—
The Age of Prokaryotes—
And episodic cratonization, Ur, and episodic glaciation—
And photosynthetic bacteria—

If I could split myself in two, one of us could do land: Artica—

And—though you can't separate land and life—the other could do life: *aerobic bacteria—*

Baltica—

The advent of an oxygen-rich atmosphere—

Atlantica—

The extinction of anaerobes—

Artica and Baltica forming Nena—

And eukaryotes—

And Rodinia, and the Riphean, and the Sturtian glaciations—

And megascopic eukaryotes and the Ediacarans—

And Gondwana—

Shells and skeletons and invertebrates—

And the Ordovician extinction—

[And now land and life together] The formation of Pangea—

Vascular land plants, amphibians—

Devonian extinctions—

Seed ferns—

The Permian extinctions—

Dinosaurs rising—

And the Triassic extinction—

And the first mammals—

And the breakup of Pangea—

Flowering plants and the Cretaceous extinction and the adaptive radiation of the mammals and—

But now it starts to get messy . . . the Ice Age—

Which we're still in—

But it always gets messy when you're coming in for a landing, everything comes up so fast: Paleolithic, Mesolithic, Neolithic—

Chalcolithic, Bronze Age, Iron Age, Middle Ages, Modern Times—

The Rock Era. The 20th anniversary of the birth of hip-hop wasn't too long ago—

Even though Bob Dylan's still touring—

The wind!

I was talking about the wind. What it said in the notes, where I second-guess the child-like simplicity of the story I tell in my text, was something like, "How does the colossal mechanism of the winds run itself?" Well, this is *exactly* what it says, I'm copying from it: "What are its causes, and what does it achieve? There is no simple answer to these questions, which involve some of the most intractable of all geophysical problems. It is not enough to say that the atmosphere is a heat engine driven between an equatorial heat source and a polar sink. It is an inefficient engine, only about one percent of the heat input appearing as kinetic energy. The north polar surface is not a heat sink on an annual basis. All that can be done is to sharpen the questions and then outline some qualitative answers."

Which leaves us . . . ?

The burden of the remark, which follows 25 pages of a classical presentation of the mechanisms of climate in the *Britannica*, is that the classical presentation is itself no more than a foundation for talk, for thinking, for, as the guy says, sharpening questions. You sharpen the questions. You answer them. And the excitement of being able to answer them *at all* gives the answers an air of conviction they probably don't merit. You explain it to someone less into the problem. It sounds definitive to them. The next thing you know, it's in the science column in the local paper. Pretty soon it's a metaphor in a poem by John Donne. (Donne sure liked those scientific metaphors.)

Or someone like Donne. What I'm trying to say is, pretty soon it's entered into everyday discourse. It's become . . . *the* explanation. Meanwhile, the guys who articulated the problem have become, if they hadn't been from the beginning, increasingly aware of the answers' shortcomings. It turns out the answers don't really answer the question so much as focus it or shape it or raise it in a different form. The whole process is violently dialectical. And it's hermeneutic too. It keeps circling back on itself to rephrase even the most fundamental phrases—

One point is, we can forget the beguiling idea of modeling human behaviors on "simple natural—or physical—systems" like the wind, (1) because the simple natural—or physical—systems aren't simple, certainly no simpler than *we* are no matter how high school physics books make them look; and (2) because we're just as natural, just as physical ourselves. The idea that "natural systems" are somehow simpler is an illusion, typically induced by a *superficial acquaintance* with the physical system coupled with an *intuitive grasp* of the complexities in the human case.

Which is not quite to deny that trade is like the wind. It's more to admit that it's really hard to think about trade which, from a biogeochemical perspective, has got to be about the flows of compounds, molecules, elements. Vernadsky once famously saw phosphorus in flight in a flock of geese on the wing. Or potassium. I forget. But what part *does* the movement of bioentities play in the chemical fluxes of the biosphere?

The cloves were carried in boats, but boats, schmoats! What difference do boats make? Did it matter to the cloves whether they were heading toward Madagascar on a leaf or in a bag? Because humans are as natural as everything else, their agency is only interesting *from their perspective*, not that of the cloves or anything else. It's like those trees we were talking about: mudslide or bulldozer, they come down the same.

What I can't help seeing is that for almost all the years they were traded, the cloves were carried by the winds. I squint a little and make the boats disappear, and the cloves are flying with the winds like any other piece of flotsam and jetsam, seeds, fronds, pods, all the stuff that occupies the realm of the aerobiota.

When I put the boats back in the picture, what I see is that humans are intervening to modulate these flows, like any other life-form, encouraging the movement of *these* substances, of cloves (strong enough to bury bad breath and so be held in the mouth during audiences with Chinese emperors) instead of some other plant in the Moluccas growing next to them. Instead of moving stuff for the first time, people are really just reaching into existing flows and shaping them, encouraging this, discouraging that. This keeps people within a biospheric tradition. It makes people less alien, so that what people do can be thought about within the same frameworks we use for thinking about other biospheric flows.

The impacts can be dramatic: in the 17th century the Dutch wiped out cloves everywhere except on Amboina and Ternate to keep the prices up, thus throttling back a human-encouraged flow, until the French in the latter part of the 18th century broke the Dutch monopoly by spreading clove culture into the Indian Ocean and out to the Americas.

Suddenly it sounds like it's all about politics and economics and colonization, which it *is*, but it's also about cloves. Though undoubtedly it's also about traders having something to move. *They* don't care whether it's cloves they move or sneakers—150 million pairs of which moved from China to the United States last year—and so what we see here is the intersection of a couple of systems, the long-term historical motion of stuff around the planet (the cloves in motion), and the need-ability of selected humans (traders) to couple with the environment by attaching themselves *to* these flows, *as flows*, guiding them, modulating them, skimming and dipping from them.

There's always been this motion. By coupling with it, we just shape it. Again, no need to explain how we invented something we didn't—and keep in mind that we too are a flow, or an eddy in a flow.

Breath of the land.

So trade *is* like a current in a flux. It is like the wind. It is *the* wind. Or *a* wind. You've got the circulation of air and rock and water; and superimposed on those, you've got the great flows of carbon and nitrogen and sulfur, and we're like eddies in those great global fluxes, and trade is like, eddies off the eddies. . . .

While this would mean trade's always been global, it's important to recall that there are flows and there are flows. Some make tight cycles. The soil nitrogen cycle can be very tight. The hydrological cycle in rain forests can be very tight. This means the water rained *on* the forest is transpired *by* the forest. (It's not water brought in from a distant ocean. It's local water.) It rises, it falls. Obviously the cycle's leaky, and so over time all the water does get caught up in the global flux, but there are a lot of tight local cycles in which stuff can get "caught" for a while. So I have to imagine there's always been a lot of local trade, little tight cycles. And I have to imagine that when human population densities were low—I mean low—and there weren't boats yet or domestic animals, that trade, if there were any trade, was local. And part-time. People hunted and gathered and made things and harvested grain and maybe had dooryard or dumpheap gardens and traded stuff. I'm not going to say, "And these had good stone and those had raw copper, and the copper moved toward those who didn't have it, as did the obsidian, being exchanged. . . . "

No, though that may be exactly how it was. Certainly it's coherent with the way I talked about niche expansion. But really I have no idea. People might have begun moving things around just to move them. Now, I'm not going to deny resource stabilization under population pressure, especially the subtle way Mark Nathan Cohen puts it, but I'm not going to demand it either. I hope it's clear I frankly don't understand the motivation structure of autonomous metabolic organizations, especially those with elaborated nervous systems (*us*), and I really don't think anyone else does either. I'm seduced by—and therefore distrust—the kind of argument that goes, "So demographic pressure led to agricultural intensification. This resulted in competition within sociocultural groupings which selected for greater centralized controls in decision making, which was largely a by-product of viable dispute arbitration," which is Robert Santley on state formation. I mean, it sounds great. It's highly plausible. It may be true. But, my god! the reifications!

What I'm confident of is that state formations arose everywhere independently, just like agriculture, *and on its heels*, another global fluctuation: *the self-domestication of the domesticators.* I'm less confident of all this bands → tribes → chiefdoms → states business, and not the least confident about where you're supposed to draw the lines. What a mess *that* literature is! But it's perfectly clear that concomitant with the rise of—no, with the *crystallization* of—these polities, there's the *crystallization* of long-distance trade.

My problem with "the rise of" is that it suggests a teleological transformation, whereas things can crystallize out of solution and dissolve back into it without that sense of *rise* and *fall*, that sense of . . . moral decay, of losing it. With less of it anyhow. I mean, *The Crystallization and Dissolution of the Roman Empire* isn't anywhere near as moralistic as *The Rise and Fall.*

It's less instructive, less cautionary. Besides, it could always *crystallize* out again whenever conditions were right.

That's one thing. But the other is, remember that that's my metaphor for the appearance of long distance trade. Remember how I talked about the self-organization of trade, like the self-organization of life, as a simultaneous reaching out of hands that grasp each other to form a knot or a ring? I talked about how little it would take to precipitate the larger trading structure, through which the Mesopotamians—or others—would be able, would be *encouraged*, to press their demands on Tidore, or wherever, however elaborately buffered; and in that way *structure up* the entire sequence of links. Suddenly the chain—the ring, the knot—would get taut.

It may be more garbage talk, but can you see it? Those incipient local links just beginning to form; but even before they got too extended, crystallizing this larger structure out of them, down through which all the "I'd-be-willing-to-give-you-these-for-those" could flow. And once there's a flow, well, but we've been here, and as Neal Stephenson said, "a person can go far by just skimming off a tiny bit of it."

Arbitrage.

The thing is, stuff is moving long distances right off the bat. To the objection that it all seems to be luxury trade: so? and no. I mean, so what if it's all luxury trade? In the first place, what's a "luxury"? There's some weird idea that some things are essential and others superfluous. Essential for what? *Life?* It's not that simple. For one thing, who needs to live? Where does that come from? What's the *necessity* for keeping this chain going? It's not only that this is just *our* perspective, but that in that kind of thinking there's no way of embracing so obvious a fact as . . . *suicide.* The problem is the idea of "necessity," which is as essentialist as it gets—

I find it hard to understand what people mean by "necessary" when most people I know think a car is a necessity. I'm told that all the time. It's never been a necessity for me. A library's a necessity for me. It depends on the pathways you follow, the way you couple with your environment—

People say "oxygen, that's essential," but anaerobic prokaryotes have lived nearly 4 billion years *despite* oxygen. But again, what's necessary about life? It may be inevitable, but it's not necessary.

Luxury comes from the Latin word for "excess," but from my perspective it's all excessive. The very existence of the universe is excessive. Wildly excessive. And I'm pretty well convinced that only once you can see that, can you really understand and enjoy this place. Everything becomes a ridiculous unnecessary extravagance, my sitting here writing, the sunlight flashing from the windshields of the cars, the. . . . It's all free. It's all for nothing. It's all wonderful.

It's all a luxury.

Can't you luxuriate in that?

But, believe it or not, that's not my point. My point is that, even taking the idea of luxuries for granted, and accepting as fact the contention that early long-distance trade *was* in such "nonessential" commodities as cloves, gold, cinnamon, lapis lazuli, copper, and iron (both of which later *became* necessities), I still think Jane Schneider made the case a couple of decades ago for the significance of the so-called luxury trade, if Robert Adams hadn't already made it before her. But she built on Adams. And that was 20 years ago. *So-called* luxury goods—silks, fine brocades, perfumes, spices, precious metals, gemstones. . . .

Their trade could structure massive alterations in technology, leadership, class structure, ideology, and not only within trading populations—in the East Indians, people in the Fertile Crescent—but within relay populations along the trade routes too, the Inner Asian polities along the Silk Roads, the great entrepôts along the cinnamon routes. These goods weren't jokes. They weren't whims or caprices. People who conceptualize these goods as status markers usually fail to consider that status isn't something an elite possesses by itself. Its maintenance requires the participation of everyone in a social system: the elite to assert it, everyone else to validate the assertion. If the marker takes the form of special clothing, for instance, the active participation of whole sectors of society is typically required to produce it, and they in turn must be supported in their production by the rest of society: with silk, for example, the tending of the trees, the nursing of the worms, the patient spinning of the silk, the weaving. . . .

A useful way to think about this is to reflect on the drug trade, which is an obvious contemporary example of a classic luxury trade. We're always amazed at how refractory it is, but that's because we refuse to see drugs as integral to the societies that produce, relay, and consume them. As long as we keep on thinking, oh well, we have society, and then on top, as a sort of . . . jewelry . . . there's cigarettes and alcohol and drugs, and we could just take them off and put them aside. . . .

As though we could do that with jewelry either.

It's really clarifying to think about luxuries this way because the drug trade shapes all our lives. Many don't want to admit this, but with the country annually spending $50 billion on prisons, and two-thirds of that for drug-related activity, it unquestionably does. Incarceration's a huge industry, just huge, and it's essentially drug supported. And that doesn't tally the costs—and the social costs—of the cops and courts and judges and attorneys. It's a staple on television. It's an essential ingredient in the nightmares, in the whole *image repertoire* of middle-class parents, whatever their own "recreational" drug use—

No, even if early long-distance trade had been in nothing but so-called luxuries, it was still powerfully formative and incredibly integrative. . . .

That's the "So?" side. The "No" side is that, in fact, there was bulk trade early on too. It wasn't just that Mesopotamia was shipping grain out through the Persian Gulf, it was importing raw wool too, beyond that provided by the 50,000 sheep it raised, to feed the looms whose product was shipped to Anatolia for—

Doesn't it sound just like the Lowlands, just like England, centuries later?

And it wasn't just Mesopotamia: similar things were going on in the valleys of the Nile, the Indus, the Ganges, the Hwang, the Yangtze—

Not everywhere. Many places, but not everywhere. Not everyone made the switch to agriculture by any means—that process still isn't completed (see Botswana, see Australia, see Canada [see Idaho!])—and among those who did, different niches returned the investment of human energy differently. So there's a whole lot of what people call uneven development, cores and peripheries, centers and margins—

This is more of that "the rise of" talk. The language is too loaded. Humans societies vary to begin with—within limits—on top of which they embody in some way, not to be deterministic about it, differences in the niches they occupy. These were always pretty patchy, reflecting their various locations in the various global fluxes and so their variant histories; and, needless to say, so were the societies that variously came to live them. Like I said, there's a range, a multiplicity, a patchiness of lifeways: hunter-gatherers, and doubtless here and there Sauerian fisherfolk, and village agriculturalists, some with economies capable of supporting "states" within 15, within 20, generations. So there are all kinds of potential gradients.

I see the agricultural economies that crystallized out of those affluent hunter–gatherer societies settled in large permanent communities beside lakes or springs or marshes in well-watered river valleys—the ones in the tributaries of the great river valleys—as having had the ability to produce sufficiently great surpluses early on to support the kind of trade that accelerates cultural transformation, writing soon enough, and the kind of transformations it supports. That is, getting hooked up—

Into networks with their ability to accelerate change. . . . And from the same logic reaching out—

Or the skimmers *pushing out* to maintain their source of livelihood—

To others reaching out. . . . It's the same logic that created the Net.

The Net's just where it's going right now: agricultural intensification, trade, boats, caravans, ships, trains, telegraph, trucks, radio, cars, telephones, planes, television, UPS, FedEx, Williams-Sonoma and the World Wide Web—

Not going anywhere, the skimmers just running their logic as far as it will take them.

What happens when everyone's skimming and no one's producing?

It crashes. It wouldn't be the first time something went extinct.

Does this mean it's alive?

The Net? I don't know. It's another structure jacked off other structures brought into being by the biosphere, another face of third-order unities. I guess it's like soil. I guess if soil's alive, there's a sense in which the Net's alive. I mean. . . .

. . . It's *all* alive.

10

The History Hidden by Paradise

A CASE STUDY

One problem for any new story is that the form of the old ones is not adventitious. This form is rooted in the ontological experience of everyone of us for whom at birth the world is literally new. Growing up we labor to construct a stable platform for our experience of this world, a platform, that is, uncontaminated by our own apparently internal and individual transformation. Erasing, by means of this construction, the most profound evidence for sources of change *within*, and lacking any sense of the processes that brought and continue to bring the world (and ourselves) into being, we come not only to accept our world as given—rock and sky, family structure and labor relations—but to locate *outside* it the source of every change: the rains fail for too many years and the family has to give up the farm; the empty lot at the end of the street where you've played for years is cleared one morning, a new house is erected, and *horrible* people move in; the school board scrambles attendance zones and you and your friends are split up; changes at the mill force your mother to take a new job; your parents divorce; your dog is hit by a car.

Or bombs drop. Or troops come ashore. Or one day a TV is installed in the village center. . . .

Indeed, what possible intervention could any of us make in any of these . . . *catastrophes*, which indeed are the model of *every* catastrophe, the model of *every* change, greater or lesser?

The form of these experiences becomes the form of . . . *experience*, and its structure is replicated in the form of the stories of loss we tell without ceasing. As we hurtle toward senescence we promise ourselves with every fresh start—new town, new business, new relationship, new home—that *this is it*, that this time (finally) we have it right; only to find (as ever, *as foretold!*) that we hadn't anticipated the flood, the fire, the . . . economic downturn. Adrift among these mini-Edens apocalyptically annihilated, we find simultaneously annihilated every awareness of ongoing processes operating across lengths of time longer than that of our lives.

Without a past we lack a platform for the construction of a past. Small wonder, then, when we fail to see in a boulder the historical forces that moved it there. Small wonder, then, when we feel put out by the noise of a football victory celebration. Small wonder, then, when we feel picked upon by a storm.

The phenomenological presentness of the experienced world compounds this ontological effect. Soon enough the two are confounded. They become impossible to disentangle, so that in the fullness of the present it seems not only that the world has *just* begun, but that it is *complete*. Since in immediate sensory experience there is nothing to call into question this world as given, in effect it is given twice: first in its ontogenic novelty, then in the completeness of its indubitable presence.

It is hard to exaggerate the difficulty this *presentness* poses for seeing the past. That boulder? It's huge. It's hard. You kick it and it hurts your foot. It's hot in the sun to touch, and when the sun is high enough it's so bright it's all but colorless. You lean against it. It doesn't move. Not only doesn't it move, it promises never to move. It's immobile. You sit down against it to eat your lunch. Its shady side is cool. It's *got* your back.

So present is this present that it interferes with *my* thinking, gets in the way of *my* writing.

It's hard—where I'm trying to write this with my head propped on my hand here on this lawn in Raleigh—to think about that boulder on the other side of the continent in Los Angeles. As close as I am to the grass, it's hard *not* to pay attention to the clover, to the wire grass, to the Bermuda, and the rye, all—in this rush-hour August sunlight—a washed-out citrony-yellow.

Below me is a strip of grass and a concrete curb and an asphalt street. In this brightness the street is almost white when you can catch sight of it between the speeding cars. Beside the strip of grass is a sidewalk and then another strip of grass. Then there's a concrete retaining wall, a steep slope, and 10 steps up to the grass where I'm lying. There are four more steps to a porch where you can sit. From there you can look across the street to the Wilmot Apartments behind *their* curb, *their* strip of grass, *their* sidewalk, *their* lawn.

There are young crepe myrtles in the tree lawn of the Wilmot and along the foot of the building there are privets and cotoneasters. There's a dogwood behind me and a spirea. To my right there's a cypress.

The Present Is Propelled by the Past

Even here where the history's not sensed as such it's present. *There's so much of it.* These grasses and shrubs were raised, sold, bought, and planted. They've been watered, fertilized, trimmed, pruned, and cut. Implicated in no more than these is an entire global commerce: plant explorers, herbaria, a history of nurseries, changing fashions in foundation planting, the *evolution* of an entire industry devoted to . . . cutting grass. The asphalt, first used 5,000 years ago to line a reservoir in Mohenjo-Daro (in the Indus River Valley, in Pakistan), was laid down on this street according to precepts laid down by John McAdam in his 1816 *Remarks on the Present System of Road-Making.* Don't get me started on the concrete.

But none of this is evident in the perceptual present. The work of the past does not reveal itself as the past. *The work of the past reveals itself as the present.* There is no awareness in this scene of the work of the past that propels everything in it into the future. Were I younger, or were it new to me, this would seem even more the case. Then the scene would really seem fulfilled in its presentness.

Even so: I *feel* the cars. They are all around me in the throbbing air, in the smell, in the taste on my tongue. Every day 30,000 brake, shift, accelerate, spew by, at this time of day about one a second. Like a stove each car radiates heat. Reactive hydrocarbons simmer in the merciless sunlight and oxides of nitrogen and sulfur. The whole toxic burgoo threatens to overflow the banks of the cut through which the street gently rises. The stew laps, in fact, on the lawn where I lie. The grit settles on my paper. It sticks to my sweaty skin.

The illusion of the given world: the *presentness* of the present. . . .

The sun beats on the grass. It hammers on the hot steel stream in the street. It mauls the face of the Wilmot where no one is sitting on the narrow metal balconies in this Carolina summer light.

A thin cornice of green tiles above the fourth floor wants to recall the Alcazar in Seville. There is green milk glass in the fringe of the awnings over the entrances. When it is not so hot and the cars are not so loud and the air is cleaner, there is something nearly elegant about the Wilmot, and you remember that when the building was new there was a streetcar out Hillsborough Street.

Hillsborough Street is Raleigh's old main drag. It runs from downtown west to Chapel Hill, to Hillsborough, along a ridge once beat bare by Tuscarora, past what used to be State College but is now North Carolina State University. Here, where an earlier generation's understanding of streets built things closer, Hillsborough Street is only three lanes wide. There's a sandwich shop to my left and an auto repair shop to my right and Piedmont Litho—Copies 5¢—and a coffee shop and The Reader's Corner and a comic book store and an aikido dojo and an Indian restaurant and a laundromat and a new Mexican grocery—La Esquina Latina Grocery Store—and the Wilmot and this old house all within spitting distance. A generation ago they didn't eat Mexican or Indian food in Raleigh or practice aikido. Then again they didn't smoke Marlboros in Mexico or drink Coke in India or use Windows in Japan.

Closer to town, closer to the *university*, Hillsborough Street is clotted with bars and clubs. *That* hasn't changed.

It was all *residential* once, and back from the street it still is. It's a wall of green behind Sotos International Auto Care and running back along the side streets the canopy soon closes.

It was all broadleaf forest once.

Once it was at the bottom of the Iapetus Ocean.

None of it's given and if you want to think about it that way, go ahead, but I *live* in it and very shortly, even in this heat—though "it's not the heat it's the humidity" we like to say—I'm going to put myself in motion and go down the stairs and off to Cup-A-Joe's for a *doppio*. The land isn't a done deal. It's a work in progress.

There *are* sidewalks, but there are cracks in them, and grass has taken root.

The Whole World Whole

The story's about the world, but about the *whole* world. It's not about some *part* of it, it's not about politics or culture or the so-called land*scape*. It's about the whole thing working as a whole, the sidewalks we walk on, but also the air we breathe, and the music that we listen to. Because although the world will never be complete, it's always been whole.

Down here on the sidewalk each car heatslaps you as it passes. The cars are all new and shiny, and like the grass and the shrubs they come here from all over the world. Most have the windows up and the AC on, but plenty don't and the bass thudding from them hits harder than the thrumming engines, is more percussive anyway, is mostly rap. Snatches—rolling samples—of Public Enemy and KRS-1 and Master P's younger brothers Silkk and Murder-C, whip their rides down the street and . . . but whoa! *wait!* what's this? "Love was out to get me / That's the way it seemed"?

That's all you get in a street sample, one shot, no loop, but I knew the next line, "Disappointment haunted all my dreams," and the next too—it's the kick into the refrain—"Then I saw her face, now I'm a believer, duh da dum da dum duh da dum," and so on; though the day was when I could have sung the whole song for you, not in 1966 when it went #1 in England and the United States, but later, in 1968, when I found it tucked into juke-boxes all over Puerto Rico.

It was weird. You'd be running your finger down the selections: Trio Los Panchos, Duo Perez Rodriquez, Julio Jaramillo, Los Condes, Emilio Quiñones, Chago Alvarado, El Jíbaro de Lares, José Manuel Calderon, Trio San Juan, El Gran Combo, La Lupe, Raphael, Ricardo Ray, Duo Los Amigos, Monchito Motta, Roberto Ledesma, The Monkees . . . *The Monkees?*

It stuck out like the boulder.

If the jukebox had just the one single, it was "I'm a Believer" with "Steppin' Stone" ("I'm Not Your—") on the other side, which had gone to #20 in its own right. If there was another it was more likely to be "Last Train to Clarksville" (with "Take a Giant Step" on the B-side) than it was to be "A Little Bit Me, a Little Bit You," though *its* flipside, "The Girl I Knew Somewhere," had also charted.[1]

A few, like the jukebox in the Bar Refresqueria Ortíz, had all three.

This isn't casual intelligence. That January—in 1968—I'd made a systematic survey of Puerto Rican jukeboxes, cutting transects across the island from Santa Isabel, Salinas, and Guayama on the Caribbean to Playa Mar Chiquita, Playa de Vega Baja, and Playa de Dorado on the Atlantic. It was an exercise for a required course in geographical field methods and I was working with a woman—well, Ingrid, soon to be my wife—who was studying rural housing. For her work we stopped every second kilometer to observe and note five characteristics of all the housing within a hectometer. For mine we stopped at every fifth place we thought might have a jukebox—bar, restaurant, *colmado*, whatever—to tally the types of songs. In every third one of these I copied the whole selection. You don't go into a bar and copy down its jukebox without playing a song (or two) and having a Corona (or two) and maybe some conversation. It was pleasant work.

Actually . . . *it was paradise.* I don't just mean it was sunny and warm and there were flowers on the trees, while back home in Massachusetts there were dog-piss stains in the slush along the curbs, though unavoidably that was part of it. I mean . . . I'd never been here before. It was new. *It was overwhelmingly present.* I had no *past* in it. *It* . . . had no past.

How *could* such a place have a past?

As is always the case with every new situation, it was as if I'd fallen, not precisely into a place out of time (there *were* the Monkees), but then into a place untouched, or at least *less* touched, by the alienated anxiety, by the mediated corruption, of the historical world in which I lived. Compared to graduate school in a "hot" program in the Northeast, I was to find Puerto Rico, especially the mountains around Barranquitas, self-present and authentic, innocent and pure. I was to find it Edenic.

It wasn't just the novelty either, just the sharpness of the distinction between where I was coming from and where I'd landed. It was also the way the place fulfilled a fantasy of "the Other." The point, after all, of this required course in the geography curriculum to which we students had committed ourselves, was to expose us to . . . *another world,* to get us to . . . *another place,* a simpler one (despite all the theory), one (despite all the theory) purer, more . . . *natural.* Flying south on the red-eye from the Pan Am terminal at Kennedy we were (despite all the theory) . . . *escaping from history.*

Or, if not from history, from most of it. In our classroom preparations Puerto Rico *had* been granted a past, but a past that began with the Spanish–American War when *we*—the United States—took over the island. We'd been lectured on our chauvinist efforts to impose English on these Spanish speakers, and on the imperialism implicit in Operation Bootstrap. Among our readings was Claude Lévi-Strauss's *Tristes Tropiques,* a book all but paralyzed by the burden of Paradises Lost. Having invoked the Eden of the Nambikwara (though even it was a paradise inescapably soiled by the anthropologist's slithering in on his belly), Lévi-Strauss concluded that "[i]t was not our world and we bear responsibility for the crime of its destruction." Even for one less susceptible than I, it would have been hard to miss the implication that in the paradise of Puerto Rico, *we* were the snakes.[2]

And though the central highlands of Puerto Rico were not the rain forests of Brazil, nonetheless stripped of history as *in their novelty* they necessarily appeared to me, they were as "Other" a world as that invoked by Lévi-Strauss, and every bit as fragile and as threatened. When it would be said "that in Puerto Rico people used to *paranda,* at Christmas they used to go from house to house singing . . . " all I could hear were the words "*used to.*" I was going to catch it . . . *just in time.*

The Jukebox Song Survey

They don't say what *kinds* of songs they are on jukeboxes anymore, but once they did because once people danced to jukeboxes, and it mattered. Right after the name of the song it used to say "Swing," "Waltz," "Cha Cha Cha." Though it was a fading fashion in 1968, there were still 40 different kinds of songs indicated on the jukeboxes I looked at.[3] There was everything from "Aguinaldo" to "Vals."

Not that *I* could tell the difference. Meaninglessness is another thing about paradise. Meaninglessness is what most connects paradise and childhood, what assures each its innocence. Meaninglessness is what innocence is all about, really.

Meaning comes with history, so with no history I had a hard time in the beginning making distinctions. Everything I heard was Puerto Rican to me, even the *merengue* which everyone said was Dominican; and I was as dazzled by the richness of this minuscule musical scene as I was by the rich strong coffee, and the almost syrupy rum, and the bright cool sunlight of the mountains.

The presentness of the present. . . .

I was looking for . . . *what*? I wasn't sure. I had come into the study of geography with some idea about the role of the environment—of the American West, of writing for western newspapers—in the shaping of a distinctly American detective story that I saw had broken away from the tradition—inaugurated by Edgar Allan Poe—that had been developed by British writers like Wilkie Collins and Arthur Conan Doyle. I wasn't altogether sure how it worked, but there was more than a "national" difference between Hercule Poirot and Sam Spade, and I had traced it back through dime novels to . . . again, I wasn't sure what, but I couldn't get away from an image of Mark Twain writing for the papers in Nevada.

While I was worrying out this thesis in my undergraduate English classes I was also sinking into a pop music scene that wore regional character like tattoos. It was the mid-1960s. There was country out of Nashville and Detroit's Motown Sound and a beach scene in California and a mainline scene in New York—not to mention the "British Invasion"—and you could hear all of it anywhere. But if you rode around the country, you could pass through regions, marked in other ways as well, where one or another of these sounds was dominant, that is, was heard more often on more radio stations and occupied more slots on more jukeboxes in more different kinds of places and—

Why? Why were they playing these heartbreaking walking-the-floor-over-you songs in West Virginia, and these let's-go-surfing-now-everyone-is-learning-how songs in California? I had no idea.[4]

I was feeling my way into it, but as a first-year graduate student one of the things I was talking about was using radio playlists to construct a pop music geography of America, and when I had to come up with a project for this field methods course the idea of a jukebox survey popped into my mind like a tune when you turn on the radio.[5]

What I Found in Paradise

It's hard to recover what those first few weeks on the island were like. Paradise is easy to lose and impossible to recover. Paradise *depends on* an unselfconsciousness within which it can't be perceived as paradise. It's only the loss of paradise in history (thanks to the snake, the debris flow [the graduate student in geography]), that makes it obvious what it was.

Afterward we know too much, we wallow in distinctions we couldn't make in the beginning. I *had* lived in Mexico, and I did recognize the *boleros* and the *rancheros*; and I did know the Monkees and the Shadows of Night (they meant "Knight"), whose "Gloria" and "Dark Side" showed up on one bizarre jukebox, but otherwise it was all just music and I only knew I loved it all. . . .

Paradise, like I said. . . .

In fact, the songs *I* could barely distinguish, and which seemed to me to comprise *a single music* of unquestionable indigenousness, had been recorded on three different continents and in twice as many countries; and in their filiations the songs caught up, once I was able to "hear" through the strands of time that bind us all together, almost all the rest of the world.

Even something as apparently insular as Puerto Rican songs turned out to be caught up in the history of the entire planet.

This was as invisible to me as the torrent of students is to the client looking in the early summer for housing near the university; or as the debris flows are to the home buyer poking into the canyons of the San Gabriels; or as the hurricanes are to someone searching for shells along the beaches of the Outer Banks in June. You stop at a *bar-restaurant* hanging off a road overlooking a lake. It's late morning. A man leans against the doorway with his head on his forearm looking out across the road at a *flamboyán* in flower. Steam rises from a *cafe con leche* in front of a man at the bar. You drop a quarter in the jukebox and punch some numbers ("Let's see what a *seis* sounds like," you're thinking), and the room fills with a sound not quite like any you've heard before. There's a stringed instrument but it's not a guitar (it's a *cuatro*, but you don't know that yet), and some rhythm instrument (it's a *güiro*, that serrated gourd you may have played in elementary school), and—

But *so much* is present! The language, the words, the lyric, the melody, the rhythm, the . . . and the man at the bar is lifting a little paper cup of rum in your direction, there's a bottle of Barrilito on the bar in front of him and—

It's like that boulder, it's like the sun on the grass where I'm trying to write this, it's like the street in Raleigh: it's *presentness* leaves no room for the years-that-propel-it-into-the-future to manifest themselves. There is no room for any awareness of the rest of the world that supports its coming-into-being.

I had *no* idea what I was listening to in the beginning—I was just trying to get a handle on the song types—but in my anxiety about the imperialist energies of the American *fox-trot* and *rock and roll*, I was managing to overlook completely the imperialist energies of Spain, of Mexico, of Argentina, of even the Dominican Republic. Here, it's January 1968. During the previous 6 months the Doors had charted with "Light My Fire," the Beatles with "All You Need Is Love," and Aretha Franklin with "Respect," yet maybe the biggest recording act *in the world* was . . . Raphael, from Madrid, from Linares actually, in Andalusia.

Who?

Exactly. Somehow I hadn't anticipated pop music in paradise.

Folk? Folk, of course. Folk was timeless. Folk was authentic. Folk was the *essential* music of paradise.

Pop, on the other hand, was precisely the music of the fallen world. Pop was *completely* historical, it was . . . *fashionable*. Its very essence was its being in time. It went without saying that it was inauthentic, corrupt. Listening to a marimba band in southern Mexico, I unhesitatingly conflated it with folk—with *ethnic* music—even when it played old Cole Porter tunes. If I'd heard of Raphael, I would have conflated him with folk too, with—from my limited perspective—some degenerate *flamenco* tradition. Raphael in fact was not only *all over* the jukeboxes in this commonwealth of the United States where I never did find even one song by the Beatles, he was also selling chart-topping numbers in *30 countries* around the world, filling soccer stadiums wherever he went, releasing popular movies, and appearing on international television (doing duets with . . . Tom Jones!).

Raphael *and* Raphael *and* Raphael. The young girls on their way home from school would come into the *colmado* in their uniforms and play "Yo Soy Aquel" or "Estuve Enamorado" on the jukebox, and when Raphael would give that little squeal he gives after the repetition of the chorus on "Estuve Enamorado," they would close their eyes and crush their books to their chests, and if they had another quarter they would play the song two more times.

Older women were pressing the buttons to listen to Roberto Ledesma—a popular Mexican with a smoky voice—float "Parce Que Fue Ayer" out over the pool tables. "Parce Que Fue Ayer" was a standard penned by Armando Manzanero, Mexico's *compositor del momento*. It was occasionally sung by Manzanero too, even on some of my jukeboxes. In his squeaky little voice it was like listening to Irving Berlin sing "White Christmas," except that Manzanero was a star in his own right, with his own records, and TV show, and movies.

The world no more announces itself as the world than the past announces itself as the past (the present is *here*), but if you knew what you were listening to you could punch up a lot of the world on the old Rock-Olas, Rowe-AMIs, Seeburgs, and Wurlitzers in these little no-name *cafetines* 2 kilometers north of Torrecillas on Route 155; or at the junction of Routes 159 and 160 on the road to Corozal; or just south of Cayey on Route 15 about 3 kilometers past Route 1. You could punch up the Spanish group Tomas de San Julian, who had had a big hit in 1967 with a pop flamenco version of "Guajira Guantanamera" and spirit Spain (*and* Cuba) into the room where the old men playing dominoes weren't paying attention anyway; or hit the buttons for Cuco Sanchez and take everyone to Mexico, northern Mexico and the Revolution just over; or press B-5 and there was Johnny Pacheco *with* the streets of New York, *los ritmos de origen claramente african* overwriting everything else that might have been there, and the kids at the pool table in their tight creased pants dancing while they played.

The world *was* present in the jukeboxes, it sang in all the bars and restaurants. *And the past was with it.* But blinded by the music's novelty, by its *presence* (to say nothing of the mystifying constructions I wore like glasses), none of this was evident to me. Or little of it was, though little by little, more and more was.

I began my survey just after *las trullas aguinalderas*, that is, just after the extended Christmas that in Puerto Rico runs from Christmas Eve through Twelfth Night. During *las trullas* either the guys who owned the jukeboxes—like Hernandez Sales and Services/Veloneras Seeburg and Wurlitzer of Cayey—or, if not them (or sometimes in addition *to* them), then the guys in the *colmados* or *bar-restaurantes* who had the keys to the jukeboxes, opened the jukeboxes up to let the local *cantantes*, the local *trovadores*, load them up with the 45s they'd had freshly minted in Hialeah, Florida, 45s pressed with their latest *aguinaldos*. Basically these were "home recordings." Their presence on the jukeboxes only strengthened my conviction that I'd fallen out of time.[6]

It was precisely their home-grown—their *folkloristic*—quality, however, that was to help me chisel the first crack I would make in the monolith of the music I was listening to. It was a crack through which *history*, through which the *world*, was to pour.

These "home recordings" were the very songs I had been encouraged by my mentors to listen for. These were the songs that were supposed to be the *authentic* music of the island, the music being threatened by commercial trash from the north. Taking this Eden story in—hook, line and sinker (and why not?)—I was soon able to distinguish these songs from the rest. It wasn't all that hard. The "*Oi, lei, lei, lei, / lo, lai, le, lai, le la,*" with which *aguinaldos* formulaically open, is distinctive. So is the *sound* with its clenched-throated vocalizations, the secco quality of the *cuatro* playing, the fluid rasp of the *güiro*. Loaded into the jukeboxes at the same time were the "commercial" *aguinaldos* of Aníbal Laureano, of Maso Rivera ("*el rey del cuatro*"), of German Rosario ("*el jíbaro del Yumac*"), of Baltazar Carrero, so I was hearing these too. These were more up-tempo. They were instrumentally virtuosic. They were reminiscent, in their way, of bluegrass.[7]

The *aguinaldos* turned out to be a kind of Christmas carol. Each verse of an *aguinaldo* is 10 lines long, and each line has six syllables. The form, a *décima*, had come to the island in the 17th century in the Spanish poetry of the time.[8] Occasionally, in 1968 (and later) you'd still hear lyrics—and not at concerts—that could be documented to the 18th century, but new ones were made up all the time, often improvised on the spot. In years to come I would hear *aguinaldos* sung a cappella (and a capriccio) by fathers coaxed to the stage at junior high school Christmas assemblies; and, accompanied by forks on graters and spoons on bottles, by neighbors in our living room. *Aguinaldos* were anything but a dead form, whatever the books lamenting the loss of paradise were mourning.[9]

Ironically, the *aguinaldos* themselves were about the loss of paradise. The professionals like Laureano, Rivera, Rosario, and Carrero may have had *aguinaldos* on jukeboxes all over the island, and on jukeboxes all over Hartford and the Bronx and Brooklyn too; yet these virtuosi wanted to pass themselves off as local singers who had *aguinaldos* only on the jukeboxes in their *barrios*, as though, that is, *they* were peasant farmers, simple *jíbaros*; and they wanted to pass off their studio productions as home recordings. This was the illusion that lay at the heart of the songs: that of a paradisical life in a prehistoric past, the *flamboyán* thick with blossom, peasant neighbors crowding the *batey* (the very syllables of this Taíno word for "patio" dripping with heightened primitivist associations), Francisco with his guitar, Rubén with his *cuatro*, you with your *güiro*, the mountains blue in the distance, the *aguinaldos*, the *seises*, the *trullas*, the *plenas* heavy with the tastes of cane and coffee and tobacco and rum.[10]

Paradise! It's exactly what every record cover promised: life as it was, life as it *always had been*, life as it *was going to be* . . . forever.

It didn't matter that like Rosario (that most *jíbaro* of all the *jíbaros*), Laureano, Rivera, Carrero, and the others played every possible venue in the migrant communities of Jersey, Philly, Lorain, Chicago, Nueva York, Connecticut, and Massachusetts. Despite this cosmopolitanism, every *aguinaldero* had still to pass himself off as a hillbilly, a hayseed, a hick—when you got down to it—a hick from hicksville. It's not that they weren't proud of where they came from, but that their pride was a form of resistance perverted: they knew exactly where they came from and knew that it *wasn't* paradise, no matter what they said in the songs.

It was so Opry it hurt.[11]

El Trio Vegabajeño sang "Donde Estaras en Este Navidad?" ("Where Will You Be This Christmas?") and "Navidad de Ayer" ("Yesterday's Christmas"). *Talk about walking the floor!*

It *was* country music. But here, where the rigors of the jukebox survey forced me to pay a kind of attention I hadn't had to pay to country music on the radio back home, I was able to hear the songs as I had come to see the boulder, as *songs* but at the same time as *signs* that pointed to the forces that brought the songs into being.[12] Listening to these songs in the little no-name *cafetines* up in the hills among *jíbaros* with mud on their boots who parked their machetes at the door before tossing back their Don Q taught me not only to hear what *was* but what *wasn't* said, not only the pleasure they took in the mountains *but their regret that the mountains weren't the ocean*. It was through this kind of listening that I was gradually brought to an awareness of the *dispossession* that generations earlier had carried the *aguinaldos,* together with the people who sang them, from the coast to the highlands.

The window the *aguinaldos* opened on this history was being cranked shut even as I was taking my survey: "If you'd come tomorrow none of these *aguinaldos* would have been here," the owner of a bar on the plaza in Comerio told me whose jukebox was serviced by Casa Wurlitzer out of San Juan; but the rest of the *criolla* would have been, and the trace, though much harder to see, would have cut the same figure.

Criolla: sometimes that's all it said next to a song. When I asked people hanging around the jukeboxes what *criolla* meant, they said *nativa*.[13] Books said that *la música criolla* was *la música del pueblo puertorriqueño*—"the music of the Puerto Rican people"—that is, *la música tradicional*.[14]

Not that this tradition was all that settled a thing. There were those, for instance, who regarded the *danza* as the "real" symbol of traditional Puerto Ricanism, of *puertorriqueñidad*.[15] Now, the *danza* had evolved—via the Cuban *habanera*—out of a *contradanza* that came to the island on a wave of *19th*-century Spanish immigration. This made the *danza* kind of a recent arrival for *deep* Puerto Ricanness, and in any case the *danza's* importance turned out to be a bourgeois notion, for the *danza* had always been . . . salon music.

From the perspective of the peasant, of the highland peasant in particular—of the *jíbaro*—it was the *seis* that was *la espina dorsal de nuestra música*—"the dorsal spine of our music"—and we've already seen that the *décimas*, in which the *seis* is sung, reached the island in the *17th* century.[16] Neither *danza* nor *seis*, however, nor any of the other *criolla*, would have been traditional for the Taínos, the Indians who had lived on what they had called *Boriquen* long before the Spanish came to work them to death.[17] What counts as *native*, as *traditional* varies. It depends on who you ask and how far back you care to go.

I categorized a song as *criolla* if the label said it was an *aguinaldo*, *bomba*, *criolla*, *danza*, *danzon*, *guajira*, *guaracha*, *jazz jíbaro*, *mapeyé* (or *mapellé*), *mazurca* (or *mazurka*), *plena*, *seis*, *seis chorreao*, or *seis controversia.*[18]

I probably should have added some of the songs labeled *chacarela*, *charanga*, *merengue*, *pasillo*, *polka*, and *vals*, but to be sure I'd have had to have listened to each of them, and it still wouldn't have changed much. As it was I lumped these in with the *boleros*, the *boleros cha cha*, the *boleros moruna*, the *boleros son*, the *boleros tango*; with the *bugaloos*, the *huapangos*, the *jala jalas*, the *mambos*, *pasodobles*, *pachangas*, *rancheros*, and *tangos*; with the rest of the big-background Hispanic pop scene.

I put songs labeled *fox*, *jazz*, and *rock* in a third category. I wonder now what I was thinking? Well, it's obvious. I was thinking these were the bad imperialist songs.[19]

I did English *titles* as a fourth cut, though I wasn't sure what I was going to make of it. I mean, the biggest Puerto Rican act on the jukeboxes was El Gran Combo.[20] When its songs weren't *in* English as its smash "Clap Your Hands" was, they were riddled with it. Look, half the population lived on the mainland, mostly in New York, and English was everywhere and to write it off as . . . as what? *not Puerto Rican?* . . . would have been as fatuous as dismissing Spanish because it wasn't what the Taínos had spoken.[21] In any case the *combos*—El Gran Combo, Sonora Matancera, Cortijo y Su Combo[22]— were part of a whole other tradition, more coastal and urban—more black— that wrapped a mainland, big band, swing sound around that of the *bomba*, around that of the *bomba*, the *son*, the *mambo*, the *merengue*, the *calypso*, around some sort of black, Caribbean–Gulf lowland, sugarcane plantation sound, one increasingly stretching from Venezuela to New York.

Salsa!

Yeah . . . well, not in 1968. *Salsa* was being born while I was playing the jukeboxes and I hadn't a clue.[23]

And the figure the *criolla* cut? There were *aguinaldos* or *seises*, mostly *aguinaldos* and *seises* but other *criolla* too, on 64 of the 93 jukeboxes I studied, and it turned out that the higher the altitude, the higher the percentage of *criolla*.[24]

I'd known this intuitively. What I mean is, I knew that I hadn't found any *aguinaldos* on jukeboxes along the coast before I'd gone to the trouble of calculating the percentage of songs that were *aguinaldos* and *seises* and plotting them on a map, and. . . .

And I knew that the further south I went into the mountains, the more likely it was that I was going to find *aguinaldos*. . . .

. . . That as the temperature dropped, that as the road began to twist and turn, that as coffee began to appear along the roadside and the slopes got steeper, that as the fields grew smaller and the mountain vistas opened up and the road began to turn back upon itself . . . the likelihood that I'd hear that clenched-throat "Oi, lei, lei, lei—" went up.

I just knew it.

The music came from the land it came from. It was as much a part of the land as the color of the soil.

And then you reached the crest of the Cordillera Central and dropped down to the Caribbean and the likelihood dropped to zero as fast as the road, and in the jukeboxes in Santa Isabel and Salinas and Guayama *aguinaldos* were as scarce as they were along the Atlantic. *Seises* were scarcer.[25]

The higher in Puerto Rico you are above sea level, the greater your chance of encountering an *aguinaldo,* as the probability declines that you'll find a field of sugarcane.

Is there a natural antipathy between *aguinaldos* and sugarcane, or an affinity between *aguinaldos* and the highlands? No, of course not, but at this point the three are locked together in an historical dance that began . . . billions (and billions) of years ago, and the only way to make sense of the pattern on the ground is to know something about this dance.

It's either that or accept it as a given: "And the Lord made the earth and he put *aguinaldos* in the jukeboxes of the highlands of Puerto Rico—"

I don't think so. . . .

Plate Tectonics, Sugar, and Song

The story—big surprise—turns out to be about the loss of paradise.

About 80 million years ago, during the Campanian stage of the Senonian Epoch, a piece of the Pacific Plate began insinuating itself between the North and South American Plates. At this point the North and

South American Plates had been moving apart from each other for nearly 120 million years. The new piece of the Pacific Plate, which today we call the Caribbean Plate, began working its way eastward a couple of centimeters a year. It's still working its way eastward a couple of centimeters a year.

This movement keeps the plate's northern margin tectonically active. The margin is a ridge. Where this ridge humps above sea level we have the four large islands of the Greater Antilles: Cuba, Jamaica, Hispaniola, Puerto Rico. Puerto Rico is the smallest and most easterly of these. It broke free of Hispaniola during the Miocene, only 20 million years ago. Volcanism, intrusion, folding, faulting, uplifting, tilting, and erosion since the Campanian have given Puerto Rico a roughly dissected topography of high relief that all but vaults out of the Caribbean while slipping gradually to a broad coastal plain along the Atlantic.[26]

A *relatively* broad coastal plain: *three-quarters* of the island is in mountains. None of these is an Everest, but Cerro de Punta rises to 4,389 feet. The thing is, it's *all* up and down: only a quarter of Puerto Rico slopes less than 15%. Another quarter slopes between 15% and 45%. All the rest is steeper. You look up, and it's as if the cattle are pasted to the hillsides.[27]

These days the island lies 18° north of the equator and 66° west of the Greenwich meridian, smack dab in the northern Trades. This gives it a tropical oceanic climate. Except for the southwest corner of the island, which lies in a rain shadow cast by the mountainous interior, the coast is ideal for raising sugarcane.

One hundred and twenty-five million years ago—that is, 45 million years before the Caribbean Plate began its push eastward—flowering plants evolved. "Shortly after," say Margulis and Sagan—I love that "*shortly* after"—"by 114 million years ago, [flowering plants] had spread to all parts of the world's land."[28] Sugarcane—which is to say, the genus *Saccharum*—is a flowering plant that most likely originated in southern Asia. Though it will grow in the subtropics, sugarcane most thrives in the hot, seasonally wet tropics. It loves the coastal lowlands of Java, India, Guyana, Cuba, and Puerto Rico.

Sugar diffused—and evolved—from Southeast Asia out through the islands of the Pacific and back again, to reach India—ultimately from the gardens of New Guinea—around 2000 B.C.[29] There is documentary evidence of the cultivation of sugar in northern India in the Sanskrit hymns of the Vedic period (c. 1500–500 B.C.), and by the seventh century sugar was well established among the Chinese[30] and the Sassanians (i.e., in Iran). Sugar moved into the Mediterranean basin with Islam. Though it wasn't a major part of the Islamic agricultural revolution, sugar was widely grown[31]; and although it had been cultivated in Moorish Spain since the late ninth century—and as far north as Valencia—other West Europeans first encountered sugarcane in the crusader states of Cyprus and Sicily. Soon they couldn't get enough of it.

It was during this *Mediterranean* phase that most of the features of plantation sugar cultivation appeared.

Sugar is a labor-intensive crop, and after the Black Death (of 1348) slaves—Greeks, Bulgars, Turks, even Tartars from around the Black Sea[32]—became essential elements of sugar production in Cyprus and Crete, as later in Morocco (where the expansion of slavery was a primary motive for the Moroccan trans-Saharan expeditions of the late 16th century). Sugar refining is fuel-intensive too. Mediterranean deforestation, particularly acute in the southern Muslim lands, encouraged the construction of refineries closer to consumers, particularly in northern Europe where there were still trees to burn (well, trees to burn, the desire for sugar, and the wealth to pay for it).[33] This shift soon made the sugar producer subservient to the sugar importer.

These developments accelerated when sugar production moved to the Atlantic islands, to the coastal fringes of the Madeiras and the Canaries, especially when it moved to São Tomé in the Gulf of Guinea. By the time sugar production shifted to the Americas, the plantation system with its African slaves and dependence on metropolitan centers was far from an experimental form of colonial land use.

Columbus was an old Madeira hand and he carried sugarcane with him on his second trip to the Caribbean. The production of sugar was to dominate life in Brazil, along the Guiana coast, in the British and French Leewards, in Jamaica, and in Saint Dominique for centuries. In the 19th century—and well into the 20th—it dominated life in Puerto Rico and Cuba.[34] Its trade structured life around the Atlantic, in New England and Europe no less than in Africa, from which 11, almost 12, million young blacks, mostly males—the same population we throw in prison in the United States today—were shipped west as slaves between 1450 and 1900.[35] The best part went to the sugar fields where they died like flies.[36]

From the beginning Spanish immigrants planted sugar along the Puerto Rican coast where they settled, but Spanish imperial trade policies restricted the island's residents—and other Spanish colonists in the Caribbean—from growing sugar commercially for the next 300 years. In the late 18th century, when Spain belatedly became interested in growing cane in Cuba and Puerto Rico—in exploiting Cuba and Puerto Rico generally— Puerto Rico had long since evolved a kind of split society, a minor "metropolitan" one centered on the military bastion of San Juan, and a much larger, racially mixed peasantry living along the coast outside San Juan— "en la isla"—dedicated to subsistence farming, cattle raising, and contraband trade with French, English, and North American pirates and merchantmen. It is said that "San Juan belonged to the crown, but Ponce, Coamo and Manatí were nurtured a good distance from—and despite—the crown. They belonged to the Puerto Rican people and were revered because they represented the developing *criollo* culture, not the Spanish import."[37]

Needless to say, these two societies had different tastes in song and dance. "En la isla" in the late 18th century they danced *cadenas*, the *fandanguillo*, the *sonduro*, the *cabayo*, the *zapateado,* and the *seis*, of which the *seis* in its various forms has survived; while in metropolitan San Juan they danced provincial echoes of whatever they were dancing in Europe: the *galop*, the *mazurka*, the *britano*, the *cotillón*, the *waltz*, the *polka*, the *minué*, the *contradanza,* and the *rigodón*.[38] Although at Christmas members of both societies sang *aguinaldos*, there were otherwise two distinct musical cultures—there had been for a couple of centuries—*both* metropolitan in origin, but sedimented, as it were, at different times. These musical cultures now embodied *antagonistic*—almost irreconcilable—values which, of course, it was part of the role of the music to do.

When Spain finally decided to develop the island, it promised huge tracts of land along the coast to immigrants for growing sugar, and then underwrote, at precisely the time other European nations were getting *out of* slavery, the massive importation of slaves to a colony in which the trade had been dead for 200 years. The number of slaves rose from around 5,000 in 1765, to 17,500 in 1812, to 51,300 in 1846. In the 19th century alone 60,300 slaves were brought to Puerto Rico.[39] Here's how Francisco Scarano describes what happened next:

> The burgeoning of cane haciendas after 1815 had a dramatic impact on the lives of the Puerto Rican lowland inhabitants. Reversing an inherent tendency of the peasant economy to fragment landholdings, the sugar plantations monopolized the ancestral lands of the hateros and estancieros. These developments hindered the further extension, and threatened the very survival, of the peasant society in the lowland plains, and as the process of land concentration evolved through the 1820s and 1830s, the inhabitants of the *bajura* [the coastal lowlands] migrated en masse to the mountainous and as yet underpopulated interior [where they would grow tobacco and coffee]. *Ousted from their lands by economic forces and administrative policies beyond their control* [my emphasis], thousands of peasants took to the *altura* [the mountainous interior] in an attempt to recreate there the conditions of independent subsistence which sugar now denied them in the lowlands.[40]

What is this if not the expulsion from the garden?

Needless to say, these Spanish peasants took their music to the highlands with them, where they nurtured it as a self-conscious expression of cultural identity, clinging to it the more ferociously the more marginal they became. Simultaneously, new sounds began to be heard along the coast outside San Juan, where the recently arrived slave populations were engaged in a more complicated, but not unrelated effort, notably in Ponce, Dorado, Salinas, Loíza, and Guayama. That is, in precisely those places that once had been strongholds of the old *criollo*, where there'd been strings there were now drums, where there'd been the *seis* there was now the *bomba*,[41] and soon enough, among the more "urban" of this Afro-Antillean element, the *plena* as well.[42]

The old opposition between the *criollo* "en la isla" and the Spanish in the fort—between the country and the metropolis, between the bumpkin and the sophisticate, between the native and the foreigner, between the "authentic" and the merely . . . *fashionable* (between the *cuatro* and the piano)—remained, but now redrawn as an opposition between the *altura* and the *bajura*, between the mountains and the coast; though the coast *itself* remained split, along the old lines too, though now these were understood to run between the *jungle* and the city, that is, between the drum and the piano, the *bomba* and the *britano*, the black and the white (the mix had fled to the mountains).

It was that simple, really: black, white . . . mixed.

That is: drums, piano . . . cuatro.

Washed then, the whole thing, in the 20th century, with a coat of whatever came with the United States: first the *jazz band*, another piece (ultimately) of that lowland, black, plantation sound; and after World War II with whatever else came with living on the mainland, with rock and roll, and later rap[43]; and whatever else came with living on the *altísimoplano* of that circular migration on Pan Am and Eastern Airlines[44]; the whole thing stirred and smoothed and extended by the miracle of radio and the man who restocked the jukeboxes, taking everybody's music everywhere, in unequal degree.

As the metropolis shifted from Madrid to New York, the salon music of the "golden age" of the plantation withered. Its fading echoes—echo, the *danza*, really, and the rare *mazurka* and *polka*—were assimilated to the *criolla* of the *altura* as further marks of that *puertorriqueñidad* the United States now threatened far more than Spain. With their smaller investment in *puertorriqueñidad*, coastal blacks assimilated what they wanted of mainland *jazz band* sounds to the *bomba*, and created the platform from which *salsa* would arise.

By the time I got to the island all the old allegiances were dying. They weren't dead yet but they were dying. Coffee and tobacco were not dead yet either, but dying too and even sugar wasn't what it had been. On the cover of his *Cosa Nuestra* album, Willie Colon stood beside the East River below the Brooklyn Bridge with his hat in his hand and his head bowed over the barefoot body of a *jíbaro* with a concrete block around his ankle, presumably murdered by the trombone Willie's carrying in the evil-looking black case tucked beneath his arm. *Adios aguinaldo!* So long *batey! La música criolla* had been "all that" but now "our thing" was *salsa.*[45]

Residence in Boston and Bridgeport and the Bronx (Willie was from the Bronx), and work in the lingerie and pharmaceutical plants Boot-strapped all over the island, and U.S. military service, and television were all eroding the "old ways." The *cuatro* had become an instrument that came out only at Christmas, and the rest of the year the kids played electric guitars or pretended they did, and a few years later there was Menudo, a bunch of kids who sold more records than all the old-timers put together. And that was before Ricky Martin.[46] It was sad, but it's what it was.

It's Not Just the Music, Because the Music Can't Be Separated from Everything Else

And it *was* sad. The sense of loss in this story is palpable. Nothing can be gained by denying it: the *cuatro* has lost its honored place, the old ways are eroding, the center cannot hold. The stink of degradation arises with it—youth being merchandised instead of talent—for the loss of the old ways is always a sinking into decay, a kind of rot. But . . . *what to expect from a fallen world?* . . . which, given our original Fall, is what it is.

As *salsa* conquered the airways, and rock, and rap, many felt forsaken: *What was happening to the world?* The songs that had sustained them were harder and harder to hear, new sounds were everywhere, and with the new sounds . . . *new everything else.*

For of course it was not the music alone that was changing, it was the world. In fact it was less that "change" was threatening the old music than that the land from which the music had grown was evolving, and it was growing a new music as part of that evolution that left less room for the old.

Disorientation came with this evolution, and loss and absence and pain. People did hurt. They *do* hurt.

But it was not Willie Colon who caused the hurt. It was not Menudo. It was not even Ricky Martin. These were all no more than people living their lives in *their* present. The sadness arose from a dead-end way of thinking about the world. It arose *from the form of the story where the pain is the price we pay for a certain kind of imagining of a former presence.*

We know this presence. It is that of the woods where you grew up playing. It is that of your family before your parents divorced. It is that Land of Youth to which none of us can return. It is that Cockaigne of family and friends, coffee and rum evoked by the songs. It is that of the Garden from which mankind was thrust. It is that of a Nature singularly marked by harmony and balance. (It is that Arcadia of innocence I knew hearing this music for the first time!)

The sadness comes to an unassuageable *yearning* for the harmony and balance of this prehistory foregone, for the ease of this Eden inaccessible, for the abundance of this Cockaigne lost, for the comforts of our vanished youth, for the irresponsibility, the meaninglessness of an innocence no longer at hand. But then . . . *it never was at hand in a world scarred by history.*

It is *history* in which Eden plays no part. And it is the *absence* of history alone that authorizes the imagining of a world whose loss could lead to such a sadness.

The *"Oi, lei, lei, lei . . . "* of the *aguinaldo* is not an echo of Eden. *Aguinaldos* were conceived, evolved, and nurtured in history. The throats in voice propel into a future sounds given shape in an agonizing past. It's just that, caught up as we usually are in the presence, in the immediacy, in the *grain* of an individual voice, of a tattoo beaten by real wooden claves, of the rattle of maracas shaken by flesh-and-blood hands, this past is not something especially easy to attend to.[47] Yet all Hunter S. Thompson heard was a howling emptiness:

> As we sat down and ordered our drinks I realized that we were the only gringos in the place. The others were locals. They made a great deal of noise singing with the jukebox, but they all seemed tired and depressed. It was not the rhythmic sadness of Mexican music but the howling emptiness of a sound I have never heard anywhere but in Puerto Rico—a combination of groaning and whining, backed up by a dreary thumping and voices bogged down in despair.[48]

Some think it smart to dismiss Thompson as a drunken buffoon, but groaning and whining are not inconceivable things to hear in the music of a colonial people, one marginal prior to leaving Spain and who, further marginalized in coastal Puerto Rico, were later shoved into its mountains by a people still more marginal than they, the lot of them marginal to the United States that proceeded to colonize the island all over again.[49]

No, the history of Puerto Rico makes it easier than is often the case to see how advantageous it can be to . . . forget history. Only a refusal, only a willful renunciation of *history*, can account for the exaltation of the *aguinaldo* as a sign of a beneficent past, for the *aguinaldo* above all else is the mark of a violence piled on a violence piled on a violence. Its deployment as the sign of a Cockaigne—that of the pre-Operation Bootstrap, highland, *jíbaro* culture—serves mutually reinforcing functions. The first is the rewriting of the *jíbaro's* utter marginality into a kind of (sentimental) centrality, this in a desperate bid to manufacture an overarching "national" identity for a population from three continents brought together in pain. It's a way of starting the clock that obviates all that.[50]

One way it manages this is by obscuring the *aguinaldo*'s 19th-century preservation as a sign of the resistance to a dispossession: that of a local economy (the mixed ground crops of the *criolla* peasantry *en la isla*) by a global one (the monoculture of lowland plantation sugar), that of a class of peasants by a class of landowners, that of freemen by slaves (themselves extracted from their homes by a still more violent dispossession).

It is by drawing the line here, by starting the clock at this point—*after* the *criolla* peasantry has lost its place on the coast and been driven into the mountains—that a "golden age" can be brought into being, even *if* a "golden age" of plantation slavery and refuge marginality. Starting the clock here not only obscures the origins of the highland communities in a flight from the coast, and those of the coastal plantations in a crossing through the Middle Passage; not only sweeps away the utter marginality of the *criollo* peasantry when it *did* inhabit the coast; but contrives to overlook entirely the disquieting fact that this tenancy itself was the consequence of a yet earlier dispossession, one still more violent, the 15th-century dispossession of an *indigenous* culture (that of the Taínos), by a colonial one (that of the Spaniard), this dispossession scarred, not by a displacement to the highlands, but by enslavement, murder, and genocide.

Of course the Taínos weren't the first to settle the island either. They had arrived sometime before 200 A.D. to displace, by war (and then by intermarriage), a population of Archaic Indians previously resident some 2,500 years.

And they? Is there, *has there ever been*, an . . . innocent occupation? Not in history.

Because It's Whole There Is Only One Story

There never was a paradise in Puerto Rico, or Boriquen, or whatever the Archaic Indians might have called it. You aim toward this paradise, searching, peeling back layer after layer, and soon enough the island itself is sinking back beneath the waves, flowering plants are furling their petals, the Pacific is shrinking. . . .

There never was a paradise in Puerto Rico or *anywhere else*. No matter how far back you go there has never been a state of innocence outside of history, never been a Nature—a Nature of Harmony and Balance, that stable state we dream about, that state of perpetual beginning, the seasons succeeding each other (but *this* spring *like the last*), around and around (magical image of the circle, of the cycle of the years), forever and ever, amen. No, instead you go back and it just gets more and more unstable and chaotic until you're in the fury of the Big Bang.

The way forward is not back.

Unless we can come to terms with this we're lost. *But a knowledge of the past is the only way to come to terms with this.* We have to let our images of paradise go. We have to let go of images of a Nature outside history. We have to let go of our images of a stable world, settled finally, done. And nothing but history is capable of revealing the story in anything like the wholeness this letting go requires.

And this is the problem, I thought as I snapped back to Hillsborough Street where I'd been standing staring at the car from whose open windows the Monkees had trailed, *making clear how deeply interconnected everything is,* including the Monkees, whose old gold radio station is laboring to construct a *musical paradise* that Silkk and Murder-C are doing their best to deconstruct, one man's paradise—as we've just seen in the case of Puerto Rico—being another man's hell.

We can twist it and turn it, follow this thread or that, pick it up here, jump around, plunge back in again, but *aguinaldo* or old gold, it is whole with everything else. Though we may choose not to tell it, there's no part that's not called for, none we can really do without. To tell the story of the *aguinaldo* without begging questions we need the sugar and the slavery, we need Spain and Africa, we need the island in the tropics with the mountains and the coast, we need the planet with the mobile crust, we need it all. *It's not that you have this land, and on top of it—just squatting on it, not integral to it—you have these people and they have this music they've brought with them from some other place on the land . . .*

It's not like that. It's all one. It's all connected. The music grows from the land like plants, is just as caught up in rock and water, is driven—like everything else—by the energy from the Sun.

You're in It Too

I still don't know what you're doing now besides reading this book, but, as I said before, if there's not a cup of coffee beside you or a soft drink, there was or there soon will be, and they tie you right into this story I've tried to unfold about the probability of hearing *aguinaldos* in the highlands of Puerto Rico. The coffee and cola imply the world no less than the *aguinaldo.* Which is the point, isn't it, that to say *anything* about the *aguinaldo* we had to invoke *everything,* the island's settlement history, its geology to raise the mountains, the history of sugar to drive the *jíbaro* into them? And since coffee and cola and *aguinaldos* each implies the world, they're bound to be entangled somewhere.

Actually it doesn't have to be that roundabout. There has to be a store where you bought the cola or the coffee (they didn't come from the aether), and then streets that connect you to the stores, and trains and boats and planes that connect the stores to the rest of the world where the coffee and the sugar came from. This means there are railroad tracks and harbors and airports. This means there are warehouses and freeways and freeway exits and great green-and-white signs looming over the roads. All of which means there have to be quarries too, and oil wells and asphalt and concrete plants and concrete trucks, orange-and-white stripped barrels, guys wearing orange-and-white stripped vests directing traffic . . . I could go on. It's all part of the story.

Somehow the coffee and sugar get to you, at which point you're mainlined into the story of the *aguinaldo,* which is another part of the story—which is unimaginable *without* the story—of the world trade in coffee and sugar. This trade implies worldwide systems of production and transportation and communication and banking. These may sound like abstractions but they're physical things: fields and farmers and ships and stevedores and telephones and linemen and offices and brokers and supermarkets and checkout clerks, all (not incidentally) with their music. They are systematically related, these things. They shape your day. They are the form *the land you live takes.* They are the form *the land takes that you live.*

Aguinaldos in the highlands, the Monkees on Hillsborough Street. . . .

In or *on* we're used to saying—live *in* the land, live *on* the land—but we don't live *in* it or *on* it. It is not apart from us. If we wish to understand anything about our situation we've got to get this. This is the new story we've got to learn how to tell. How does Mickey put it at the end of *In the Night Kitchen?* "I'm in the milk and the milk's in me. God Bless Milk and God Bless Me."[51]

We are the land and the land is us.

What we need, then, is a sketch—a *history*—of how it all came to have the form it has, one without paradise and golden ages and harmony and balance.

This is it. Or this is a stab at it, a first draft, a first draft of a stab at a sketch of the new story we have to learn how to tell.

From first principles, really, from the Big Bang to the moment not too long ago when that car rolled past me down the street blaring "I'm a Believer."

It's all salt from now on. No more sugar.

Notes

DON'T SKIP THIS

1. That lecture became the paper "Culture Naturale: Some Words about Gardening" that appeared in *Landscape Journal*, *11*(1), Spring 1992, pp. 58–66. It worked its way into this book's Chapter 8.
2. That lecture became the paper "The Spell of the Land," which was published in George Thompson, ed., *Landscape in America* (University of Texas Press, Austin, 1995, pp. 3–13). It too worked its way into Chapter 8.
3. This is "Popular Culture," *Encyclopedia of Global Change* (Oxford University Press, New York, 2001, pp. 269–272). It's been incorporated into Chapter 2 of this book.

CHAPTER 2

1. For a rich analysis of how misconceptualization of the Columbian encounter keeps us from thinking straight about many things, see James W. Loewen's *The Truth about Columbus: A Subversively True Poster Book for a Dubiously Celebratory Occasion* (New Press, New York, 1992).
2. For this story I'm wholly indebted to John McPhee's "Los Angeles against the Mountains," in his *The Control of Nature* (Farrar Straus Giroux, New York, 1989, pp. 183–272). Brilliant geography!
3. This myth is cogently analyzed in the context of the loss of the English countryside by Raymond Williams in his *The Country and the City* (Oxford University Press, New York, 1973). In its exhilarating second chapter we step onto an "historical escalator" that sweeps us back through one period of mourning for the loss of the countryside after another, from a couple of such episodes in the last century, to the perception of loss in 1860 (as in George Eliot), to George Crabbe's sense of loss in 1783 (*The Village*), to Bastard's *Chrestoleros* of 1598, to

William Langland's *Piers Plowman* of the 1370s. In search of that "timeless" stable time, Williams casts his eyes back to the Doomsday Book, to the "free" Saxon days, and to the Celtic days before those: "Where indeed shall we go, before the escalator stops?" In each case Williams sees that "a moral protest was based on a temporary stability: as again and in the subsequent history of rural complaint. It is authentic and moving yet it is in other ways unreal," unreal because the perception of stability—which never existed—was in every case generated as a form of the reaction to an undesired (but ongoing) transformation of the land, authentic and moving because people were losing their land, their homes, their villages.

4. This describes the logical relationship between developers and conservationists. Since they're using a common logic to no more than opposite ends, only power can distinguish between them. The logic of conservation is indistinguishable from that of development, and precisely as sterile.

5. In the classic millenarian vision this hoped-for future is precisely the longed-for paradise lost. History here is no more than a momentary interruption in an otherwise timeless existence.

6. By using the word "sign" I do not wish to be read as taking sides in the debate Tim Ingold rehearses between Claude Lévi-Strauss and Gregory Bateson as to whether the mind recovers the world by *decoding*—as Ingold reads Lévi-Strauss—or through a process of revelation—as Ingold reads Bateson, a distinction Ingold catches up in a distinction between "cipher" and "clue." While signs are operationalized by codes, the process of decoding rarely requires a cipher. Instead, the signs are usually taken as clues, and the whole system is "read" before its parts. But see Tim Ingold, *The Perception of the Environment: Essays on Livelihood, Dwelling and Skill* (Routledge, London, 2000, pp. 13–26, and throughout). This is a book to which I shall return often.

7. Much of the following is drawn from my article "Popular Culture" in Andrew Goudie, ed., *Encyclopedia of Global Change* (Oxford University Press, New York, 2001, pp. 269–272). I in turn leaned heavily on, among others, Elana Gomel, "Mystery, Apocalypse and Utopia: The Case of the Ontological Detective Story" (*Science-Fiction Studies*, 22, 1995, pp. 343–356); and Arthur B. Evans and Ron Miller, "Jules Verne, Misunderstood Visionary" (*Scientific American*, April 1997, pp. 92–97). For more detail, see the list of further reading at the end of my article.

8. This story draws on widespread antecedent accounts of worldwide flooding in, among others, the Sumerian epic of Gilgamesh and the Greek story of Pyrrha and Deucalion, though indeed most peoples have told such tales.

9. Although the form Shelley gave it was new, the theme of the recovery of culture by nature was anything but. See, for example, the chapter "Nature and Culture," in Edward Snow's *Inside Bruegel: The Play of Images in Children's Games* (North Point, New York, 1997, pp. 83–91); or Erwin Panofsky's *Early Netherlandish Paintings, I* (Harvard University Press, Cambridge, MA, 1953), especially, pp. 142–144, where he unfolds the dialectics of grace and nature in a discussion of a wall, in ruin, overrun by . . . nature.

10. For further analysis of the form of these tales see my "No Place for a Kid: Critical Commentary on *The Last Starfighter*" (*Journal of Popular Film and Television*, 14(2), Summer 1986, pp. 52–63).

11. Indeed it marks the allegiance of Tolkien's thought to what Mircea Eliade referred to as "cyclic time," the mythological time of traditional ways of thought. See Eliade's exposition in *The Myth of the Eternal Return* (Bollingen Series 46, Pantheon Books, New York, 1954). Incidentally *Return* is a book Eliade thought to give the title "Introduction to a Philosophy of History." Eliade's was, needless to say, a conception of our relation to time in complete opposition to that of his contemporary existentialists for whom humans *are* precisely insofar as they makes themselves in history.

12. In his *J. R. R. Tolkien: Author of the Century* (HarperCollins, New York, 1999), T. A. Shippey makes the point that fantasy is the dominant literary mode of the past century—"the natural way to respond to the traumatic events of the twentieth century," in the words of Stephen Medcalf's illuminating review of Shippey (*Times Literary Supplement*, September 22, 2000, pp. 24–25). I'm not saying that Tolkien had become an existentialist in the *Ring*, but there was sure no going back to Eden this time.

13. Though perhaps we should admit that it is not notably more intractable than geology or chemistry.

14. I don't mean to minimize the effect of Willie Horton, but the dirty water in Boston Harbor excised the environment as an issue in the contest, to Bush's advantage.

15. Old-fashioned westerns are all about threatened Edens, whether these Edens are family farms (Ma, Pa, and the Kid) threatened by cattleman (and their odious gunmen) or cattle spreads (the Ponderosa) threatened by incursions of farmers (with their odious fences).

CHAPTER 3

1. Fred Hoyle used the term "big bang" first, in a derisive reference to the theory whose singular past his "steady state" theory rejected, on a radio program in 1950. Hoyle was an effective and popular writer as well as a leading cosmologist, and his *The Nature of the Universe* (Blackwell, Oxford, 1950), while dated, remains an engaging introduction to astrophysics, steady state cosmology notwithstanding. For the failure of subsequent efforts to come up with a "better" name, see Timothy Ferris, *The Whole Shebang* (Simon & Schuster, New York, 1997, footnote 10, p. 323).

2. Indeed which even *cosmologists* still tell. Thus Steven Weinberg opens *The First Three Minutes: A Modern View of the Origin of the Universe* with the Nordic origin story of the giant and the cow from the *Younger Edda* (Basic Books, New York, 1977, pp. 3–4), while Stephen Hawking begins *A Brief History of Time: From the Big Bang to Black Holes* with a version of the oriental image of the world supported on the back of a giant tortoise (Bantam Books, New York, 1988, p. 1). Weinberg actually closes his account with the Nordic story of Ragnorak (p. 153) and even cites the *Younger Edda* in his further reading list with the note: "For another view of the beginning and end of the universe" (p. 180).

3. And why not finesse it? It's hard to do anything *with* it. Listen to Steven Weinberg recently twisting his tongue into all sorts of contortions: "I remarked in a recent article in *The New York Review of Books* that for me as a physicist the laws

of nature are real in the same sense (whatever that is) as the rocks on the ground. A few months after the publication of my article I was attacked for this remark by Richard Rorty. He accused me of thinking that as a physicist I can easily clear up questions about reality and truth that have engaged philosophers for millennia. But that is not my position. I know that it is terribly hard to say precisely what we mean when we use words like 'real' and 'true.' That is why I said that the laws of nature and the rocks on the grounds are real in the same sense. I added in parentheses 'whatever that is.' " (See Steven Weinberg, "The Revolution That Didn't Happen" (*New York Review of Books*, October 8, 1998, p. 52).

4. Though, according to Hoyle, Big Bang cosmology is *all* about belief: "Big-bang cosmology is a form of religious fundamentalism, as is the furor over black holes, and this is why these peculiar states of mind have flourished so strongly over the past quarter of a century. It is in the nature of fundamentalism that it should contain a powerful streak of irrationality and that it should not relate, in a verifiable, practical way, to the everyday world. It is also necessary for a fundamentalist belief that it should permit the emergence of gurus, whose pronouncements can be widely reported and pondered on endlessly—endlessly for the reason that they contain nothing of substance, so that it would take an eternity of time to distill even one drop of sense from them. Big-bang cosmology refers to an epoch that cannot be reached by any form of astronomy, and, in more than three decades, it has not produced a single successful prediction." This jeremiad is from Hoyle's funny, incendiary, exhausting, exhilarating autobiography, *Home Is Where the Wind Blows* (University Science Books, Mill Valley, CA, 1994, pp. 413–414). Remember: Hoyle was a *highly* respected cosmologist.

5. T. Padmanabhan, *Structure Formation in the Universe* (Cambridge University Press, Cambridge, UK, 1993, p. xiv).

6. This is from Steven Shapin and Simon Schaffer's truly great *Leviathan and the Air-Pump: Hobbes, Boyle and the Experimental Life* (Princeton University Press, Princeton, NJ, 1985, p. 25). While harboring no doubts about the validity of Boyle's gas laws, Shapin and Schaffer establish it as a "matter of fact" that Boyle constructed science as a machine for the social construction of assent. Shapin and Schaffer show how Boyle utilized the material technology of the gas-pump, a literary technology for spreading the story of the pump, and a social technology for adjudicating competing or novel knowledge claims. This is a *long* book, but well worth the slog. For a more laser-like approach, try Elizabeth Potter's much slimmer *Gender and Boyle's Law of Gases* (Indiana University Press, Bloomington, 2001). Potter acknowledges Boyle as a *great* scientist, but one whose science was nonetheless caught up in his humanness, that is, raveled up with his construction of gender (no less than his Christianity). Potter stresses communities as primary epistemic agents, and takes the demise of foundationalism for granted.

7. Dennis Cosgrove has recently published an interesting "genealogy," as he calls it, of the idea of the spherical Earth in Western thought. It illustrates the extent to which every *bit* of the story I think we should be telling has been hard won. See his *Apollo's Eye: A Cartographic Genealogy of the Earth in the Western Imagination* (Johns Hopkins University Press, Baltimore, 2001).

8. From Feyerabend's posthumous *Conquest of Abundance: A Tale of Abstraction versus the Richness of Being* (University of Chicago Press, Chicago, 1999). Feyerabend was the preeminent philosopher of science of the last quarter of the 20th century. He hadn't finished much of *Conquest of Abundance* when he died, and two-thirds of the book consists of previously published essays on the book's themes, but if you're looking for intellectually bracing iconoclasm of the highest order, look no further. It's demanding, though. Less so is his wonderfully heady autobiography, *Killing Time* (University of Chicago Press, Chicago, 1995); more so is his widely influential *Against Method* (3rd ed., Verso, London, 1993). His whole approach can be summed up in a sentence from the latter: "Science is an essentially anarchic enterprise" (p. 5).

9. Speaking of science, Feyerabend says, "We can tell many interesting *stories* [emphasis his]. We cannot explain, however, how the chosen approach is related to the world and why it is successful, in terms of the world. This would mean knowing the result of all possible approaches or, what amounts to the same, we would know the history of the world before the world has come to an end" (*Conquest of Abundance*, p. 145). The next sentence, however, reads: "And yet we cannot do without scientific know-how." There is nothing anti-science about trying to understand its social construction. Quite the contrary.

10. Richard Feynman, *The Character of Physical Law* (MIT Press, Cambridge, MA, 1965, pp. 22–23). Feynman was the preeminent physicist of the second half of the 20th century. This is a transcription of the Messenger Lectures Feynman gave in 1964 and captures better than the transcriptions of any of his other lectures the vibrancy of his thought in speech—a sweet little book. The "Mr. Roemer" referred to is the 17th-century Danish astronomer Olaus Roemer.

11. This is from the epigraph to Gerald Holton's *Thematic Origins of Scientific Thought: Kepler to Einstein* (Harvard University Press, Cambridge, MA, 1973, p. 5).

12. Giorgio de Santillana and Hertha von Dechend, *Hamlet's Mill: An Essay on Myth and the Frame of Time* (Gambit, Boston, 1969). This is a heartrending book about the loss of the archaic world that is as filled with feeling as it is learned, and it is very learned indeed. One of those improbable books you keep returning to, worrying it, wondering. . . .

13. Despite which it must be evident that what I really buy into is the dictum with which Martin Klein opens *Slavery and Colonial Rule in French West Africa:* "Most histories have neither a beginning nor an end. They are part of a seamless web" (Cambridge University Press, Cambridge, UK, 1998, p. xiii).

14. Joseph Silk, *A Short History of the Universe* (Scientific American Library, New York, 1994, p. 66).

15. This is all just standard physics. See any textbook or even any decent encyclopedia. The Schwarzchild radius is also known as the "gravitational radius."

16. And in which, presumably, gravity was unified with the three quantum forces. "Seeing" into this earlier superunified universe is one of the goals of superstring theory. The introduction and collection edited by P. C. W. Davies and Julian Brown, *Superstrings: A Theory of Everything?* (Cambridge University Press, Cambridge, UK, 1988) is still a great place to start for those with no math skills; and a great place to move to is Brian Greene's *The Elegant Universe* (Norton, New York, 1999). Greene is an accessible writer as well as an impor-

tant contributor to construction of the theory. Those with no math skills will also appreciate Richard Feynman's acidic observation about string theorists that "the mathematics is far too difficult for the individuals who are doing it" (quoted on p. 194 of the Davies and Brown book).

17. The concluding line of the *Tractatus Logico-Philosophicus*, "Wovon man nicht sprechen kann, darüber muss man schweigen," in the original translation (Kegan Paul, Trench, Trubner, London, 1922, p. 189). In the newer translation of D. F. Pears and B. F. McGuinness (2nd ed., Routledge & Kegan Paul, London, 1971, p. 151) it's rendered, "What we cannot speak about we must pass over in silence," causing me to mutter, "What one cannot improve, one ought to goddamn leave well enough alone."

18. This is controversial even among those who buy into the larger picture, even among those who created it. Stephen Hawking, who with Roger Penrose proved that the universe began in a space–time singularity (S. W. Hawking and R. Penrose, "The Singularities of Gravitational Collapse and Cosmology" [*Proceedings of the Royal Society of London, A314*, 1969, pp. 529–542]), later worked out a model without the singularity, that is, a model without a boundary condition. This, in his words, means the universe "would have neither beginning nor end: it would simply BE" (Hawking, op. cit., p. 141).

19. It was then that gravity first expressed itself as such (what words to use to say this?), released by the cooling of the expanding universe from its unification with the electromagnetic, weak, and strong forces in a spontaneous shattering of their symmetry. Something like that. The other three forces remain unified, since at very high energy density levels their effects are indistinguishable. See Silk (op. cit., pp. 75–76), but also the really helpful article by Andrei Linde, "The Universe: Inflation Out of Chaos" (*New Scientist, 105*[1446], March 7, 1985, pp. 14–18). Linde was one of the creators of the "new" inflationary universe I'll get to in a second.

20. Alan Guth is the theorist who came up with the "original" inflationary universe. In his wonderful memoir *The Inflationary Universe: The Quest for a New Theory of Cosmic Origins* (Addison-Wesley, Reading, MA, 1997) he claims that "to my knowledge, the first serious suggestion that the creation of the universe from nothing could be discussed in scientific terms was the 1973 paper by Edward Tryon, 'Is the Universe a Vacuum Fluctuation?' " (p. 271). Tryon's paper, short and disarmingly straightforward and simple, is worth reading. Of course, it takes a lot for granted: "The spontaneous temporary emergence of particles from a vacuum is called a vacuum fluctuation, and it is utterly commonplace in quantum field theory." Why? Because in "quantum field theory . . . every phenomenon that could happen in principle actually does happen occasionally in practice, on a statistically random basis." Having established that the universe does have a zero net value for all conserved quantities— which we'll skip over here—Tryon concludes: "If it is true that our Universe has a zero net value for all conserved quantities, then it may simply be a fluctuation of the vacuum, the vacuum of some larger space in which our Universe is imbedded. In answer to the question of why it happened, I offer the modest proposal that our Universe is simply one of those things which happen from time to time" ("Is the Universe a Vacuum Fluctuation?" [*Nature, 246*(5433), December 14, 1973, pp. 396–397]—I mean, that's the length of the entire pa-

per). Incidentally, the trivial transformation of "simply" into "just" in the last line turns Edward Tryon into Cole Porter. This may mean, as I've felt almost since I can remember, that the whole thing is just one of those things, you know, one of those crazy flings, a trip to the moon on gossamer wings, great fun, but . . . you know, just one of those things. (For a treatment that assumes less, try Henning Genz's *Nothingness: The Science of Empty Space* [Perseus Books, Reading, MA, 1999]).

21. Nicely caught in the dictum of John Archibald Wheeler to the effect that "spacetime grips mass, telling it how to move; mass grips spacetime, telling it how to curve" (*A Journey into Gravity and Spacetime* [Scientific American Library, New York, 1990, pp. 11–12]). My friend Gordie Hinzmann introduced me to Wheeler, for which I have always been grateful.

22. In the language of gauge theories, their symmetry remained unbroken, still *unified*, still all jumbled up. To quote again from Linde (op. cit.): "Without any scalar field, there is no difference between, say, the weak nuclear force and electromagnetism. But when an appropriate scalar field fills the Universe, it has the effect of breaking the symmetry between these two forces making them go their separate ways. In the modern unified theories, there are many scalar fields, each one responsible for breaking one of the original symmetries of this kind. At a state of very high energy (such as the moment of creation) the scalar fields play no part and the forces are unified; but as the energy density falls (as the Universe expands and cools) each scalar field 'switches on' in turn." Linde's article is reprinted in John Leslie, ed., *Physical Cosmology and Philosophy* (Macmillan, New York, 1990, in which reprint this quote appears on p. 242).

23. That is, for the symmetry between the strong force (on the one hand) and the weak and electromagnetic forces (on the other) to break.

24. Thinking about this as a phase transition is extremely common. Again, "each scalar field has a state of minimum potential energy, and each will tend to 'roll' down into that state as soon as the temperature (or energy density) of the Universe falls far enough. . . . When the temperature of the Universe was very high, the minimum energy state of the scalar field corresponded to zero energy which is why there was at that time no distinction between the forces of nature. It is only as the Universe cooled that the scalar fields with non-zero energy appeared and caused the symmetry breaking. The change from a symmetric state to a state of broken symmetry is a phase transition, and the best analogy to make is with the change that occurs when a liquid crystallizes" (Linde, ibid.). Each of these guys writes with complete authority, as if with that of the early universe itself, but then, as Lev Landau somewhere wrote, "Cosmologists are often in error, but never in doubt."

25. Guth (op. cit.) defines the vacuum as "roughly speaking . . . a space devoid of matter," going on to note that "this definition suffers from the ambiguity of the word 'matter.' *Particle physicists therefore define the vacuum as the state of lowest possible energy density* [emphasis mine]. The vacuum is not simple since the inherently probabilistic nature of quantum theory implies that unpredictable events, such as the chance materialization of an electron and its antiparticle, the positron, can occur at any time [as Tryon observed]. Such pairs have a fleeting existence of perhaps 10^{-21} second, and then they annihilate into noth-

ingness" (p. 343). It's precisely the presence of this matter, however, no matter how fleeting, that mandates a definition of the vacuum in terms of minimum energy density. Yet *this* definition suffers from an ambiguity surrounding the word "minimum": "Originally the phrase 'false vacuum' was used to describe a region in which one or more fields have a set of values that do not minimize the energy density [as the vacuum does], but which are in a local valley of the energy density diagram [a local minimization of the energy density that is vacuum-like in every other way]. Classically such a state would be absolutely stable, since there would be no energy available for the fields to jump over the hills that surround the valley. By the rules of quantum theory, however, the fields can tunnel through the hill of the energy density diagram and settle eventually in the true vacuum, the state of lowest energy density" (p. 331). Such hills, valleys, and tunnels are metaphorical, and refer to graphs of energy density (see the diagrams Guth provides, especially the one on p. 168). In physics and chemistry, whenever an action, interaction, or reaction requires energy, it is thought of as confronting an energy barrier, or hill. Richard Feynman says: "If we have one substance and another very similar substance, the one does not just turn into the other, because the two forms are usually separated by an energy barrier or 'hill.' Consider this analogy: If we wanted to take an object from one place to another, at the same level but on the other side of a hill, we could push it over the top, but to do so requires the addition of some energy. Thus most chemical reactions do not occur, because there is what is called an activation energy in the way" (*Six Easy Pieces* [Addison-Wesley, Reading, MA, 1995, p. 52]). It's as if the false vacuum were a pool of water on the floor of a mountain valley. Classically this would be a stable situation (i.e., a vacuum), because energy would be required to get the water up over the mountains before it could run off to the lowlands and the ocean on the other side (at an evidently lower energy density); but in the quantum world—our world at infinitesimal scales—it's possible to "tunnel" through the mountains, and so realize, in a manner of speaking, the "potential energy" of the drop (manifested as virtual particles that annihilate each other). Keep in mind that this is a *metaphorical description* of a *diagram* of variations in energy density. When used as above, "classically" means no account is taken of the Heisenberg uncertainty relation. For a rigorous treatment of tunneling events between vacuum states with different topological quantum numbers interpreted as instatons, see Ta-Pei Cheng and Ling-Fong Li, *Gauge Theory of Elementary Particle Physics* (Oxford University Press, New York, 1984, pp. 482–487). For a more visual—but nonetheless rigorous—treatment of tunneling and metastable states, pick up Siegmund Brandt and Hans Dieter Dahmen, *The Picture Book of Quantum Mechanics* (2nd ed., Springer-Verlag, New York, 1995); see especially pp. 83–90. Don't be misled by the title. The pictures are all graphs and there are 486 of them. Is any of this "right"? Who knows? As Michael Hawkins says, " 'Scientific truths' is simply another way of saying, 'the fittest, most beautiful and most elegant survivors of scientific debate and testing' " (*Hunting Down the Universe* [Addison-Wesley, Reading, MA, 1997, p. 7]). Don't forget: *useful principle!*

26. Symmetry breaking, transition from the false to the true vacuum, the materialization of particles, and the inflation of the rate of expansion of the uni-

verse seem to have been simultaneous, and *may* be variant ways of knowing a common phenomenon, causally related, or serendipitously associated (fat chance). Padmanabhan concludes his review of inflationary models by stating that "no single model for inflation suggested so far, can be considered completely satisfactory" (op. cit., p. 364). Of course he's approaching it from the perspective of the density perturbations required to explain the present large-scale structure of the universe. Still, writing more popularly a year later, Silk put it like this: "The release of energy at symmetry breaking corresponds to a change of the quantum vacuum: the quantum vacuum of the symmetric state is the so-called false vacuum, equivalent to the vacuum of 'negative energy' seen in the Casimir effect. At the time of symmetry breaking, it changes to the true vacuum of the present epoch, which exists along with the very disparate forces of the universe today. The energy released accelerates the expansion rate of the universe. Indeed, the universe inflates at an exponentially increasing rate until the transition to the true vacuum is completed" (op. cit., pp. 77–78).

27. Which is one way of thinking about the conclusion of this epoch. Here's another: "The negative energy of the false vacuum was soon swamped by the thermal energy of the ordinary matter and radiation, and the inflationary expansion abruptly ceased" (Silk [op. cit., p. 81]). The problem is *so much* is going on.

28. Here's Ferris distinguishing between virtual and real: "Owing to what is called wave–particle duality, quantum physics sees nature as if through two eyes, one of which beholds particles and the other waves. Look through the particle eye and we find that for every 'real' (meaning long-lived) electron there are countless 'virtual' electrons and positrons. Look through the wave eye and we find quantum fields roiling the vacuum like winds across water" (op. cit., p. 236). *Well.* I know this all sounds like a fairy tale, but stone-sober college texts have subheads that read "During inflation, all of the mass and energy in the universe burst forth from the vacuum of space" and "As the primordial fireball cooled off, most of the matter and antimatter in the early universe annihilated each other." See William Kaufmann's best-selling astronomy text *Universe* (Freeman, New York, 1994, pp. 556, 558). Or try the article on the cosmos in the 15th edition of the *Britannica* where it straightforwardly says things like "Notably, the false vacuum had a negative pressure capable of behaving as a cosmological constant . . . " and so on (*The New Encyclopaedia Britannica*, Chicago, 1987, Vol. 16, p. 847]). The point is that this is all absolutely taken for granted. Again, this was Alan Guth's baby and his *The Inflationary Universe* (op. cit.) is where you want to start. If you don't have time for Guth's book, try the review article he wrote for Norriss Hetherington, ed., *Encyclopedia of Cosmology* (Garland Books, New York, 1993, pp. 301–322). A. D. Linde, and independently Andreas Albrecht and Paul Steinhardt, fixed up the shortcomings of Guth's model to make it useful. The historical introduction to Linde's *Inflation and Quantum Cosmology* is surprisingly simple and really clear (Academic Press, New York, 1990, say, pp. 1–16).

29. Feynman (1965, op. cit., pp. 20–21).

30. And while this is true, and while it is also true that this simple fact of daily life is all but invariably included in accounts of this story at precisely this point, it

is also true that Craig Horton, Robert Kirshner, and Nicholas Suntzeff warn that "though often ascribed to the Doppler effect—the phenomenon responsible for changing the pitch of a passing train whistle or car horn—the cosmological redshift is more correctly thought of as a result of the ongoing expansion of the universe. Emissions from more distant objects, having traveled for a greater time, become more redshifted than radiation from nearby sources" ("Surveying Space–Time with Supernovae" [*Scientific American*, January, 1999, p. 47]).

31. "I do not want to give the impression that everyone agrees with this interpretation of the red shift," Steven Weinberg wrote in 1977. "We do not actually observe galaxies rushing away from us; all we are sure of is that the lines in their spectra are shifted to the red, i.e., toward longer wavelength. There are eminent astronomers who doubt that the red shifts have anything to do with Doppler shifts or with an expansion of the universe" (Weinberg, op. cit., pp. 28–29). He therefore moves to provide evidence not dependent on the redshifts.

32. "The difficulty of dealing with the dynamics of an infinite medium pretty well paralyzed further progress until the advent of general relativity" is how Weinberg puts it. See his discussion of Newton's confrontation of the problem (Weinberg, op. cit., pp. 31–32).

33. Alexander Friedmannn was a Soviet meteorologist who, in the words of Jeremy Bernstein and Gerald Feinberg, "got caught up in the wave of excitement that followed the 1919 confirmation of the bending of light by the sun's gravitational field by the amount Einstein's theory predicted. He not only taught himself the theory but began improving on it." See the "Introduction" to their *Cosmological Constants: Papers in Modern Cosmology* (Columbia University Press, New York, 1986, p. 11), a collection of reprints that includes English translations of both Friedmann's important papers along with comments by Einstein (pp. 49–67). Friedmann's terribly short life is the subject of E. A. Tropp, V. Y. Frenkel, and A. D. Chernin's *Alexander Friedmann: The Man Who Made the Universe Expand* (Cambridge University Press, Cambridge, UK, 1993) which was not, but should have been, written by his former student, George Gamow, who really knew how to write. Speaking of Gamow, Helge Kragh crowns him, not Friedmann, "father of the Big Bang," noting that Gamow—with his associates—did all the work (Helge Kragh, *Cosmology and Controversy: The Historical Development of Two Theories of the Universe* [Princeton University Press, Princeton, NJ, 1996]; this is actually a *detailed* history of the rivalry of the expanding and steady state universes). But who *cares* who was first?

34. Lemaître's and Friedmann's views were, however, in no way identical. Lemaître championed a "primeval atom" model in which the original stuff of the universe was matter, not radiation, though this is not an aspect of his 1931 paper "A Homogeneous Universe of Constant Mass and Increasing Radius Accounting for the Radial Velocity of Extra-Galactic Nebulae" (translated in Bernstein and Feinberg, eds., op. cit., pp. 92–101). For more on the primeval atom model, see Ernest Sternglass's conceptually challenging but narrative and chatty exposition of a modern version of Lemaître's idea in *Before the Big Bang: The Origins of the Universe* (Four Walls Eight Windows, New York, 1997). There are respected cosmologists at work on every variety of the story.

35. Actually there's nothing inevitable about it at all. In 1948, Thomas Gold, Hermann Bondi, and Fred Hoyle, Cambridge physicists all (of course we've met Fred), proposed a steady state cosmology. Here expansion is offset by the continuous creation of new matter at a rate that maintains the large-scale features of the universe over time. This idea is immensely appealing on philosophical, even emotional, grounds to those who have no interest in doubting the mainstream interpretation of the redshift, but are uninterested in images of creation or cataclysm. As a young teenager I found the theory as presented in Hoyle's *The Nature of the Universe* (op. cit.) completely satisfying. The rise and fall of steady state cosmology is well told in Kragh (op. cit., or see Kragh's review article in Hetherington, ed., op. cit., pp. 629–636), or can be followed through the Leslie reprints (op. cit.). Incidentally, there is *no* overlap between Leslie's collection and that edited by Bernstein and Feinberg (op. cit.), who would, I assume, imagine their collection to represent the mainstream, Leslie's the fringe.

36. The young theorist was P. J. E. Peebles, then working under Robert Dicke—champion of an *oscillating* universe—who had suggested both that Peebles carry out the calculation, and that two other researchers, P. G. Roll and D. T. Wilkinson, look for the radiation. In the event, Penzias and Wilson had already found it, and two companion letters were published: A. A. Penzias and R. W. Wilson, "A Measurement of Excess Antenna Temperature at 4080 Mc/s," pp. 419–421; and R. H. Dicke, P. J. E. Peebles, P. G. Roll, and D. T. Wilkinson, "Cosmic Black-Body Radiation," pp. 414–419 (*Astrophysical Journal, 142,* 1965). Both are eminently readable. Peebles describes the experience in an interview collected by Alan Lightman and Roberta Brawer in *Origins: The Lives and World of Modern Cosmologists* (Harvard University Press, Cambridge, MA, 1990, pp. 214–231, especially pp. 218–219).

37. This is Weinberg again (op. cit., p. 50). His account is at once rigorous and wonderfully chatty. And out of date!

38. I've skipped over a complicated history of particle and antiparticle creation and annihilation that went on from the time protons and neutrons appeared around 10^{-5} second until the beginning of nucleosynthesis at about minute 3.

39. "Tuned into" is precisely wrong. Penzias and Wilson found the microwave background radiation entirely by happenstance. Weinberg devotes an entire chapter to the question: "Why Did It Have to Be by Accident? Or to Put It Another Way, Why Was There No Systematic Search for This Radiation, Years before 1965?" He advances many reasons but concludes: "most importantly, the 'big bang' theory did not lead to a search for the $3°K$ microwave background because it was extraordinarily difficult for physicists to take seriously *any* theory of the early universe. (I speak here in part from recollections of my own attitude before 1965.) Every one of the difficulties mentioned above could have been overcome with a little effort. However, the first three minutes are so remote from us in time, the conditions of temperature and density are so unfamiliar, that we feel uncomfortable in applying our ordinary theories of statistical mechanics and nuclear physics" (op. cit., p. 131). For the rest of us it's got to be infinitely worse.

40. This is a ludicrous conceit, but the point is that until this moment photons—light particles (particles of electromagnetic radiation)—would no sooner start

moving than they'd wham into an electron (etc.). Now at last they could keep moving, could propagate, well, as long as the universe continued to expand. This is when the microwave radiation picked up by Penzias and Wilson started heading our way. It had been propagating for nigh on 15 billion years. This is what is meant by the universe becoming "transparent": something happens, and then light moves away from that something into the universe. Before, whatever was happening *and* the light—the radiation—were so mashed together there wasn't any question of the light carrying anything *away* from the happening, it *was* the happening (it was *part* of the happening anyway). Yes? No? Clear? I'm trying. Look, if you just want the *feel* of this without the rest of it, the best thing you can read is Masamune Shirow's *Orion* (Dark Horse Comics, Milwaukie, OR, 1994). This is a *manga* masterpiece filled with footnoted drawings of immense wit and intelligence. Here, I've opened at random (yes, actually): "In the first stage of creation, there was yin and yang," Susano says. "The very beginning?" asks Fuzen. "Right. Before that even nothingness did not exist—there was a virtual vacuum without time or meaning. But there was yin and yang." Of course Susano is a crazy long-haired blond boy-god and Fuzen is the father of the beautiful heroine and it's very funny, but for those without the math it better conveys the *sense* of modern cosmology than almost *any*thing . . . Try it. It is a *great* comic book.

41. Where do the initial inhomogeneities come from that lead to this clumpiness? Big debate. Hawking argues that under his "no boundary" proposal it would have arisen from the initial quantum fluctuations in the positions and velocities of particles: "In an expanding universe in which the density of matter varied slightly from place to place, gravity would have caused the denser regions to slow down their expansion and start contracting. This would lead to the formation of galaxies, stars, and eventually even insignificant creatures like ourselves. Thus all the complicated structures that we see in the universe might be explained by the no boundary condition for the universe together with the uncertainty principle of quantum mechanics" (op. cit., p. 140). But see Padmanabhan (op. cit.) for more on the large-scale structure of the universe.

42. Or maybe it didn't take this long. Here's a line I really love. After 11 chapters dealing with the *really* early universe, Silk finally gets to the question of galaxy formation: "Unfortunately," he writes, "this is where the physics becomes messy and poorly understood." I howled when I first read that. Absurdly enough, it's true: "Now we have to contend with dark matter that evolves collisionlessly, conserving its energy, and with gas that evolves by dissipating its energy. We have to incorporate star formation, a process that is not well understood in the vicinity of the sun, let alone in a remote, forming galaxy. We will need to extrapolate poorly understood processes over periods of billions of years and over distances of millions of parsecs" (op cit., p. 214). And we *hadn't* been doing that in the first 11 chapters? Ah, well. . . .

43. Two hundred and six years to be exact. Pierre-Simon, Marquis de Laplace, published his *Exposition du système du monde*—"an acknowledged model of French prose"—in 1796. The model suffered continuous vicissitudes as its inadequacies surfaced, but as refined it remains the model of choice.

44. James Kaler's *Stars* (Scientific American Library, New York, 1992) is an accessible, beautiful, and continuously clear account of what we know about stars,

whose core concern is the Sun. Need I say the subject is evolving and fraught with controversy?

45. During the main-sequence life of a star the "inward tug of gravity balances the outward force exerted by the higher central pressure, and the star is in hydrostatic equilibrium" (Silk, op. cit., p. 18). Eventually the fuel gives out and the star dies according to one of a variety of scenarios ranging from supernovas to black holes that are a function of its size, composition, and so on. See the diagram plotting mass against radius for all objects in the universe from which every fate can be read in Simon Mitton, ed., *The Cambridge Encyclopedia of Astronomy* (Crown, New York, 1977, p. 107).

CHAPTER 4

1. This is a wonderful feature of Nigel Calder's *Timescale: An Atlas of the Fourth Dimension* (Viking Books, New York, 1983), a running head on nearly every other page (starting with p. 99): "We were quarks," "We were hydrogen," "We were carbon," "We were tar," "We were bacteria," and so on, that inexplicably stops shy of, say, "We were idiots." But, yes, we *are* an eddy in an expanding and cooling gas.

2. In *The Tree of Knowledge* (rev. ed., Shambhala, Boston, 1992), Humberto Maturana and Francisco Varela define history this way: "Each time in a system a state arises as a modification of a previous state, we have a *historical phenomenon*" (p. 56).

3. Stars, as we've seen, are maintained in hydrostatic equilibrium by the equivalence of gravity and the radiation pressure produced by the fusion of hydrogen into helium. Sooner or later, however, stars run out of hydrogen. With the decline of fusion, the radiation pressure drops, and gravity causes their cores to contract. This jacks up the temperature to the point where helium can begin to fuse, ultimately producing carbon and oxygen. The same thing happens when stars run out of helium: the pressure drops, the stars contract, and the temperature rises until the carbon and oxygen start fusing to produce neon, sodium, magnesium, and silicon. The process continues, depending on the size and composition of the star. Stars larger than eight solar masses evolve through a complete series of cycles of nuclear "burning" as shells successively form of hydrogen, helium, carbon, nitrogen, oxygen, and silicon. There are many intermediate isotopic products. At the end of the star's life is iron. Iron is the most tightly bound of all atomic nuclei. No energy can be gained from its fusion and the star implodes to form a neutron star, generating a supernova in the act. In Kaler's words: "The core contracts faster and faster, then goes into a catastrophic collapse. The star has lived for 10 million years. In less than a tenth of a second the iron core flies inward at a quarter the speed of light to a sphere only 100 kilometers across. The gravitational energy released is beyond imagining. In that blink of an eye the star expends over 10^{46} joules. . . . 100 times more than the Sun has radiated in its lifetime. . . . The sudden implosion creates a shock wave that propagates outward. . . . In the envelope's nuclear fury vast numbers of neutrons are created, which rapidly attach themselves to highly radioactive isotopes. This r-process can build heavy

isotopes that the s-process of the giants cannot reach. Several solar masses of stellar material, enriched in the elements created in both the supergiant and the supernova phases, are then suddenly returned to interstellar space at several thousand kilometers per second" (op. cit., p. 190). Whew! Also see Kaufmann (op cit., pp. 128–138); Silk (op. cit., pp. 214–226); and Kaler (op. cit., pp. 119–203, et passim). This synthesis has been in place for some time now (see, e.g., the chapters "The Distribution and Origin of the Chemical Elements" and "Between the Stars," in Mitton [op. cit., pp. 120–125 and 258–277]), but this should not be taken to imply that it is not problematic.

4. Silk (op. cit., p. 222).

5. This metaphor is beautifully developed in Roman Smoluchowski, *The Solar System: The Sun, Planets and Life* (Scientific American Library, New York, 1983, pp. 25–26), a beautifully illustrated, handsomely produced, and luminously calm (if dated) description of the solar system. (For a more up-to-date equivalent, get hold of David Morrison's *Exploring Planetary Worlds* [Scientific American Library, New York, 1993].) Stuart Taylor's *Solar System Evolution: A New Perspective* (Cambridge University Press, Cambridge, UK, 1992) is all but its opposite, a sort of aggressively ugly book, that sets out to doubt every assertion ever made about the solar system and so is unnervingly critical. Taylor has read everything by everybody, and while you might want to take Smoluchowski to bed to bliss out on the Sun and planets, you want to wake up with Taylor.

6. With respect to the inner planets (Mercury, Venus, Earth, and Mars), repeated simulations with a wide range of initial starting conditions ultimately end up in the same place: accretion continues for 100 million years and usually forms four terrestrial planets with orbits between 0.3 and 1.6 AU from the Sun. See the summary in Kaufmann (op. cit., pp. 138–140). (*Terrestrial planets*, which are small and rocky, are distinguished from *Jovian planets*, which are huge and gaseous. An AU is an astronomical unit, i.e., the mean distance from the Earth to the Sun, equivalent to 1.496×10^8 meters, or, as you might have grown up hearing, 93 million miles. Astronomers have developed scads of these astronomical quantities.)

7. Here's another of those confirming convergences of evidentiary chains: the 100 million years hypothesized for the formation of the solar system based on the dynamics of a slowly rotating gas, "more or less conforms with the approximately 100 million years from the completion of the nucleosynthesis to the formation of the earth and meteorites that was calculated from the existence of iodine-129," in the words of geochemist Minoru Ozima in his *The Earth: Its Birth and Growth* (Cambridge University Press, Cambridge, UK, 1981, p. 16). So this means we're on the right track, there is no going back now. . . .

8. What's hell on earth is the nomenclature for geological time, with its ceaselessly evolving eons (and æons), its eras, its periods or systems, its epochs or stages or ages. Subdivisions are proposed by appropriate International Union of Geological Sciences stratigraphic subcommissions and . . . *well*, you can imagine. I had originally planned to write a long note tracing its history and attempting to sort it out; but I realized that task would actually take a book, and decided to no more than bewail the widespread inconsistency. But you know what? Though one author's "Precambrian" may be another's

"Prephanerozoic," ones "eon" another's "era," practically nobody gets confused. It's not like there isn't a *lot* of context. For a guide to what's up, see W. Brian Harland et al., *A Geologic Time Scale 1989* (Cambridge University Press, Cambridge, UK, 1990).

9. With the composition of the early Earth we begin to read geophysicists, geochemists, geologists, paleoclimatologists, and biologists instead of cosmologists, astrophysicists, and astronomers. My Virgil through this literature is Harold Morowitz, whose *Beginnings of Cellular Life: Metabolism Recapitulates Biogenesis* (Yale University Press, New Haven, CT, 1992) is as clear an exposition of what a theory of biogenesis has to satisfy—and why—as I can imagine.

10. In the words of Sherwood Chang, "Early catastrophic outgassing is supported by isotopic systematics of radiogenic and primordial noble gases in terrestrial materials" ("The Planetary Setting of Prebiotic Evolution," in Stefan Bengston, ed., *Early Life on Earth* [Columbia University Press, New York, 1994, p. 11]). For more detail, see especially C. J. Allegre, T. Staudacher, and P. Sarda, "Rare Gas Systematics: Formation of the Atmosphere, Evolution and Structure of the Earth's Mantle" (*Earth and Planetary Science Letters, 81*, 1987, pp. 127–150). Believe it or not, "burp" is a technical term. The period is even sometimes called "the Big Burp."

11. Thomas Graedel and Paul Crutzen, *Atmosphere, Climate and Change* (Scientific American Library, New York, 1995, p. 61). Note that there is plenty of disagreement about the source and quantity of hydrogen.

12. Mitton (op cit., pp. 169–173).

13. Though "the oxidation state at the time when life originated remains a matter of dispute," according to J. F. Kasting and S. Chang in their summary "Formation of the Earth and Origin of Life," in J. William Schopf and Cornelius Klein, eds., *The Proterozoic Biosphere: A Multidisciplinary Study* (Cambridge University Press, Cambridge, UK, 1992, p. 10). This, and its companion predecessor, *Earth's Earliest Biosphere: Its Origin and Evolution* (Princeton University Press, Princeton, NJ, 1983), which Schopf also edited, are canonical, all 1,889 aggregate pages. They are ineradicable traces of a rapidly evolving field of study, and people who should know regard them as indispensable. But this stuff does date unbelievably rapidly. Therefore also keep at hand Bengston (op. cit.), the proceedings of a symposium at which most of the authors in Schopf (op. cit.) and Schopf and Klein (op. cit.) gave papers (though a third of Bengston's volume is concerned with the Phanerozoic). And whatever else on the topic was published yesterday. . . .

14. This is based on a reading of the lunar record. See E. M. Shoemaker's, "The Collision of Solid Bodies," in J. K. Beatty, B. O'Leary, and A. Chaikin, eds., *The New Solar System* (Cambridge University Press, Cambridge, UK, 1982, pp. 33–44); as well as Shoemaker's "Large Body Impacts through Geologic Time," in H. D. Holland and A. F. Trendall, eds., *Patterns of Change in Earth Evolution* (Springer-Verlag, Berlin, 1984, pp. 15–40). For consideration of these data with respect to the evolving biosphere, see N. H. Sleep, K. J. Zahnle, J. F. Kasting, and H. J. Morowitz's "Annihilation of Ecosystems by Large Asteroid Impacts on the Early Earth" (*Nature, 342*, 1989, pp. 139–142).

15. Which is different from saying they wouldn't have *formed*. During the Hadean, terrestrial chemistry undoubtedly exhibited both self-reference and auto-

catalysis, but at these temperatures, it was easy come, easy go. As I said earlier, ours is the story of an expanding and cooling gas. As it cools, as energy levels drop, increasingly complicated forms amalgamate. In the *really* high energy levels of the early universe it was too much for even the strong force to hold quarks together (and as particle physicists who try to smash them know, those strong force bonds are tough). Only when the energy level dropped could protons and neutrons form. Only when it dropped still further could they form atoms. Only at still lower energy levels could stable molecules form. And only at still lower levels could we form. I know I keep repeating this. I want you to think about it as the refrain to a song. . . .

16. In "The Planetary Setting of Prebiotic Evolution," Sherwood Chang puts it this way: "Only a fragmentary record of Earth's earliest surface environments is preserved in the 3.8 Ga [billion years ago] metasediments of Isua, Greenland. These rocks indicate surface temperatures below 100°C; an extensive body of liquid water; carbon dioxide, water vapor, and presumably nitrogen in the atmosphere; higher heat flow and more intense volcanism than now; the beginnings of continental growth; weathering and, by inference, hydrologic and carbon geochemical cycles" (Bengston [op. cit., p. 10]). Talk about sermons in stones!

17. Though as Chang notes, "Most relevant for the origin of life may have been the ubiquitous, geophysically active regions at the ocean–atmosphere interface and at the ocean–crust interface in marine hydrothermal systems. These realms contain phase boundaries between gas, liquid, and solid states where disequilibrium resulting from gradients in physical and chemical properties are maintained by physical and chemical energy fluxes. Within these environments, small-scale interfaces are provided by aerosols, volcanic and cometary dust, hydrothermal minerals, chemical precipitates, and vesicle-like structures of organic or mineral composition. Inasmuch as life itself must have emerged as a phase-bounded system, the formation, dissipation, and reformation of small-scale interfaces must have been a prerequisite for the origin of life" (ibid., pp. 17–18). When Chang says "life itself must have emerged as a phase-bounded system," what he means is that the living thing, let's say it was in water, wouldn't have been homogenous with that water, but would have been separated from it by some kind of membrane (phase separation).

18. Lynn Margulis and Dorian Sagan, *Microcosmos: Four Billion Years of Microbial Evolution* (Summit, New York, 1986, p. 51). This is a terrific book.

19. Mitton (op. cit., p. 271). Note the possibility of these compounds polymerizing, and further, the fact that they seem to have done so, at least as evidenced by the infrared and mass spectrography of comets and the analysis of meteorites. See, inter alia, D. A. Allen and D. T. Wickramasinghe, "Discovery of Organic Grains in Comet Wilson" (*Nature, 329,* 1987, pp. 615–616); and Stanley Miller's review, "Which Organic Compounds Could Have Occurred on the Prebiotic Earth?" (*Cold Spring Harbor Symposia on Quantitative Biology, 52,* 1987, pp. 17–27).

20. It's interesting. She says it to Fred Astaire in Robert Mamoulian's *Silk Stockings* where she plays a Soviet commissar to his bourgeois gentilhomme (well, film producer). They're in Paris. They're in love. She unfolds her dialectical materialist interpretation of their attraction in "It's a Chemical Reaction, That's

All." Astaire retorts that whatever it is it has nothing to do with electromagnets, and segues into "All of You" (both written by Cole Porter). The problem with construing this as a parody of dialectical materialism is that Cole Porter is essentially a materialist himself ("I've Got You Under My Skin," "Night and Day," "Why Can't You Behave," "You Do Something to Me"). Anyhow, what's interesting is that the only scientists during the first half of the century to take the problem of the origin of life seriously—Soviet biochemist Alexander Oparin, British geneticist J. B. S. Haldane, and British crystallographer John D. Bernal—were all Marxists—"militant defenders of dialectical materialism," in the words of Belgian–American evolutionary molecular biologist Christian DeDuve (*Blueprint for a Cell: The Nature and Origin of Life* [Neil Patterson, Burlington, NC, 1991, p. 109]). Haldane published a number of his essays on the origin of life in the *Daily Worker* (collected in *On Being the Right Size and Other Essays* [Oxford University Press, Oxford, UK, 1985]). In *Life: Its Nature, Origin and Development* (Academic Press, New York, 1962), Oparin quotes Engels (on the definition of life, p. 8, et passim), Lenin, and even Lysenko (trivially, pp. 178 and 184). *Life* superseded Oparin's *The Origin of Life on Earth* (Academic Press, New York, 1957) which he'd originally published in 1924. Oparin is tough: "Life is a form of the motion of matter." Period.

21. Morowitz (op. cit., p. 32). It's the evidence of the Isua supracrustals from western Greenland that convinces Morowitz of this.

22. Ibid., pp. 36–37.

23. Urey was an atomic physicist (a Nobel laureate for his discovery of deuterium in 1931) whose work with this heavy isotope of oxygen and a related technique for determining temperatures of the early ocean propelled him into the study of the elemental composition of the early Earth and the origin of the planets, in the course of which he came to take the work of Oparin and Haldane seriously. Working in Urey's lab, Miller showed that when exposed to ultraviolet radiation, the primitive atmosphere Urey proposed would react to form a variety of organic compounds. Miller's paper is a model of straightforwardness: "A Production of Amino Acids under Possible Primitive Earth Conditions" (*Science, 117*, 1953, pp. 528–529).

24. Stuart Kauffman, *At Home in the Universe: The Search for the Laws of Self-Organization and Complexity* (Oxford University Press, New York, 1995, p. 35). Kauffman's a complexity guru who contends that complexity itself triggers self-organization; that if enough molecules pass a certain threshold of complexity, they begin to self-organize into a living cell; that far from improbable, life is inevitable. Since it *is*, that sort of goes without saying. It's all a bit breathless. There's not a lot of chemistry in Kauffman's world, but there is lots of information.

25. This is DeDuve (op. cit., pp. 109–110), who credits Wöhler with freeing the organic world from the vital spirit, as later Miller credits the *prebiotic*.

26. "The list has become so long that emphasis is now put on those substances that still await 'prebiotic' generation," writes DeDuve (ibid., p. 123).

27. "It is becoming clear that however life began on earth, the usually conceived notion that life emerged from an oceanic soup of organic chemicals is a most implausible hypothesis. We may therefore with fairness call this scenario 'the myth of the prebiotic soup' " (C. B. Thaxton, W. L. Bradley, and R. L. Olsen,

The Mystery of Life's Origins: Reassessing Current Theories [Philosophical Library, New York, 1984, p. 66]). While their inability to find plausibility in any of the theories then current drove this chemist, marine scientist, and geochemist to a creationist conclusion, their critique of the primeval soup hypothesis has been widely accepted.

28. These two paragraphs are a summary of Morowitz. In *A Case against Accident and Self-Organization* (Rowman & Littlefield, Lanham, MD, 1997), Dean Overman summarizes Morowitz's position as follows: "Harold Morowitz, having performed the probability calculations described above, rejects the chance origin of life scenario and proposes a scenario based on his belief in the self-ordering power of the elements in the periodic table. Morowitz does not consider the primeval soup to be a plausible paradigm and *speculates that life is a natural extension of the laws of physics and chemistry*" (p. 82, my emphasis) and so on for another six pages. Overman's reduction, if hostile, is acute, and if you haven't time for Morowitz—whose book is elegant and concise—get your hands on Overman. He summarizes the case against accident thoroughly, including even the contributions of our friend Fred Hoyle who, ever the iconoclast, was led by its improbability to propose extraterrestrial origins (best argued in the book Hoyle coauthored with Chandra Wickramasinghe, *Evolution from Space* [Dent, London, 1981]). But hasn't the argument from accident always been no more than a way of giving a name to our lack of consensus about the certain chain of events? For Overman, however, self-organization is scant improvement. (Need I add that Overman has even greater problems with the quantum fluctuation origin of the universe arguments of Silk, Tyron, Linde, and Company, which he nonetheless summarizes with unequaled lucidity, especially on pages 154–159?) In the end, Overman has no counterargument. His objections reduce to the "that can't be" variety. By the way, he's a lawyer, which may or may not explain the simplicity and precision of his précis.

29. This is the essential burden of DeDuve's book (op. cit.).

30. A. G. Cairns-Smith, *Genetic Takeover and the Mineral Origins of Life* (Cambridge University Press, Cambridge, UK, 1982).

31. G. Wächtershäuser, "Before Enzymes and Templates: Theory of Surface Metabolism" (*Microbiology Review, 52*, 1988, pp. 452–484).

32. Margulis and Sagan (op. cit., p. 54).

33. This is from Morowitz's *Cosmic Joy and Local Pain: Musings of a Mystic Scientist* (Scribner's, New York, 1987, p. 234).

34. Gerald Weissmann, in the "Introduction" to G. Weissmann and R. Claiborne, eds., *Cell Membranes: Biochemistry, Cell Biology and Pathology* (Hospital Practicing Publishing, New York, 1975, p. xiv).

35. F. M. Harold, *The Vital Force: A Study of Bioenergetics* (Freeman, New York, 1986, p. 168). An energetics perspective almost forces an emphasis on the membrane.

36. Margulis and Sagan (op. cit., p. 55). Margulis herself seems to be moving toward the primacy-of-the-membrane camp however. Twelve years later in her *Symbiotic Planet: A New View of Evolution* (Basic Books, New York, 1998), she writes: "Yet although we cannot create cells from chemicals, cell-like membranous enclosures form as naturally as bubbles when oil is shaken with water. In the earliest days of the still lifeless Earth, such bubble enclosures separated in-

side from outside. As Harold J. Morowitz, distinguished professor at George Mason University, Fairfax, Virginia, and director of the Krasnow Institute for the Study of Evolution of Consciousness, argues in his amusing mayonnaise book, we think that prelife, with a suitable source of energy inside a greasy membrane, grew chemically complex. These lipidic bags grew and developed self-maintenance. They, through exchange of parts, maintained their structure in a more or less increasingly faithful way," and so on (pp. 71–72). The amusing mayonnaise book is Morowitz's *Mayonnaise and the Origin of Life: Thoughts of Minds and Molecules* (Scribner's, New York, 1985).

37. A dissipative structure, according to Ilya Prigogine, who won a Nobel Prize for studying them, "corresponds to a giant fluctuation stabilized by the exchange of energy with the outside world"; see his *From Being to Becoming: Time and Complexity in the Physical Sciences* (Freeman, San Francisco, 1980, p. 90). This giant fluctuation is a far-from-equilibrium structure, like (famously) the scrolls in the Belousov–Zhabotinskii reaction, like a convection cell, a whirlpool, a tornado, like an . . . autocatalyzing system of self-bounding macromolecular reactions (i.e., . . . a *cell*) which maintains its structure—is stable—as long as energy is available to it from the encysting system. As the examples may indicate, such structures are self-organizing and far from random. Having discussed the oscillatory aspect of chemical reactions in the glycolytic (metabolic) cycle, Prigogine concludes: "It seems that most biological mechanisms of action show that life involves far-from-equilibrium conditions beyond the stability threshold of the thermodynamic branch. It is therefore very tempting to suggest that the origin of life may be related to successive instabilities somewhat analogous to the successive bifurcations that have led to a state of matter of increasing coherence" (p. 123). Supposedly intended for the general reader ("with some background in physical chemistry and thermodynamics"), Prigogine's book is actually as abstract and difficult as they come.

38. There's no point in rehearsing the endless controversy over the fossiliferousness of the first findings from the Early Archean Warrawoona Group of Western Australia since incontrovertible fossils from the group have been subsequently discovered, but dated ~3400 to 3500 Ma (million years) ago instead of the ~3600 to 3700 Ma originally claimed. "It is evident that as early as ~3500 Ma ago microbial communities were extant, morphologically varied, and possibly physiologically advanced, and that Archean microfossils, both filamentous and colonial, are notably similar in morphological detail to extant prokaryotes," according to J. William Schopf, in "Paleobiology of the Archean" (in Schopf and Klein [op. cit., p. 39]), where all the data is meticulously reviewed.

39. Since all its members are prokaryotes, increasingly the Kingdom Monera is being called the Kingdom Prokaryote.

40. "People and other eukaryotes are like solids frozen in a specific genetic mold, whereas the mobile, interchanging suite of bacterial genes is akin to a liquid or a gas" is how Margulis and Sagan put it (op. cit., p. 89). Margulis is the chief cheerleader of the Kingdom Prokaryote.

41. Steven M. Stanley, *Earth and Life through Time* (2nd ed., Freeman, New York, 1989, p. 258).

42. These paragraphs draw very heavily on Lynn Margulis, either from Margulis

and Sagan (op. cit.) or from Lynn Margulis and Karlene Schwartz, *Five King-doms: A Illustrated Guide to the Phyla of Life on Earth* (3rd ed., Freeman, New York, 1997). The 99.99% figure is from Margulis and Sagan (p. 66). Note, however, that while most species have gone extinct, no major metabolic type seems to have.

43. Ibid., p. 74.
44. See the straightforward presentation in H. Robert Horton et al., *Principles of Biochemistry* (Prentice-Hall, Englewood Cliffs, NJ, 1993, pp. 16.3–16.7).
45. Humberto Maturana and Francisco Varela, *The Tree of Knowledge: The Biological Roots of Human Understanding* (rev. ed., Shambhala, Boston, 1992, pp. 76–77). This is the best book about life ever written.
46. Recall the dictum of John Archibald Wheeler quoted earlier: "Spacetime grips mass, telling it how to move; mass grips spacetime, telling it how to curve" (Wheeler, op. cit., pp. 11–12). It is precisely this kind of relationship.
47. Contrasting the traditional view of the relationship between an organism and its environment with that of "the new biology," Francisco Varela says, "We have changed our point of view from an externally instructed unit with an independent environment linked to a privileged observer, to an autonomous unit with an environment whose features are inseparable from the history of coupling with that unit, and thus with no privileged perspective" ("Laying Down a Path in Walking," in William Irwin Thompson, ed., *Gaia: A Way of Knowing: Political Implications of the New Biology* [Lindisfarne Press, Great Barrington, MA, 1987, p. 53]).

CHAPTER 5

1. For one thing, "When the two of us were college students in the 1960s, almost nothing was known of the chemistry of the atmosphere and the resulting effects on the climate of our planet," Thomas Graedel and Paul Crutzen observe. They add that "it was our great good fortune to help launch these areas as fields of study and then to participate in the explosive growth of understanding as increasing numbers of scientists addressed these topics" (*Atmosphere, Climate, and Change* [Scientific American Library, New York, 1995, p. vii]). Today it practically goes without saying that even the kitchen sink is integral to some larger and *dynamic* system. This, for instance, is Ehrhard Raschke describing the climate: "Our climate system, consisting of the atmosphere, oceans with varying sea-ice cover, land-surfaces with their complex vegetative cover and orography, and of the cryosphere (ice sheets over Greenland and Antarctica, continental glaciers), is steadily in motion within a wide spectrum of spatial and temporal scales which are ranging from millimeters to 10,000 km and from milliseconds to several or hundreds of years, respectively"; see his "Energy and Water Cycles in the Climate System," in Ehrhard Raschke, ed., *Radiation and Water in the Climate System, NATO ASI Series I: Global and Environmental Change*, Vol. 45 (Springer, Boston, 1996, p. 3). All this makes the climate hard to model. At the same time, it means the models are actually beginning to work.

2. This is from Peter Westbroek's *Life as a Geological Force: Dynamics of the Earth* (Norton, New York, 1991, pp. 140–146). One of the world's experts on coccoliths, he tells his story with both clarity and panache, but indeed this is a book all of which is very much worth reading—and a pleasure besides.

3. For a general introduction to the biospheric cycling of the elements, see Vaclav Smil, *Cycles of Life: Civilization and the Biosphere* (Scientific American Library, New York, 1997, p. 8).

4. For, hardly accidentally, the elements it circulates most thoroughly are those most broadly involved in metabolism: hydrogen, oxygen, carbon, nitrogen, and sulfur.

5. Well, three or four. We've already seen Ehrhard Raschke, a few notes back, talk about the "the *atmosphere, oceans* with varying sea-ice cover, *land-surfaces* with their complex vegetative cover and orography, and of the *cryosphere* (ice sheets over Greenland and Antarctica, continental glaciers)," which is *four* anyway you read it. Graedel and Crutzen are perfectly straightforward: "Temperature, air motions, and climate are influenced by five different Earth system regimes with widely varying impacts and time scales. The atmosphere is one of the five; the others are the biosphere, the hydrosphere, the cryosphere, and the pedosphere" (op. cit., p. 3). Note the assertion that there is only *one* system, with five different *regimes*, and the reduction of the lithosphere (the solid, the rocky Earth) to the pedosphere ("the solid portion of the Earth's *surface*," op. cit., p. 4, emphasis mine). Recent technical literature—that is, of the 1990s—almost uniformly adopts the five-regime (or system) model (five, that is, with the biosphere). This comes to Gaia without saying so.

6. One of the three fundamental hypotheses undergirding the work of Valentine P. Dymnikov and Aleksander N. Filatov is that climate models belong to the class of nonlinear dissipative systems, as laid out in their *Mathematics of Climate Modeling* (Birkhauser, Boston, 1997).

7. Or metastable state, through, in any case a bifurcation point into a trajectory in phase space subject to a new attractor. The language of "phase space," "bifurcations," and "attractors," so pervasive in contemporary discussions of dynamic systems, can be readily mastered, even by those with no—or very little—mathematical background ("we assume nothing in the way of prior mathematical training, beyond vectors in three dimensions and complex numbers") through an acquaintance with Ralph Abraham and Christopher Shaw's series of volumes: *Visual Mathematical Library* (Aerial Press, Santa Cruz, CA, 1983/1984). These are serious math books without symbols: "Math symbols have been kept to a minimum. In fact, they are almost completely suppressed. Our purpose is to make the book work for readers who are not practiced in symbolic representation. We rely exclusively on visual representations, with brief verbal explanations" ("Introduction" to *Dynamics: The Geometry of Behavior: Part One: Periodic Behavior* [1984, p. ix], the first volume in the *Visual Mathematical Library* series). The drawings are cartoony and easy to get. Don't be put off by the need to understand vectors and complex numbers. You may not recognize it, but you already know what you need.

8. The *heat*—whatever happens to the *air* which is recirculated by more heat from the radiator—ultimately passes into, and out of, the house, into the atmo-

sphere, and is finally lost to space. This is the fate of all the energy absorbed by the Earth: it's reradiated out into space. Before it gets there, though, it can sure spin a lot of wheels.

9. For a simple overview of this variation, see the treatment of insolation in Arthur and Alan Strahler, *Elements of Physical Geography* (Wiley, New York, 1989, pp. 37–45, et passim). This is a classic text.

10. Smil (op cit., p. 8).

11. It doesn't rise, however, in a sheet or wall, in a single, planet-girdling uplift, but in endless individual columns of rising air, incidentally—or not so incidentally—filled with towering clouds that produce showers and thunderstorms.

12. This longwave radiation is emitted from both land *and* water surfaces. The use of the word "ground" here reflects the parochial orientation of a land-dweller.

13. It's expanding as it rises and therefore cooling, since the pressure of the air column decreases with altitude. The energy it's expending in rising against gravity is not, however, being replaced by the short-wave photons streaming through it.

14. Named after George Hadley (1685–1768), the first to theorize the trades, in his paper of 1735. See the neat historical introduction provided by Richard Grotjahn in his *Global Atmospheric Circulation: Observations and Theories* (Oxford University Press, New York, 1993, pp. 6–14). The maps and diagrams, which Grotjahn reproduces, of (Edmund) Halley, Hadley, (Matthew F.) Maury, and William Ferrel (see below) didn't have a lot of data beneath them, but floated on insight.

15. Named after William Ferrel (1817–1891), whose paper of 1856 was the first to make much sense out of the westerlies.

16. "Theories for Zonal Fields," the sixth chapter of Grotjahn's book (op. cit.) builds, on a contemporary history of Ferrel cell research, an extended brief for his argument that "eddies in fact do drive the Ferrel circulation," as he puts it (p. 240). The fact is, as Grotjahn makes perfectly clear, that the theory of the circulation of this part of the globe remains ... *incomplete*. Still, Grotjahn's is a rich book, deep, well argued, and full of provocative insight.

17. What I'm describing as seething caps over the poles are often characterized as Hadley cells, though as weak ones; and, at least from the perspective of 60° N and S, they sort of have that form, though it's a more "frontal" sloping of warmer air up over colder, rather than the columnar rising of air seen along the ITC. The singularity over the poles, where the "cooling" air descends, throws another monkey wrench into the classic description of the Hadley cell. The fact is that "cellular" models break down over the poles.

18. Using language echoing Prigogine's (op. cit.), W. Dieminger and G. K. Hartman write: "The earth's atmosphere is an open system far from equilibrium. It is irradiated by the short-wave radiation of the sun, emits radiation in the infrared and continuously exchanges matter, energy and momentum with the cryosphere, the hydrosphere, the biosphere and the lithosphere. Systems far from equilibrium can spontaneously form spatial–temporal patterns, in contrast to closed systems in thermal equilibrium whose macroscopic states are spatially homogenous and time-independent ... via self organization, i.e., without any specific interference from outside"; see their "Introduction to the Earth's Atmosphere," in W. Dieminger, G. K. Hartman, and R. Leitinger, eds.,

The Upper Atmosphere (Springer, Berlin, 1996, p. 19). For an infinitely more accessible treatment, try Fritjof Capra on the bathtub whirlpool as a dissipative structure, where he attempts to wed the work of Maturana and Varela to that of Ilya Prigogine, in his own *The Web of Life* (Anchor Books, New York, 1996, pp. 168–172; see also all of chap. 8).

19. There's another of these over the subtropical high, but that's a whole other story.

20. "It has been calculated that five well-developed mid-latitude cyclones can effect the heat transfer needed to balance the Earth–atmosphere heat budget, because of the equatorward component to cold-air flow and the poleward component to the warm-air flow," say Peter Brimblecombe and Trevor Davies, in their "Atmosphere, Water and Weather," in David Smith, ed., *The Cambridge Encyclopedia of Earth Sciences* (Crown/Cambridge University Press, New York, 1981, pp. 285–286).

21. My reading is most generally informed by Brimblecombe and Davies (op. cit., pp. 276–296), but I can't resist quoting Grotjahn (op. cit.), one more time: "Eddies are fundamental to the system" is how he emphatically puts it (p. 233). The strongest statement, however, appears in the *Britannica* where the author of the general article on "Climate and Weather" asserts: "In fact it is now realized that the vast suite of disturbances described above as storms are the driving mechanisms of the general circulation of the atmosphere. They are not embroidery on the basic mechanism of the zonal wind systems but the mechanism's very heart" (*The New Encyclopaedia Britannica* [15th ed., Chicago, 1987, Vol. 16, p. 498]). Note that despite this, he still refers to them as "disturbances." Climate modeling is beset with problems similar to those we'll see in the case of Puerto Rican geology where a convincing global picture is confounded by local anomalies. There too we find global models, local data. See the interesting confrontation Jesús Idlefonso Díaz has set up as editor of *The Mathematics of Models for Climatology and Environment: NATO ASI Series I: Global and Environmental Change, Vol. 48* (Springer, Berlin, 1997), the global models (pp. 3–364) problematized by local anomalies (pp. 367–471). It's dense, not well written, but interesting.

22. Of course it *does* wobble. Here's what it says about *nutation* in the *Britannica*: "In astronomy, a small irregularity in the precession of the equinoxes. Precession is the slow, toplike wobbling of the spinning Earth, with a period of about 26,000 years. Nutation (Latin, *nutare*, "to nod") superimposes a small oscillation, with a period of 18.6 years and an amplitude of 9.2 seconds of arc, upon this great slow movement. The cause of nutation lies chiefly in the fact that the plane of the Moon's orbit around the Earth is tilted about 5° from the plane of the Earth's orbit around the Sun. The Moon's orbital plane precesses around the earth in 18.6 years, and the effect of the Moon on the precession of the equinoxes varies with this same period" (op. cit., Vol. 8, p. 836).

23. Although it's pointless to overemphasize the recency of this awareness. At a local level, the relationship between deforestation and rainfall was long understood almost everywhere. As early as the 18th century it began to be current in European scientific thinking. "Forest reserves in wood for rain" were established on Tobago as early as 1764. (These "rain reserves" continue to exist.) Such thinking rapidly began global, and as early as 1858 J. Spotswood Wilson

argued that the "general and gradual desiccation of the earth and the atmosphere" was due to changes in the proportion of oxygen and carbonic acid in the atmosphere. Spotswood went so far as to announce our extinction as a consequence. See Richard Grove's "Origins of Western Environmentalism" (*Scientific American*, July, 1992, pp. 42–47). Nevertheless it's the case that this kind of thinking remained limited to a few scientists and policymakers. It's still not generally accepted.

24. Having said this, and having laid out the story we currently tell about the cause and form of the atmospheric circulation, which is the story you will find in any introductory meteorology, climatology, or physical geography text, I have to hit you with this from that most normative of normative sources, *The Encyclopaedia Britannica*. This follows 25 pages of classical exposition: "How does the colossal mechanism of the winds run itself? What are its causes, and what does it achieve? There is no simple answer to these questions, which involve some of the most intractable of all geophysical problems. It is not enough to say that the atmosphere is a heat engine driven between an equatorial heat source and a polar sink. It is an inefficient engine, only about one percent of the heat input appearing as kinetic energy. The north polar surface is not a heat sink on an annual basis. All that can be done is to sharpen the questions and then outline some qualitative answers" (op. cit., vol. 16, pp. 497–498). It's sobering to accept that the most fundamental aspects of global behavior remain open to question in their most fundamental form.

25. This is largely due to the requirement of the conservation of angular momentum, an issue I evaded in my presentation of the atmospheric circulation system but one every bit as involved in the evolution and behavior of these systems as the Earth's radiation budget, especially when it comes to the role played by midlatitude eddy motions (see, inter alia, Grotjahn [op cit., especially chapters 5 and 6]). The papers in which this understanding was initially worked out for the oceans, including the classics of H. Sverdrup and H. Stommel, are collected in Allan Robinson, ed., *Wind-Driven Ocean Circulation: A Collection of Theoretical Studies* (Blaisdell, New York, 1963).

26. Though, as we'll see in a couple of paragraphs, what look here like surface currents—and *are* surface currents—are also, and perhaps more importantly, part of the thermohaline circulation, which is . . . basically about deep water. Especially in the North Atlantic we need to think about this phase of the gyre as part of a meridional overturning akin to that we saw in the Hadley cells. Here, for instance, is Jochem Marotzke speaking of the Gulf Stream: "We will concentrate entirely on the thermohaline circulation, because the meridional overturning accounts for a much greater part of the large total heat transport in the Atlantic, at 25°N, than does the horizontal gyre circulation." He adds: "It is not possible to decompose ocean circulation in purely wind-driven and purely thermohaline flows," though, naturally, for the sake of simplicity, he largely omits discussion of wind forcing. See his "Ocean Models in Climate Problems," in Paola Malanotte-Rizzoli and Allan Robinson, eds., *Ocean Processes in Climate Dynamics: Global and Mediterranean Examples, NATO ASI Series C: Mathematical and Physical Sciences, Vol. 419* (Kluwer, Dordrecht, The Netherlands, 1994, p. 85).

27. Note that in discussing these currents, especially the intense warm currents that are so well studied, no mention is made of the "broad slow flow in the oceanic *interior* (the region to the east of the boundary currents)" that Joseph Pedlosky takes as his point of departure in *Ocean Circulation Theory* (Springer, Berlin, 1996, p. 2), to say nothing of the almost jet stream-like Equatorial Undercurrent. In conceptualizing ocean flows as restricted to the surface currents and the thermohaline circulation, we actually miss most of the water, even most of the surface water. To the surprise of no one who's been reading these notes, this water is populated by small- and mesoscale eddies of variable intensity and nonuniform geographical distribution. These can be made to disappear by choosing a suitable (global) scale of analysis. Pedlosky does this (pp. 15–17) in order to concentrate on the dynamics of the interior mass—the Sverdrup interior—and its relationship to the western boundary current, and ultimately on the relationship of this integrated system to the Equatorial Currents, where, at least on the global scale considered by Pedlosky, the Coriolis parameter vanishes.

28. Though, to forestall any vociferous objections at this point, note that I grant completely the answer Marotzke (op. cit., p. 87) gives to his own rhetorical question, "Does the ocean drive the atmosphere, or the atmosphere the ocean?": "The question seems inappropriate because oceans and atmosphere are a coupled system, so one can only talk about interactions, and not of one subsystem driving the other," notwithstanding which he proceeds to the construction of uncoupled models, as alluded to above.

29. Indeed, Sverdrup (and others) working on the general theory of the ocean circulation have explicitly treated the effects of heating and cooling as negligible, concluding that wind stress alone was responsible for the form of the permanent ocean currents. See his "Wind-Driven Currents in a Baroclinic Ocean" (in Robinson [op. cit., pp. 3–10]). Note that this entire tradition of work relates solely to the surface, not the thermohaline, circulation. We'll get to their interaction in a second.

30. Named after V. Walfrid Ekman (1874–1954), the Swedish oceanographer who first described it. This is an idealization, sorely confounded by pressure-driven geostrophic flows, and hard to find in the world as such. Water *would* do this *if* . . . and so it's a major contributor to the complicated reality we actually encounter. For a general introduction, see Strahler and Strahler (op. cit., especially pp. 94–99); for more detail consult Christopher E. Vincent, "The Oceans" (in Smith [op. cit., pp. 311–324]).

31. About 1/23rd of it in fact. Obviously we're not talking about the Equatorial Undercurrent here which is whipping along, eastward, against the trades at speeds equivalent to those of the Gulf Stream or Kuroshio. See the awesome cross sections at 155°W between Hawaii and Tahiti in Pedlosky (op. cit., pp. 323–324).

32. This varies with latitude. This wind-stirred layer can reach 1,600 feet in the tropics where it exists year-round, diminishing in depth and duration poleward, to disappear entirely in arctic latitudes. The 600-foot figure is a global average. *Nothing* is as simple as it seems, but everything partakes of this common terrestrial disposition.

33. As Marotzke notes, "The ocean has a vastly larger thermal inertia than the atmosphere whose entire heat capacity is equivalent to the upper 2.5 meters of ocean. The mixed layer's turnaround time for heat is roughly a factor of 20 larger than the atmosphere's whose entire heat content is recycled in less than a year, and the deep ocean's turnaround time is another factor of a 100 larger" (op cit., p. 79). This enormous thermal inertia means that thermal states established in the ocean respond slowly, like a train trying to stop, to changes in the thermal state of the atmosphere, which is more like a bicycle in terms of its responsiveness.

34. It was the presence of this cold water in the tropical ocean that nearly 200 years ago led Benjamin Thompson (1753–1814), at the time Count Rumford, to predict the existence of a global, deep circulation. The logic, as recapitulated by Pedlosky, is simple: "Waters of such low temperatures can be formed by the cooling of the ocean by the atmosphere only in polar regions. The presence at other latitudes of such cold water implies a large-scale deep circulation, the *abyssal circulation*, which carries the water formed in polar regions to the rest of the ocean. Cold water flows from the polar regions to fill deep ocean basins, from which the water must eventually rise to the surface and, heated to the observed surface temperatures, must then flow poleward to replace the water which has sunk to the bottom, forming an endless global cell of motion" (Pedlosky [op. cit., p. 379]). What I can't help observing is the way this *abyssal* circulation ends up . . . *on the surface*, where, not at all incidentally, it is the open (mixed) boundary conditions that give rise to multiple equilibria in virtually every model of the thermohaline circulation (i.e., to the catastrophic changes, into and out of ice ages, for example, that have so galvanized the popular press). What *I see* is a sort of oceanic equivalent of a convection cell—okay, call it a conveyor belt—whose surface manifestations bear the same relationship to this convectional circulation as the surface *winds* do to the Hadley and Ferrel cells. *Now* let's ask Marotzke's question, "Does the ocean drive the atmosphere, or the atmosphere the ocean?" P.S.: What's the relationship between the deep currents flowing in these endnotes and the surface read in the main text? Hmm. . . .

35. As a consequence, the vertical structure of deep water is unnervingly complicated, with long tongues of water from varying sources overlaying each other, and sooner or later, all of them rising and falling like blobs in a lava lamp, all of them mixing. See the illustration in Vincent (op. cit., p. 321) that shows the main water masses of the Atlantic, with Antarctic bottom water (AABW) puddling below North Atlantic deep water (NADW), which is overlaid by Antarctic intermediate (AAIW) and Mediterranean water. The main thermocline rides on top of all this deep water with above it the layer of wind-stirred water. Now try thinking about it dynamically, in three dimensions. Yeah, right!

36. This fate of this water is beautifully caught by Robert Kunzig, who begins the story thus: "In our mind's eye we can almost see it whole, the round-the-world journey that seawater takes. We can imagine taking the trip ourselves. It begins north of Iceland, a hundred miles off the coast of Greenland, say, and on a black winter's night. The west wind has been screaming off the ice cap for days now, driving us to ferocious foaming breakers, sucking every last ounce

of heat from us, stealing it for Scandinavia. We are freezing now and spent, and burdened by the only memory we still have of our northward passage through the tropics: a heavy load of salt. It weighs on us now, tempts us to give up, as the harsh cold itself does. Finally comes that night when, so dense and cold we are almost ready to flash into ice, we can no longer resist: we start to sink. Slowly at first, but with gathering speed as more of us join in, and as it becomes clear that there is nothing to catch us—no water below that is denser than we are. We fall freely through the tranquil dark until we hit bottom, more than a mile and a half down" ("In Deep Water" [*Discover*, December, 1996, p. 86]). This, of course, is a major reason the returning cold water *surface* currents along the eastern edge of the ocean basins are so sluggish: the cold water they would be returning to the equator has dropped into the abyss and is heading south a mile or more below the surface.

37. William Calvin, "The Great Climate Flip-flop" (*Atlantic Monthly*, January, 1998, p. 50).

38. Graedel and Crutzen (op. cit., p. 24).

39. Kunzig (op. cit., p. 88). See the cross-sections across the South Atlantic at 30°S from South America to the Mid-Atlantic Ridge showing at around 6,000 feet the southward flowing North Atlantic Deep Water in Pedlosky (op. cit., pp. 383–384). These loony metaphors are fully justified by the reality.

40. There's so little salt in the North Pacific that even water cooled to the freezing point doesn't sink—overturn convectively—but instead freezes to form sea ice. The North Pacific and North Atlantic are completely different in this regard. See B. A. Warren's article, "Why Is No Deep Water Formed in the North Pacific?" (*Journal of Marine Research, 41*, 1983, pp. 327–347).

41. I know it's all so convincing, but as Elizabeth K. and Robert A. Berner point out, "Deep water circulation is not well documented and maps [showing the circulation I've been describing] are based largely on theoretical models" (*Global Environment: Water, Air and Geochemical Cycles* [Prentice-Hall, Upper Saddle River, NJ, 1996, p. 23]). Carl Wunsch, one of the modelers, admits that "ocean modelers, in the formal sense of the word, attempting to describe the ocean circulation have paid comparatively little attention in the past to the problems of working with real data," a situation he aims, to a greater or lesser degree, to deal with in his *The Ocean Circulation Inverse Problem* (Cambridge University Press, Cambridge, UK, 1996, p. 1).

42. Which is pretty much how William Kaufmann and Larry Smarr put it in *Supercomputing and the Transformation of Science* (Scientific American Library, New York, 1993, p. 173). They're making the point that the prevalence of systems thinking today is dependent on the availability of teraflop machines on which to model the systems. With respect to the oceans, Wunsch would add that "technologies of neutrally buoyant floats, long-lived current meters, chemical tracer observations, satellite altimeters, acoustical methods, etc." have "made it possible to seriously consider estimating the global circulation in ways that were visionary only a decade ago" (op. cit., p. 1). Nonlinear thinking, supercomputers, new data . . . *it's a new world*, literally; not that this one isn't as rife with what Wunsch calls myths as the old one, or that oceanography isn't characterized by the existence of what he sees as two oceans: "the

ocean as observers understand it, and the ocean as the theoreticians describe it" (p. 4). Still, the myths are very different myths, and even the ocean understood by the observer is different from what it was only a couple of decades ago. Incidentally, when he's speaking English, Wunsch is sufficiently interesting to make his book worth perusing even for those without the stomach for the math.

43. Ozima (op. cit., pp. 1–2).

44. This is the qualitative heart of the mantle plume model generally attributed to W. J. Morgan (see, e.g., his "Plate Motion and Deep Mantle Convection" [*Geological Society of America Memoirs, 132*, 1972, pp. 7–22]). It beats at the heart of every qualitative description of the driving force of plate tectonics, though it probably plays only a secondary role. For an up-to-the-minute, authoritative, but unbelievably readable review, see Geoffrey Davies, "Plates, Plumes, Mantle Convection, and Mantle Evolution," in Ian Jackson, ed., *The Earth's Mantle: Composition, Structure and Evolution* (Cambridge University Press, Cambridge, UK, 1998, pp. 228–258). In this view, there is a plate mode of convection and a plume mode of convection operating relatively independently. The plumes transport only about 10% of the heat involved to the base of the lithosphere.

45. Saying that magma wells up, descends along the ridge flanks, and migrates toward the plate margins evades the ludicrously complicated physics that you have to imagine are controlling the motions of all this matter. Force–balance models consider more than the Earth's heat budget, taking into account the vagaries of pressure gradients (the ridge-push force), shear forces exerted by the relative motions of the plate and mantle (the mantle-drag force), deviotic stress (the slab-pull force), and so on (and *so* on). It's probably true, as Frank Stacey insisted, that "we can expect to comprehend the rheological behavior of the Earth only by taking account of the fundamental atomic processes involved" (in his *Physics of the Earth* [2nd ed., Wiley, New York, 1977, p. 284]), but . . . not here, not now. For an account of how these other forces contribute to the shaping of the earth, see Michael Gurnis's "Sculpting the Earth from Inside Out" [*Scientific American*, March 2001, pp. 40–47]). The subtitle of the piece reads: "Powerful motions deep inside the planet do not merely shove fragments of the rocky shell horizontally around the globe—they also lift and lower entire continents," where the emphasis is on lift and lower. It's an active area of research.

46. The 14 long articles in Smith (op cit., pp. 36–275) provide a standard treatment, with lovely, provocative graphics. Eldridge Moores and Robert Twiss's *Tectonics* (Freeman, New York, 1995) is more straightforward and completely authoritative.

47. James Lovelock, *Healing Gaia: Practical Medicine for the Planet* (Harmony, New York, 1991, p. 47). This is a passionate, heavily illustrated, *elementary* introduction to Gaia as a self-regulating system by one of the originators of the idea.

48. "Granitic" and "basaltic"—even felsic and mafic—are frequently used by geo*physicists* without reference to their composition per se as a shorthand for the gross properties of material marked by different velocities of elastic waves in the respective layers above and below the Mohorovicic discontinuity (the lower boundary of the crust), unlike geologists who always have differences in

composition in mind. The relationship between basalt and granite is vexed (has been for 200 years). In the closing chapter of his *The Nature and Origin of Granite* (2nd ed., Chapman & Hall, London, 1997), entitled—emphasis mine— "A *kind* of conclusion: A search for order among *multifactorial processes* and *multifarious interactions*," Wallace Pitcher notes that "granite *is* born of basalt, but only by a complex series of partial meltings, first of mantle to provide a basaltic underplate, and then of the latter to supply the primitive granitic melts that are subsequently differentiated," though in so saying he is not writing off granitization of other source material (pp. 340–341). Pitcher refers to "a veritable deluge of modern literature" on the origins of basalts, but then how to describe the literature on granite? There are 900 items in Pitcher's bibliography, almost all from the last 30 years.

49. Nicholas Pinter and Mark T. Brandon, "How Erosion Builds Mountains" (*Scientific American*, April, 1997, p. 74–79, quote on p. 78). These guys study what they call "active tectonics" or "tectonic geomorphology." They have already published a textbook, Edward Keller and Nicholas Pinter's *Active Tectonics: Earthquakes, Uplift and Landscape* (Prentice-Hall, Upper Saddle River, NJ, 1996), which I cannot recommend too highly. It's hardcore, but things are presented so plainly it's a breeze. There are illustrations on every page that are informative, dramatic, or both. The heavy emphasis on Southern California almost makes it a field guide.

50. Pinter and Brandon (op. cit., p. 75).

51. "In 1943, Gutenberg discovered the existence of mountain roots, confirming the theory of isostasy," write Moores and Twiss (op. cit., p. 256), though Gutenberg himself speaks only of confirming the (1855) hypothesis of B. G. Airy, and cites a 1933 date for having done so. See his treatments of gravity anomalies and "The 'Roots of Mountains' " in Beno Gutenberg, *Physics of the Earth's Interior* (Academic Press, New York, 1959, pp. 46–59). Gutenberg was the influential seismologist (1889–1960) credited with recognizing the core–mantle structure of the earth (in 1914), and after whom the Gutenberg discontinuity—between the core and mantle—is named.

52. Admiral Frost, adrift on his iceberg, comes from Keller and Pinter (op. cit., p. 309).

53. The wording is from Pinter and Brandon (op. cit., p. 76), but I want to give the impression neither that the idea of isostasy, nor its role in raising mountains, is either theirs or new. In addition to B. G. Airy, J. H. Pratt published on the reality and role of isostasy in 1855 (though neither used the term itself), and indeed the idea can be dated to the 15th century (among others see E. N. Lyustikh, *Isostasy and Isostatic Hypotheses* [American Geophysical Union, New York, 1960]) for a succinct (and sort of contentious Soviet) pre-plate tectonic revolution review). What is new here is the conception of a mountain as a system of feedbacks in which isostasy is entrained by erosion which is entrained by mountain building, over similar time scales and at similar rates.

54. The Andean and Himalayan examples are discussed in the active tectonic framework by B. L. Isacks in " 'Long-Term' Land Surface Processes: Erosion, Tectonics and Climate History in Mountain Belts," in Paul Mather, ed., *TERRA-1: Understanding The Terrestrial Environment: The Role of Earth Observations from Space* (Taylor and Francis, London, 1992, pp. 3–34).

CHAPTER 6

1. It is precisely this scenario that has folks like Calvin writing "The Great Climate Flip-flop" (op. cit.) for the *Atlantic Monthly*, and Kunzig writing "In Deep Water" (op. cit.) for *Discovery*. Of course they'd been reading Wally Broecker (Wallace S. Broecker "Chaotic Climate" [*Scientific American*, November, 1995, pp. 62–68]; Wallace S. Broecker and George Denton, "What Drives Glacial Cycles" [*Scientific American*, January, 1990, pp. 48–56]; Wallace S. Broecker . . . *ad infinitum* (there's a *lot* of Broecker]); Steven Stanley (*Extinction* [Scientific American Library, New York, 1987] and *Children of the Ice Age* [Harmony Books, New York, 1996]); and the Imbries (John and K. P. Imbrie, *Ice Ages: Solving the Mystery* [Enslow, Short Hills, NJ, 1979; Harvard University Press, Cambridge, MA, 1986])—all of whom support an understanding of climate marked by large, rapid changes.
2. Though recent events could hardly be less encouraging. A report in this week's *Science News* (February 13, 1999) documents marked increases in the temperature, and decreases in the salinity, of Arctic waters, and a notable shrinking and thinning of the ice cover ("Sea Change in the Arctic: An Oceanful of Clues to Climate Warming in the Far North," pp. 104–106).
3. Named after the German oceanographer Hartmut Heinrich, a Heinrich event is a regular deposition in the North Atlantic, every 7,000 to 10,000 years or so during the last ice age, of limestone from the Hudson Bay area scraped up by moving ice sheets and rafted out to the North Atlantic in icebergs. In addition to giving up their limestone, the melting icebergs give up enough fresh water to reduce saline levels sufficiently to shut down the thermohaline circulation, and so . . . and so on. See Broecker, "Chaotic Climate" (op. cit.), and D. R. MacAyeal, "A Low-Order Model of the Heinrich Event Cycle" (*Paleoceanography*, December, 1993, pp. 767–773).
4. Danish glacialogist Willi Dansgaard had discovered swings in the ratio of isotopes O^{16} and O^{18} in Greenland ice cores indicative of swings in temperature that suggested transitions between quasi-stationary modes of atmospheric circulation. See W. Dansgaard, S. J. Johnson, H. B. Clausen, and C. C. Langway Jr., "Climatic Record Revealed by the Camp Century Ice Core," in Karl Turekian, ed., *Late Cenozoic Glacial Ages* (Yale University Press, New Haven, CT, 1971, pp. 37–56). (Incidentally, this is a Stillman Lectures publication, so the writing is, yes, still technical, but at the same time unusually accessible and clear. This is also true of other articles in this volume.) The Swiss geochemist Hans Oeschger, analyzing bubbles of ancient air trapped in the ice core, found that concentrations of CO_2 had risen and fallen with the swings in temperature. This suggested "a bi-stable climatic system" that "oscillated between a cold and a warm state, probably strongly influenced by different ocean circulations and ice cover"; see his "The Contribution of Ice Core Studies to the Understanding of Environmental Processes," in C. C. Langway Jr., H. Oeschger, and W. Dansgaard, eds., *Greenland Ice Core: Geophysics, Geochemistry and the Environment: Geophysical Monograph 33* (American Geophysical Union, Washington, DC, 1985, pp. 9–17). Further detail on the CO_2 concentrations appears in B. Stauffer, A. Neftel, H. Oeschger, and J. Schwander "CO_2 Concentration in Air Extracted from Greenland Ice Samples" (ibid, pp. 85–89). These and

other data of Oeschger, Dansgaard, and their colleagues involved with this Greenland ice were widely reported in *Nature, Science,* and the like. Bi-stability is a tidy idea, but how do you do estate planning with the world ready to flip into an ice age without a lot of notice? And that was before Wally Broecker got his hands on the data....

5. Milutin Milankovitch, a Serbian mathematician/astronomer, while an Astro–Hungarian prisoner during World War I, worked out the perturbations of the Earth's relationship to the Sun caused by the interactions among (1) the eccentricity of the orbit (cyclic changes in the *shape* of the Earth's orbit around the Sun caused by its interactions with the other planets), (2) the eccentricity of the elliptic (cyclic changes in the obliquity of the Earth's axis relative to the plane of its orbit), and (3) precession of the equinoxes. These cycles interact to produce continuously varying but broadly repetitive changes in the intensity of summer sunshine in high latitudes which seem to be causatively related to episodes of glaciation. Broecker and Denton conclude that "over the past 800,000 years, the global ice volume has peaked every 100,000 years, matching the period of the eccentricity variation. In addition, 'wrinkles' superimposed on each cycle—small decreases or surges in ice volume—have come at intervals of roughly 23,000 and 41,000 years, in keeping with the precession and tilt frequencies" (op. cit., p. 50), though they do not believe this happens by direct action on the ice sheets, but by flipping a bimodal climatic system from warm to cold as mediated by the thermohaline circulation. The most readily available technical treatment is W. Schwarzacher's *Cyclostratigraphy and the Milankovitch Theory* (Elsevier, Amsterdam, 1993), especially chap. 3 which summarizes Milankovitch's complete argument. The Imbries (op. cit.) have a less technical and warmly historical approach, beginning with Louis Agassiz's articulation of the "glacial theory," and covering James Croll's pioneering work on the astronomical causes of the ice ages (he failed to consider the effect of the eccentricity of the elliptic), before treating Milankovitch and the revival in the 1960s of his theory. Stephen Schneider and Randi Londer offer an unusually broad-handed (but somewhat dated) view of the matter in their wonderfully thorough *The Coevolution of Climate and Life* (Sierra Club, San Francisco, 1984), especially chap. 7. The most elementary presentation is in R. V. Fodor's delightful *Frozen Earth: Explaining the Ice Ages* (Enslow, Short Hills, NJ, 1981). Among the most focused discussions is John Terborgh's *Diversity and the Tropical Rain Forest* (Scientific American Library, New York, 1992) where the Milankovitch cycles are seen to drive a speciation pump (pp. 139–151). There are dissenters to this majoritarian view. See, for example, Fred Hoyle's (*yes! again!*) *Ice: The Ultimate Human Catastrophe* (Continuum, New York, 1981), and indeed of course *all* these cycles could be no more than non-linear properties of the climate system itself....

6. Cornelius Klein et al. "Conclusions and Unsolved Problems," in Schopf and Klein (op. cit., pp. 173–174).

7. The "small, steep-sided felsic protocontinents" is quoted from Stanley, *Earth and Life through Time* (op. cit., p. 252).

8. Though perhaps obvious in this context, "spine of time" comes here courtesy of Saul Williams's great yard monologue in *Slam*, Marc Levin's terrific 1998 film about being young and black in Washington, DC.

Notes to Chapter 6

9. There's nothing "official" about these names, nor even—at this time—that much uniform about them, or, for that matter, about the periodization. What I've been doing with this list of epochs is to try to catch the broadest outline with the greatest consensus and attach the most common and accessible names to them. The spine I'm most reliant on is that in Silk's *A Short History of the Universe* (op. cit.), summarized in "A Chronology of the Universe" (p. 87). But this is vague in spots, incomplete in others, and there's nothing sacred about it. In his *Evolution of Matter and Energy on a Cosmic and Planetary Scale* (Springer-Verlag, New York, 1985), the physicist M. Taube denominates as follows: "Era of superunified force (Planckian Era or Very Hot Era); Era of grand unified force (Hot Era); Era of Unified Force (Lukewarm Era); the Cold Era: Evolution in the 'Hadron Epoch'; the Universe a few seconds old: Lepton Epoch; the Photon Epoch; the Present Very Cold Era" (p. viii), and says, "The best established model of the evolution of the Universe includes the following eras (see Table 2.1): very hot, directly after the Big Bang, also called the Planckian era; hot; lukewarm; cold; [and] very cold, the era in which we are now living" (p. 21), which all has the advantage of stressing the ongoing decline in temperature. His Table 2.1 (p. 22) is similar to Silk's "Chronology" with somewhat different breaks and names. Others I've relied on include Ferris's *The Whole Shebang* (op. cit.), Hawking's *A Brief History of Time* (op. cit.), and Weinberg's *The First Three Minutes* (op. cit.). They, Silk, and Taube are about as consistent with their use of "epoch" and "era" and "age" as geologists; and of course are but five members of a legion.

10. This is pure Silk.

11. Of course we *live* in an Hadronic Epoch, hadrons being nothing other than the protons (hence Epoch of Protons) and neutrons that make up atoms. Made up of quarks, these nucleons are composite particles that can only form at temperatures below 300 MeV, reached about 10^{-5} second. During the ensuing phase transition, quarks start glumping up, become *confined* inside these nucleons (hence Epoch of Confinement). But the epoch doesn't end . . . we're still in it. It's just that subsequently . . . other things happen.

12. Leptons are the light constituents of our universe, electrons (and hence Era of Electrons), neutrinos, photons. After the hadrons chill out of the mix, there's a complicated period—like one of the Chinese dynasties dominated by civil war—during which the weak interaction rate continues to drop until the temperature reaches about 1 MeV (10 billion degrees Kelvin [just to put a little perspective on what is meant by cool]), at which point it's effectively halted, ending nuclear reactions and fixing the number of neutrons, and permitting them to interact with protons to form helium nuclei, which is nucleosynthesis (and hence Epoch of Nucleosynthesis). Calder (op. cit.) names them *eras*: quark era, proton era, electron era, gamma-ray era, et cetera (pp. 98–101).

13. During this time negatively charged electrons begin combining with positively charged protons to form neutral atoms. Once this has happened, photons, which interact intensely with free electrons and nucleons, become (relatively) free to propagate, the universe becomes transparent, the "Fireball" has died out. This is the dawning of the cold, transparent, low-density universe we know. I don't know why I feel compelled to go over this again here, but I do.

14. This is the nomenclature used by Morowitz, *Beginnings of Cellular Life* (op. cit.,

p. 29), in which he follows J. E. Harrison and Z. E. Peterman in their "North American Commission on Stratigraphic Nomenclature: Note 52—A Preliminary Proposal for a Chronometric Time Scale for the Precambrian of the United States and Mexico" (*Geological Society of America Bulletin, Part 1, 91,* 1980, pp. 377–380).

15. Davies (op. cit., pp. 228–229 and 255). His treatment of the thermal evolution of the mantle and tectonic evolution of the earth (pp. 248–255), is, all things considered, amazingly accessible. It's also not—hot news!—universally accepted. Indeed, contra Davies's suggestion that plate tectonics might not have worked under the higher mantle temperatures of the Archean Earth, we have Kent Condie's unequivocal "Although plate tectonics undoubtedly operated in the Archean," though he goes on to acknowledge the problems of excess upper boundary heat, the possibility of layered convection, and the likelihood that whole-mantle convection didn't begin until no more than 1.3 billion years ago. See his *Plate Tectonics and Crustal Evolution* (4th ed., Butterworth/ Heinemann, Oxford, UK, 1997), especially chap. 5 (quote on p. 176).

16. Stanley, *Earth and Life through Time* (op. cit., p. 252). The structure of my presentation here is deeply indebted to Stanley.

17. See A. Kröner and P. W. Layer, "Crust Formation in the Early Archean" (*Science, 256,* June 5, 1992, pp. 1405–1411); and Derek York, "The Earliest History of the Earth" (*Scientific American,* January, 1993, pp. 90–96). In this reading the Kaapvaal Craton has been moving too, and about as fast as modern continents, for at least the last 3.5 billion years. For a synoptic treatment of the larger issue, also check out the wonderful discussion of early continent formation in Jonathan I. Lunine's *Earth: Evolution of a Habitable World* (Cambridge University Press, Cambridge, UK, 1999, pp. 196–210). You know, if you're looking for a really authoritative but really readable account of *Earth* history, look no further: Lunine is it! Lunine isn't as detailed as Stanley, but lord! is he clear!

18. From the biological point of view, sex is simply the recombination of genes from more than one source (from which perspective Monica Lewinsky and Bill Clinton did *not* have sex). This whole issue is of immense significance to the biological history of Margulis and Sagan and they've written at least three passionate and illuminating books about it, each of which, despite extensive overlap with the others, is worth reading: *Origins of Sex: Three Billion Years of Genetic Evolution* (Yale University Press, New Haven, CT, 1986), *Mystery Dance: On the Evolution of Human Sexuality* (Summit, New York, 1991), and *What Is Sex?* (Simon & Schuster, New York, 1997).

19. Margulis, *Symbiotic Planet* (op cit., p. 24).

20. "Particle zoo" is what they started calling the particles when they started multiplying and it seemed there would be no end to the taus and bosons and muons and gluons; but I prefer Gordon Kane's "particle garden," which is the name of the most accessible yet least patronizing book I know on the topic: *The Particle Garden: Our Universe as Understood by Particle Physicists* (Addison-Wesley, Reading, MA, 1995).

21. Margulis, *Symbiotic Planet* (op. cit., p. 24).

22. Ozima (op cit., pp. 106–107). I've omitted a final paragraph on the history of Earth's magnetism.

23. Mark Ridley, *Evolution* (Blackwell, Boston, 1993, p. 5).
24. On the other hand, Arthur T. Winfree says, "In a sense everything happens by accident. Any deliberate event is guided by a purpose that selects among accidents, ignoring some and turning others into opportunities, but the outcome is always colored by the particulars of those accidents"; see the "Preface" to his absolutely amazing *The Timing of Biological Clocks* (Scientific American Library, New York, 1987, p. ix).
25. Lovelock (op. cit., p. 83).
26. You can read all kinds of figures, and obviously the soil you dig your hand into makes all the difference. "A cubic centimeter [of which there would be several in a handful] of grassland soil typically contains hundreds of millions of bacteria, tens of thousands of protozoa, hundreds of meters of fungal hyphae, several hundred nematodes, mites and insects, and a myriad of other microbes and larger organisms," according to K. Ritz, J. Dighton, and K. E. Giller, eds., in *Beyond the Biomass: Compositional and Functional Analysis of Soil Microbial Communities* (Wiley, New York, 1994, p. ix). I find it interesting that books like this aren't shelved with books on soils but with books on microbes
27. This comparison is due to Peter Volk, *Gaia's Body: Toward a Physiology of Earth* (Copernicus, New York, 1998, p. 108).
28. "The word 'soil' like other commonly used words like love and home, means different things to different people," writes G. J. Retallack in *Soils of the Past: An Introduction to Paleopedology* (Unwin Hyman, Boston, 1990, p. 9). For a farmer, it's fertile, tillable ground. For an engineer, it's any material that can be excavated without quarrying or blasting. For many soil scientists, it's the medium in which vascular plants take root. And so, of course, what to call . . . it . . . early on in Earth history?
29. All *kinds* of life. There are a bunch of wonderful terms that point toward its ineffable richness. Embraced in the *rhizosphere*, which is that soil, what? . . . under the influence of roots, is the *endorhizosphere*, those layers of root cells colonized by soil microorganisms (*yes!* it's a two-way street); the *ectorhizosphere*, the biochemically tumultuous region surrounding the root; and with the development of mycorrhizal fungal associations, the *mycorrhizosphere*, which can extend the influence of the root substantial distances. Then there's the *spermosphere*, that soil under the influence of a germinating cell. The world in a grain of sand? *The world in a root!* See, inter alia, J. M. Lynch, ed., *The Rhizosphere* (Wiley, New York, 1990), especially his introductory essay.
30. Known as the aerobiota which are the subject of aerobiology. See P. H. Gregory's *The Microbiology of the Atmosphere* (Wiley, New York, 1973; or if possible the unusually handsome first edition, Leonard Hill, London, 1961), which was the first to treat the subject on a global basis; Robert Edmunds, ed., *Aerobiology: The Ecological Systems Approach* (Dowden, Hutchinson & Ross, Stroudsburg, PA, 1979), which reports on the fruits of the International Biological Program of the late 1960s and early 1970s; or S. N. Agashe, ed., *Aerobiology: 5th International Conference, Bangalore, 1994* (Science, Enfield, NJ, 1997), which contains 68 papers of often local (Indian) interest, but is indicative of the field, including its concern with allergies and indoor air (called "intramural" in aerobiology). This is not the study of granite, which has been go-

ing on for a *long time*, but Gregory's 1973 edition already had a 900-item bibliography.

31. Margulis and Sagan (op. cit., p. 97).

32. From their perspective, we're just another perturbation of the environment. The chunks of DNA that bacteria exchange are known generically as replicons. They show up in cells (and the literature) as replicons, plasmids, episomes, prophages, phages, and viruses. In transduction the replicons are basically moved about by viruses; in transformation bacteria just pick up strands of DNA that may be (that are) lying around; and in conjugation a little tube grows between two bacteria through which the DNA moves. In his *The Outer Reaches of Life* (Cambridge University Press, Cambridge, UK, 1994), John Postgate carries out some simple calculations about the genetic mutation and transference taking place in *E. coli* in the human gut: "Every gene of the gene pool of *E. coli* traversing the human intestine mutates at least 2.5 billion times daily. . . . The upshot is that among the terrestrial population of *E. coli*, every possible mutation is occurring, an enormous number of times a day" (p. 242).

33. Again, from the biological point of view, sex is simply the recombination of genes from more than one source, and this recombination is the form of the adaptation the microcosm makes to the dynamics of the rest of the environment.

34. Margulis and Sagan, *Microcosmos* (op. cit., pp. 100–101). There are many descriptions of the rise of atmospheric oxygen. This of Margulis and Sagan—the whole of their chap. 6—is far and away the most accessible and colorful. A more technical—that is, more chemical—but still accessible treatment is provided by James and Carol Gould in the "Introduction" to their *Life at the Edge* (Freeman, New York, 1989, pp. 4–5). This is a collection of readings from *Scientific American*.

35. Postgate (op. cit.) is a wholly accessible book (without scholarly apparatus of any kind!) entirely devoted to the lives of microbes in extreme environments. For the bacteria underneath glaciers, see "Bacteria under Ice" in *Science News* (February 13, 1999, p. 101).

36. Margulis and Sagan, *Microcosmos* (op. cit., p. 109). The energy released from the oxidation of carbohydrates, fatty acids, and the like is stored in the phosphorylization of ADP, adenosine *di*phosphate, to ATP, adenosine *tri*phosphate, the so-called energy currency of life. The energy released from the cleavage of the bonds between the phophoryl groups of ATP can be coupled, later and elsewhere, to endergonic (energy-requiring) processes: muscle contraction, for example, or the active transport of molecules, or biosynthesis. What cells *do* is make, and expend, ATP in order to keep on making, and expending, ATP. Everything else is run off this. For my money, no one explains this better than Christian DeDuve in the two volumes of his magisterial *A Guided Tour of the Living Cell* (Scientific American Library, New York, 1984), though the subject is covered by every biology and biochemistry text from high school up.

37. Fermentation is the most primitive metabolic pathway that leads to the generation of ATP. Along it a single molecule of glucose is split into two of pyruvate, accompanied by the net conversion of two molecules of ADP into

ATP. In the absence of oxygen—that is, anaerobically—yeast cells convert the pyruvate to ethanol, as similarly in muscle cells it's converted to lactate. In the presence of oxygen—that is, aerobically—pyruvate can be fully oxidized to CO_2 and H_2O. In this case the fermentation is only the front end, the first half, the initial stage, of aerobic respiration, where it is better known as glycolysis. The full oxidative metabolism of glucose generates almost 20 times the ATP available to a cell through glycolysis, that is, fermentation, alone. Hence the terrific advantage aerobic cyanobacteria had over fermentors, at least in oxygen-rich environments. See Horton et al. (op. cit., pp. 12-1–12-26 and 14-1–14-21); or DeDuve, *A Guided Tour of the Living Cell* (op. cit.).

38. Of course many of these niches had been occupied by organisms that were killed by, or that had fled, cyanobacterial oxygen. More niches were opened up when, as described below, the ozone layer the oxygen would be responsible for started absorbing the ultraviolet. This opened the entire surface of the planet to life-forms that could tolerate, or use, oxygen.

39. A straightforward but relatively comprehensive treatment of ozone chemistry appears in J. E. Andrews, P. Brimblecombe, T. D. Jickells, and P. S. Liss's *An Introduction to Environmental Chemistry* (Blackwell, Oxford, UK, 1996, pp. 30–31, 192–198).

40. Margulis and Sagan, *Microcosmos* (op. cit., p. 110).

41. This idea is rigorously pursued in J. William Schopf's "Metabolic Memories of the Earth's Earliest Biosphere," which he concludes with this remark: "Thus the seeds that ultimately flowered in human intelligence can be found in the ecological structure of the Earth's earliest biosphere. Housed in our cells, in the chemistry that keeps us alive, we each contain metabolic memories of the unimaginably long evolutionary journey" (in Charles Marshall and J. William Schopf, eds., *Evolution and the Molecular Revolution* [Jones & Bartlett, Sudbury, MA, 1996, p. 105]). It's hardly an original idea, but this is a hard-eyed, fact-filled treatment. The whole book's terrific.

42. These mats are both fascinating and the subject of an academic growth industry. Lynn Margulis is completely romanced by them (see Margulis, *Symbiotic Planet* [op. cit., pp. 69–70 in particular]), but so are many others since the mats seem to have persisted for over 3 billion years. But see, about this and other things, George Zavarzin, "Cyanobacterial Mats in General Biology," in Lucas Stal and Pierre Caumette, eds., *Microbial Mats: Structure, Development and Environmental Significance, NATO ASI Series G: Ecological Sciences, Vol. 35* (Springer-Verlag, Berlin, 1994, pp. 443–452, especially p. 448), where Zavarzin cautions against extrapolating from contemporary to Archean mats across the unidirectional changes that have taken place over the past 3 billion years (including changes in weathering–sedimentation patterns, biota, and the atmosphere, especially those biologically induced). Zavarzin is also concerned to distinguish cyanobacterial mats and stromatolites. Why do the former not lithify? Why did the later become extinct—or nearly extinct—near the end of Proterozoic? Or are these the same question? (i.e., do they both point toward a change in biogenic carbonate precipitation?)

43. This chronology is very much Morowitz and Margulis. Others are more conservative, everything happening 400 or 500 million years later. See J. William

Schopf's chronology in his "The Evolution of the Earliest Cells" (in Gould and Gould [op. cit., pp. 7–23]). His timeline (p. 22) is unusually comprehensive (i.e., correlates events in multiple domains), though 20 years old. If you're up for it, his summary effort in his "Times of Origin and Earliest Evidence of Biological Groups," in Schopf and Klein (op. cit., pp. 588–589), is the definitive place to go (but see below).

44. Margulis and Sagan, *Microcosmos* (op cit., p. 115).

45. Ibid., p. 117; Stanley, *Earth and Life through Time* (op. cit., pp. 276–278); though again, the evidence is actually in Schopf, "Times of Origin" (op. cit.) and the 500+ pages he summarizes there. Note that he is bending over backward to be even-handed and inclusive. The dissension is over the significance of size, a break occurring in the fossil record, as all agree, between 1.5 and 1.4 billion years ago, smaller earlier, larger later and only later. If larger = eukaryotes, then . . . but if that's not definitively diagnostic, then. . . . His timeline is littered with entries like "Abundant 'bizarre' microfossils," 2.0 billion years ago; and "Oldest millimetric (eukaryotic) microfossils," 1.8 billion years ago. Certainly he's showing single-celled algae ~2.2 billion years ago. About all of this he says: "There can be little doubt, however, that the earliest 'complete eukaryotes' (viz., mitochondrion-, chloroplast-, and nucleus-containing walled cells capable of mitosis) were neither morphologically complex nor megascopic in size. Indeed they almost certainly were simple, probably spheroidal, mitotic haploid unicells, perhaps morphologically similar to modern, small-celled, rhodophycean or chlorophycean micro-algae and thus essentially indistinguishable in cell size and shape from coexisting coccoid cyanobacteria. If so, and in the absence of other evidence by which to distinguish between fossils of prokaryotes and early-evolving eukaryotes . . . the actual time of origin of the eukaryotic cell may remain shrouded in mystery. Thus, at present, the available fossil record can be expected to provide evidence only of the existence of 'complete' eukaryotic cells (rather than the eukaryotic genome, the origin of which evidently occurred in anaerobic heterotrophs and predated, perhaps substantially, the endosymbiotic assembly of the 'complete eukaryote'), and only after such eukaryotes had evolved sufficiently to become morphologically distinguishable from coexisting prokaryotes. For the reasons discussed in Section 5.5.2A, single-celled phytoplankters (sphaeromorph acritarchs) having a cell size larger than all extant prokaryotic unicells (viz., 'mesosphaeromorphs' 60 to 200 μm in maximum diameter; and 'megasphaeromorphs' 200 μm to more than 7.5 mm in maximum diameter), are in this sense regarded here as assuredly eukaryotic. Although a few scattered reports exist of 'mesosphaeromorphs' from units as old as ~2250 Ma, the oldest well-documented occurrences of 'meso-' and 'megasphaeromorphs' are from ~1850 Ma-old sediments, the age inferred for the oldest clear-cut evidence of single-celled eukaryotic algae" (p. 592). Yet writing only two years later, in 1992, Bruce Runnegar writes that "living eukaryotes are derived from archaebacterial prokaryotes that existed after the origin of methanogenesis (≥ 2.7 Ga ago). Before their appearance in the fossil record, eukaryotes had acquired their cytological characteristics and their organelles (chloroplasts and mitochondria). The oldest fossil though to be a eukaryote is a [recently discov-

ered] probably megascopic alga (Grypania) from 2.1 Ga old strata in Michigan" ("Proterozoic Eukaryotes: Evidence from Biology and Geology" [in Bengston, ed., op cit., p. 297]).

46. In *Symbiotic Planet* (op. cit.), Margulis tells the story of the 1969 rejection of her first book on serial endosymbiosis because "of extremely negative peer review," going on to detail and lament its dogmatic triumph: "Today I am amazed to see a watered-down version of [serial endosymbiosis theory] taught as revealed truth in high school and college texts. I find, to my dismay if not to my surprise, that the exposition is dogmatic, misleading, not logically argued, and often frankly incorrect. Unlike the science itself, SET now is uncritically accepted. So it goes" (pp. 30–31). The year after its rejection by Academic Press, Yale University Press brought her book out as *Origin of Eukaryotic Cells* (New Haven, CT, 1970). It's not written in her popular vein, but with, as G. Evelyn Hutchinson says in his "Introduction," "infectious enthusiasm" nonetheless.

47. Without the bacteria the animals die, yes, though at the scale of the biosphere this is true for *all* animals. None of us would be here but for the activity, among others, of the decomposing bacteria who turn last year's green leaves and ticklish flesh into the elements or molecules needed for releafing the trees and growing the baby up into the babe.

48. Ibid., pp. 33, emphasis mine. With respect to the conclusion of her sentence, the fact is that her ideas have generated so much research, so much of which has been so much more supportive than anyone would have dared to imagine, that Margulis has fewer and fewer detractors, and more and more adherents. Not that I want to leave you with the impression that Margulis originated the idea of the symbiogenesis of eukaryotic cells. This was probably due to K. S. Merezhkovsky (many alternate spellings and initials appear in the literature) as long ago as 1909. But his, and subsequent work on the part of I. E. Wallin and others, were all but reviled until Margulis—at the time married to Carl Sagan and writing under his last name—force-fed the idea to the modern scientific community. See L. N. Khakhina's *Concepts of Biogenesis: A Historical and Critical Study of the Research of Russian Botanists* (Yale University Press, New Haven, CT, 1992 [but originally published in Russian in 1979]) and the almost incendiary—plenty of Kropotkin—*Evolution by Association: A History of Symbiosis* by Jan Sapp (Oxford University Press, New York, 1994). Many of the, at least Anglophone, papers are reprinted in B. D. Dyer and R. Obar, eds., *The Origin of Eukaryotic Cells* (Van Nostrand Reinhold, New York, 1985). There's a whole history of suppressed science here, which brings us to Margulis's second major contribution, for in addition to her theory of serial endosymbiosis (and her trumpeting of the place and cause of bacteria generally), she's the author, with James Lovelock, of the idea called Gaia that the whole planet is what's alive. In turn this brings us to another Russian, the biogeochemist Vladimir Vernadsky, whose ghost hovers over . . . *the whole enterprise*, not just of Gaia but of microbial mat research and geochemical fluxes and non-Darwinian (*not* anti-Darwinian) ideas of global evolution, and whose *The Biosphere*, originally published in 1926, received its first complete English translation only recently (Vladimir Vernadsky, *The Biosphere* [Copernicus, New York, 1998]). It's a beautiful edition of a beautiful book.

49. Margulis had thought she was the first to suggest this, but in 1975 she learned from Armen Takhtadzhyan that the idea had been broached by B. M. Kozo-Polyansky in 1924. See the "Introduction" she coauthored with Mark McMenamin in Khakhina (op. cit., p. xix).

50. It doesn't sound quite so . . . 1960s when the details are filled in. Here's a longer version of "Later an aerobic bacterium moved in to become in time a mitochondrion" from Margulis's first book (*Origin of Eukaryotic Cells* [op. cit.]: "The history of the modern eukaryotic line began when a pleiomorphic microbe, capable only of anaerobic fermentation of glucose to pyruvate, symbiotically harbored a smaller prokaryote. The endosymbiont microbe was an aerobe, a relative of extant aerobic eubacteria, having the biosynthetic ability to form cytochromes and to oxidize all of its foodstuffs to CO_2. Flavins, ubiquinone, cytochromes, were intermediate electron carriers in the total oxidation of carbohydrates by the small aerobic symbiont. This association, one of many bizarre and different types, many of which exist today, led to the formation of primitive amoebas. The host became the ground nucleus and cytoplasm, by definition the *protoeukaryote*. It harbored the eubacterial aerobe (by definition, the *protomitochondrion*) that later became mitochondria. From this association amoebas that contained extensive cytoplasmic membrane systems and formed food vacuoles subsequently evolved. Whole-cell predation became common. The aerobic symbiotic bacteria requiring more surface membrane for oxidation-reduction reactions differentiated into the cristae-containing mitochondria of today" (pp. 57–59). Each of these sentences, still in English, can be similarly unpacked, until they read like, "Uninterrupted aerobic glycolysis requires that cytosolic NADH produced in the reaction catalyzed by glyceraldehyde 3-phosphate dehydrogenase be oxidized," which can, have no fear, be further unpacked. Not that this proves anything. This is the form of every act of mystification, but still. . . .

51. Margulis *Symbiotic Planet* (op. cit., pp. 36–37).

52. As Andrew Knoll says, "Accumulating geological data are deepening our appreciation of this era as a period of major tectonic, biogeochemical, climatic, and atmospheric change"; see his "Biological and Biogeochemical Preludes to the Ediacaran Revolution," in Jere Lipps and Philip Signor, eds., *Origin and Early Evolution of the Metazoa* (Plenum Press, New York, 1992, p. 76).

53. This is Knoll again: "The last 300 My of the Proterozoic was also a time of strong climatic fluctuation, with as many as five distinct ice ages alternating with relatively ice-free periods" (ibid., p. 76), which he describes in the context of a weak thermohaline circulation and high rates of carbon burial. For a review of other potential mechanisms driving this glaciation, see Joseph L. Kirschivink, "Late Proterozoic Low-Latitude Global Glaciation: The Snowball Earth," in Schopf and Klein (op. cit., pp. 51–52). For the (most recent) last word, see W. B. Harland's, "The Proterozoic Glacial Record" (in G. L. Medaris, C. W. Byers, D. M. Mickleson, and W. C. Shanks, eds., *Proterozoic Geology: Selected Papers from an International Proterozoic Symposium, Geological Society of America Memoir 161*, 1983, pp. 279–288).

54. Speculation, but nonetheless provocative, are these comments by Kirschivink: "In closing, it is perhaps worth noting again that this late Proterozoic glacial episode marks a major turning point in the evolution of life. Although pre-

ceded by abundant evidence of the presence of protists and prokaryotes, it is followed by the first clear record of metazoan animals (the Ediacara Faunas) and shortly thereafter by the appearance of mineralized fauna in the Cambrian. It is tempting to extend the snowball earth speculation to suggest that these evolutionary changes were made possible by the glaciations—the periodic removal of all life from higher latitudes would create a series of post-glacial sweepstakes, perhaps allowing novel forms to establish themselves, free from the competition of a preexisting biota" (op. cit., p. 52). At the same time, of course, there was a dramatic increase in atmospheric oxygen, to a point capable of supporting large animals (minimally estimated at 6–10% of current levels). There was all sorts of stuff going on—terrific phosphogenesis, for instance (with a consequent increase in oceanic fertility)—and . . . well, anyone who thinks we've got this eon, or even this Vendian tail-end of this eon, nailed, is nuts.

55. The Ediacaran fossils, named after the site where they were first found in the Flinders Mountains of southern Australia, may or may not be ancestral to any extant life-forms. They may or may not even be faunal. For this reason Mark McMenamin prefers to call them Ediacaran *biota*. For far more than an introduction (it's a wacky first-person tour of a lot of recent paleontology), see his *The Garden of Ediacara: Discovering the First Complex Life* (Columbia University Press, New York, 1998).

56. Harold Morowitz, "From Soup to Solid-State," in his collection *Mayonnaise and the Origin of Life* (Scribner's, New York, 1985, p. 214).

57. Well, for example, about the rain forests. "The rain forests we have now in the Amazon Basin, West Africa, and elsewhere depend on relatively *low* global temperatures to reduce seasonality in equatorial regions, on a configuration of continents and terrain conducive to heavy rainfall in the tropics, and on substantial time for the species making up complex communities to evolve. Fred Ziegler, a paleogeographer and climatologist at the University of Chicago, has estimated that, during the 350 million years since diverse land floras first evolved, tropical rain forests have flourished for only about a quarter of the time," observes David Raup in *Extinction: Bad Genes or Bad Luck?* (Norton, New York, 1991, pp. 134–135). My point is not that we therefore have the license to behave as wantonly and stupidly as we can—which is what we're doing—but that if we do manage to off them, once we've eliminated ourselves from the picture—which we will do—the rain forests . . . *will come back* (given supportive climate, continental configuration, and the rest of it). In any case, eating Rain Forest Crunch isn't going to help, so if you prefer Totally Nuts . . . *by all means.*

58. "Estimates of meteorite destructive power are based almost entirely on theory or computer simulations based on theory. For the impact of even a one-kilometer object (ten football fields), the estimated energy release is truly incredible, exceeding by a large multiple that released by the simultaneous detonation of every nuclear weapon in existence. Luis Alvarez quoted one estimate for the impact of a 10-kilometer object: one hundred million megatons of TNT. Even if this number is too high, the destructive power defies the imagination," writes Raup (op. cit., p. 161). Perhaps not incidentally, this is the book in which the notorious paleontologist insists that the mass extinctions he

and Jack Sepkoski have argued cycle every 26 million years—from all of which the biosphere has recovered—may have all been caused by meteorite impacts. *Very* refreshing reading!

59. Heinz Haber, *The Walt Disney Story of Our Friend the Atom* (Simon & Schuster, New York, 1956, pp. 24–27). This was an amazing book that holds up very well a half century later.

60. According to Lawrence Joseph, "Lynn Margulis believes that life cannot be stopped by any earthly means. That is the essence of her belief in the Gaia theory. She knows that neither she nor Lovelock can prove it yet, but it has long been their shared conviction that rather than thinking of nature as something 'exquisitely sensitive to the depredations of man,' as her ex-husband Sagan is fond of putting it, humanity's power is trifling when compared to the unfathomable resiliency of the plant, animal, and particularly, microbial world. Of course, human civilization may well destroy those parts of nature that are beautiful or essential to our own welfare, extinguishing species that by any moral or aesthetic measure deserve to flourish unmolested. But the ability to vandalize is not at all the same as the ability to control or conquer. Compared to the three-and-a-half-billion-year momentum of the microcosm, Margulis cannot help but snicker at the vainglory of our little species" (Lawrence Joseph, *Gaia: The Growth of an Idea* [St. Martin's Press, New York, 1990, p. 51]).

61. Again, 99.9% of all species that have lived on Earth are extinct, and if the growing consensus has it at all right, most were probably wiped out as a consequence of the impacts of asteroids. Again, see Raup (op. cit.) for interesting arguments on all sides of this issue.

CHAPTER 7

1. I'm quoting here from John J. W. Rogers, "A History of Continents in the Past Three Billion Years" (*Journal of Geology, 104,* 1996, pp. 91–92). My history of the continents comes from his paper; from the discussion and summary in McMenamin (op. cit., pp. 173–188); from Condie's overview and his exceptionally clear maps (op. cit., pp. 28–33); and from Brian F. Windley's *The Evolving Continents* (3rd ed., Wiley, New York, 1995), especially his "The Earth Evolution Paradigm" (pp. 457–465). The most straightforward description of the history of the continents is McMenamin's rehashing of Rogers (pp. 180–185). If you want to see not only how all the land fit together, but where on the globe it was, consult the series of beautiful maps in Rob Van der Voo's *Paleomagnetism of the Atlantic, Tethys and Iapetus Oceans* (Cambridge University Press, Cambridge, UK, 1993), especially the series on pp. 253–265. (Van der Voo's someone who not only quotes Robert Pirsig, but Charles Bukowski.) There are plenty of alternative models for almost every aspect of this history, but, considering that Rodinia wasn't even named until 1990, pretty amazing consensus about the big picture.

2. Or Eurasia is hard at working uniting with the Indian plate. Again, it's a matter of perspective, but by dramatically strengthening the monsoon, the initial uplift of the Tibetan Plateau radically increased erosion (increasing the sediment flux by a factor as high as 13), so increasing crustal buoyancy and in-

ducing a surge of uplift that *draws* (or at least doesn't resist) the Indian plate, which therefore continues to move northward at an awesome 5 centimeters a year (Pinter and Brandon [op. cit., pp. 78–79]). And then, given this, how to deal with the story in this morning's paper headed "Wait for Aid Drags On in Quake-Shattered Northern India"? That is, what sense to make of the daily in the context of the eonian?

3. Windley (op. cit., pp. 151–153, for details on the monsoon, and p. 464, for the quotation).

4. The C4 (or Hatch–Slack) pathway Windley refers to enables photosynthesis at much lower CO_2 levels than required in the older C3 (or Calvin cycle) pathway. Trees and broadleafed plants use the C3 pathways; grasses, most of them (including cereals), the C4. It is this sort of response of the biosphere to changes in the environmental levels of critical chemicals that calms my beating heart about the impact of *Homo sapiens* on the world at large.

5. A surge in chemical weathering attendant on rapid continental growth could have reduced CO_2 levels in the atmosphere, and so led to cooling that induced glaciation. See Condie (op. cit., p. 205) for an overview.

6. For example, see M. V. Caputo and J. C. Crowell, "Migration of Glacial Centers across Gondwana during Paleozoic Era" (*Geological Society of America Bulletin, 96*, 1985, pp. 1020–1036).

7. Condie (op. cit., p. 206).

8. A popular idea, increasingly associated with Steven Stanley. He provides simple descriptions of the putative mechanisms in his *Earth and Life through Time* (op. cit., pp. 572–574, 598–602).

9. For an overview, see Condie (op. cit., pp. 202–204).

10. That is, the mechanisms shutting down the thermohaline circulation invoking Milankovitch cycles to drive Heinrich and Dansgaard events may be no more than local triggers superimposed on climatic changes driven by supercontinent cycles. R. L. Larson thinks these may ultimately be related to mantle plume activity. He's argued, among other things, that the required changes in CO_2 levels, far above those accommodated by feedback mechanisms in the carbonate–silicate cycles, could be introduced by increased volcanism associated with superplume events. But see his own subtle "Latest Pulse of Earth: Evidence for a Mid-Cretaceous Superplume" (*Geology, 19,* 1991, pp. 547–550) and "Geological Consequences of Superplumes" (*Geology, 19*, 1991, pp. 963–966). Anyhow. . . .

11. The most notorious of these mechanisms is James Lovelock's Daisyworld, best described in his *The Ages of Gaia: A Biography of Our Living Earth* (Norton, New York, 1988, especially pp. 42–64). This is another of those books, like Westbroek's *Life as a Geologic Force* and Margulis and Sagan's *Microcosmos* from which you want to quote almost the entire volume.

12. Or some such dates. These are Condie's (op. cit., p. 218). The 410 date is for vascular plants, with mossoid forms on the land as early as 470 million year ago (in the Ordovician). Margulis and Sagan have plant spores on land 460 million years ago and animals coming ashore 425 million years ago (op. cit., p. 169). And so on. It all depends on the sense you make of the data that are really not in dispute. Frankly, I like David Norman's description: "The emergence of life from the water and onto the land was not a concerted effort

over a short period by some heroic early forms of life, but took place in fits and starts over a vast time span—probably in excess of 3,000 million years." This is based on the idea that the stromatolitic cyanobacteria periodically exposed at low water were, however temporarily, land dwellers (*Prehistoric Life: The Rise of the Vertebrates* (Macmillan, New York, 1994, p. 85). And why not?

13. I like John Sibbick's pictures in David Norman (op. cit.), though the suite of illustrations in "The Expanding Microcosm: A Pictorial Preview," in Margulis and Sagan, *Microcosmos* (op. cit., pp. 23–38), and those illustrating most chapters of Stanley's *Earth and Life through Time* (op. cit.), are probably "more accurate" in terms of representative numbers and mix of species.

14. The Burgess Shale is hot, thanks to Stephen Jay Gould's wonderful, but provocative, *Wonderful Life: The Burgess Shale and the Nature of History* (Norton, New York, 1989). This is a thesis book, advancing Gould's view of history as contingent, and it's driven plenty of paleontologists around the bend. For example, Simon Conway Morris's recent *The Crucible of Creation: The Burgess Shale and the Rise of Animals* (Oxford University Press, New York, 1998) is little more than an extended, and surprisingly ad hominem, attack on Gould. See the bizarre paragraphs, for instance, on pp. 11–12, where Morris accuses Gould of advancing not only an "underlying ideological agenda" sympathetic "to the greatest of twentieth-century pseudo-religions, Marxism," but of an interest in—gasp!—nomothetic science!!! *Well.* Of course, Morris, Derek Briggs, and Harry Whittington, as the most important contemporary students of the Burgess Shale, were extensively "subjects" of Gould's revisionism in *Wonderful Life* . . . but still. . . . For a determinedly "even-handed" treatment and by far the most comprehensive collection of very beautiful photographs of the Burgess Shale organisms, see Derek Briggs, Douglas Erwin, and Frederick Collier's *The Fossils of the Burgess Shale* (Smithsonian, Washington, DC, 1994). The photographs, it's worth mentioning, are by Chip Clark.

15. The dates I give, those of Harland et al. (op. cit.), have never been universally observed. It is quite common, for example, to find the onset of the Paleozoic (which is to say the Cambrian) given as ~540 million years ago (and ending at 505 million years ago) as in, say, Mark and Diana McMenamin's *Hypersea: Life on Land* (Columbia University Press, New York, 1994, p. 2). This boundary in particular is the subject of continuous tinkering, though as David Raup has observed of the Jurassic, "In the five major geologic time scales published in the past ten years, the end of the Jurassic has varied from 130 to 145.6 ma BP [million years before the present]," so it's no big deal. The quote is from Raup (op. cit., p. 177). You know, this is another terrific book.

16. Inaugurating, presumably, a new, as yet unnamed era. When this is done the *periods* of the Cenozoic are given as the Paleogene (66 to 24 million years ago) and the Neogene (24 to 1.8 million years ago), as in McMenamin and McMenamin (op. cit.).

17. During Cambrian times stromatolites continued to blanket enormous areas of the seafloor. So what happened? "The types of algae that form stromatolites occur widely in modern seas, but only in supratidal areas and in hypersaline lagoons do they prosper well enough to form conspicuous stromatolitic structures. Marine animals are largely absent from both these kinds of habitats. In more normal marine environments, animals burrow

through algal mats and also eat them; this destruction is so severe that the mats do not survive long enough to form stromatolites. . . . It seems evident that the great adaptive radiation of Ordovician life produced a variety of animals that tended to prevent stromatolites from developing in all but unusual habitats. As a result, the character of shallow seafloors was forever altered" (Stanley, *Earth and Life* [op. cit., pp. 327–329]).

18. For a series of maps showing, for example, late Paleozoic cratonic shelf areas, see Charles and June Ross, "Late Paleozoic Transgressive-Regressive Deposition," in Cheryl Wilgus et al., eds., *Sea-Level Changes: An Integrated Approach, Society of Economic Paleontologists and Mineralogists, Special Publication No. 42* (Tulsa, OK, 1988, pp. 227–248). The Rosses identify over 60 transgressive-regressive sequences in the Carboniferous–Permian alone, which they consider the major causes of the distinctiveness of the large numbers of warm, shallow, marine provinces.

19. These insights are largely the work of Thomas Worsley and colleagues. See T. R. Worsley, R. D. Nance, and J. B. Moody, "Tectonic Cycles and the History of the Earth's Biogeochemical and Paleoceanographic Record" (*Paleoceanography, 1,* 1986, pp. 233–263); R. Damian Nance, Thomas R. Worsley, and Judith B. Moody, "The Supercontinent Cycle" (*Scientific American,* July, 1988, pp. 72–79); and T. R. Worsley, R. D. Nance, and J. B. Moody, "Tectonics, Carbon, Life and Climate for the Last Three Billion Years: A Unified System?," in Stephen H. Schneider and Penelope J. Boston, *Scientists on Gaia* (MIT Press, Cambridge, MA, 1991, pp. 200–210). Their arguments keep getting more refined. The diagram of the stages of the cycle in the last item cited (p. 204) is the most comprehensive and subtle.

20. Atmospheric CO_2 is drawn down by a silicate–carbonate buffering system presently responsible for burying 80% of outgassed CO_2 by reacting it with silicate rock to form carbonate. The other 20% is buried by a biospheric buffer as organic matter (and hence the problem with our rereleasing it by burning coal and oil). See the discussion in Worsley (op. cit., 1991), but also that in Westbroek's *Life as a Geologic Force* (op. cit.). If I had to say Westbroek's book were about one thing, it would have to be the carbon cycle. Westbroek is good at dealing with what Stephen Schneider says is a problem with the carbon cycle: "If you do geology, you see the big slow cycles of mountain and oceans. If you do biology, you see the little, fast cycles of a microbe, a plant, or a whole population" (in Schneider's "Introduction" to Schneider and Boston [op. cit., p. xix]). Westbroek is *great* on the gearing.

21. For a taste, see no more than the papers by scientists more or less in the same ballpark in the Schneider and Boston volume just cited.

22. This contention is reviewed in Philip Signor and Jere Lipps's very helpful, "Origin and Early Radiation of the Metazoa" (in Jere Lipps and Philip Signor [op. cit., pp. 3–23]). They list and briefly evaluate the following "hypotheses" or issues involved in the early metazoan radiation: hidden evolution; skeletons and animals; oxygen and animals; predators and prey; evolution of large size; carbonate, phosphate, and oceanic chemistry; glaciations, sea level, and diversity; tectonics; genetic mechanisms; and mechanical efficiency. While they conclude that the causes of the radiation remain mysterious, they don't

have a lot of time for changes in chemistry, glaciation, sea levels, or tectonics. Which doesn't mean they're right. . . .

23. Margulis and Sagan, *Microcosmos* (op. cit., pp. 184–185).

24. Ibid., p. 186

25. Again, see Condie (op. cit., pp. 195–197) for a review.

26. See Westbroek (op. cit., pp. 69–88) for an introduction to the whole issue of sea-level change in the contexts of both Earth history and the history of geology. Westbroek includes good reproductions of the "Vail curve," Peter Vail's powerful image of sea-level change over time.

27. Margulis and Sagan, *Microcosmos* (op. cit., pp. 170–171).

28. Ibid., pp. 175–187.

29. There's a whole vocabulary that goes along with this too. "Establishing a beachhead" and "Conquering the land," for example, are common caption or chapter titles for material related to the evolution of terrestrial life. There's this whole military thing. . . . It's not just progressive. Everything becomes very martial and Roman.

30. Raup, *Extinction* (op. cit., p. 29).

31. Ibid., pp. 29–30.

32. Actually, with respect to plants, Steven Stanley has termed our time—the Neogene—"the Age of Herbs" (in *Earth and Life through Time* [op. cit., p. 558]), where herbs are "small, non-woody plants that die back to the ground after releasing their seeds" (including sunflowers, lettuces, and most weeds), but his discussion at this point is concerned no less with the success of grasses.

33. It was while reading about the amazing amount of the world man has planted in corn that the idea came to me that corn had domesticated humans. I flipped out to find that Jack Harlan, at the time professor emeritus in plant genetics at the University of Illinois and one of *the* authorities on the origins of agriculture, had had the same thought (I parade his credentials to make it clear he's not a kook): "The human species has become so completely dependent for survival on a few plant species that one could well ask which are the domesticated. Did people domesticate plants or did plants domesticate people? Are we not all in the same household? Our domesticated cereals cannot survive without us and we cannot survive without them. The symbiosis is complete; one cannot live without the other" (*The Living Fields: Our Agricultural Heritage* [Cambridge University Press, Cambridge, UK, 1995, p. 240]). This is a book so free of cant and received wisdom it's breath-taking.

34. The number of rodent and bat species comes from Raup (ibid., p. 52); the quote's from *Microcosmos* (p. 191).

35. McMenamin and McMenamin (op. cit., p. 3). The community also includes all its viral and bacterial symbionts and parasites.

36. The language of the McMenamins is incredibly powerful, evocative, provocative. In this sentence I've ripped off some of theirs from p. 25.

37. More reworking of the McMenamins, these from p. 18.

38. Ibid., p. 210.

39. Ibid., p. 216

40. In fact, as Margulis and Sagan have insisted often, "Once upon a time, we think, eating and mating were the same." This is a crucial part of their long

argument about the origin of meiotic sex and the evolution of death (which sounds like a Daniel Manus Pinkwater title). I've quoted from their *What Is Life?* (Simon & Schuster, New York, 1995, p. 114), but I've cited their wonderful books about sex above.

41. H. Robert Horton et al. (op. cit., p. 16-1).

42. As I've said, Westbroek (op. cit.), is a good place to learn about the carbon cycle, though, as I've also said, the subject is spread throughout the book. The presentation in Smil (op. cit.) is more concentrated (pp. 42–60 and 78–109), but still aimed at a general reader. A great introduction to numerical models is W. T. Holser, M. Schidlowski, F. T. Mackenzie, and J. B. Maynard, "Geochemical Cycles of Carbon and Sulfur," in C. Bryan Gregor, Robert M. Garrels, Fred T. Mackenzie, and J. Barry Maynard, eds., *Chemical Cycles in the Evolution of the Earth* (Wiley, New York, 1988, pp. 105–173). If I've given the names of all the editors, it's because Garrels is, with Mackenzie, one of the creators of the present way we have for thinking about these things. Indeed, their *Evolution of Sedimentary Rocks* (Norton, New York, 1971) was seminal. So was R. M. Garrels and E. A. Perry Jr.'s "Cycling of Carbon, Sulfur and Oxygen through Geologic Time," in E. D. Goldberg, ed., *The Sea* (Wiley, New York, 1974, vol. 5, pp. 303–336). Or you can just read *about* all this in Westbroek.

43. Again, see Smil (op. cit.), but, you know, another interesting place to read about these is in G. Tyler Miller's pretty amazing *Environmental Science: Working with the Earth* (5th ed., Wadsworth, Belmont, CA, 1995, pp. 76–82). When I say pretty amazing I mean, among other things, that "Two trees have been planted in a tropical rain forest for every tree used to make this book, courtesy of G. Tyler Miller and Wadsworth Publishing Company. The author also sees that 50 trees are planted to compensate for the paper he uses and that several hectares of tropical rain forest are protected" (on the copyright page).

44. And not just in the linguine. This from Holser et al. (op. cit.): "The linkage between tectonic events and the behavior of carbon and sulfur is likely through phosphate" (p. 157). I spare you the argument.

45. Lawrence Durrell, *Clea* (Dutton, New York, 1960, p. 98).

46. Though species don't die, as Raup reminds us, individuals do (op. cit., p. 6, et passim). As Margulis and Sagan put it: "Fatefully for the future history of life forms such as ourselves, in protoctists sexuality became inextricably linked to death. Bacteria can be killed but they do not naturally die. . . . Aging and death, in which living cells disintegrate with predictable timing, first evolved in sexual protoctists. 'Programmed' death as the final stop of a lifelong metabolism was absent at the origin of life—and for a very long time thereafter. Unlike us, bacteria are immortal; they will live until external conditions prevent autopoiesis. By contrast, like us, many protoctists age and die at the end of a regular interval. Aging and dying is an internal process, thanosis, and it arose in our microbial ancestors at some time during the evolution of sexual individuals. Strange to say, death itself evolved. Indeed, it was the first—and still is the most serious—sexually transmitted 'disease' "; see their *What Is Life?* (op. cit., p. 113), but also see almost any of their books, including the three cited earlier about sex per se. But read Raup, especially the

first four chapters, for no-nonsense talk about extinction at the species and higher levels, including discussion of the extent to which the "big five" are analytic artifacts (in fact artifacts of his own earlier analyses). If I haven't said this before, the best *general* book about extinction is Steven Stanley's *Extinction* (op. cit.). The dates and magnitudes I've given for the "big five" come from Virginia Morell's "The Sixth Extinction" (*National Geographic*, February, 1999, pp. 48–49).

47. The phrase is Calder's (op cit., p. 131).

48. Condie attributes the long-term cooling of the past 50 million years to the Alpine–Himalayan orogenies caused by the northward drift of Africa and India. These resulted in a massive drawdown of atmospheric CO_2 as a consequence of the increased weathering and erosion, with a consequent decline in temperatures (op. cit., p. 207).

49. Stanley, *Extinction* (op. cit., pp. 187–188). Apropos the Eocene and Oligocene extinctions, Stanley notes that "the various extinctions apparently took place in pulses, as cool temperatures were transmitted to lower latitudes by winds, by ocean currents that pulled waters from the circumarctic gyre, and by upwelling of cold waters from the deep sea. Through these events, the climatic history of the earth experienced profound and lasting change that . . . culminated in the recent glacial age that has episodically spread ice sheets over large areas of the Northern Hemisphere" (pp. 188–189).

50. These are big craters, 24 and 3.4 kilometers across, respectively. The bolides that made them were more than a couple of football fields long. See Raup (op. cit., p. 176).

51. Though continental ice sheets wouldn't form for another 10 million years, that is . . . *until the Ice Age*. For a review of Miocene glaciation, see Stanley, *Extinction* (op. cit., pp. 193–196). Note that glaciation begins in Antarctica at the Oligocene–Miocene transition, some 22 million years ago. What happens 14 million years ago is a sharp increase in their movement with increased spalling of icebergs so that glacial debris is rafted much farther from the continent than heretofore.

52. Neil Roberts, "Climatic Change in the Past," in Steve Jones, Robert Martin, and David Pilbeam, eds., *The Cambridge Encyclopedia of Human Evolution* (Cambridge University Press, Cambridge, UK, 1992, p. 178).

53. Bjørn G. Andersen and Harold W. Borns, *The Ice Age World: An Introduction to Quaternary History and Research with Emphasis on North American and Northern Europe during the Last 2.5 Million Years* (Scandinavian University Press, Oslo, 1994, p. 25). They continue by saying that "causes for the steps have been actively debated [how unusual!]. However, an obviously important factor was the drastic reorganization of the ocean current system during some of these periods. Oceanic gateways were opened and closed as a result of plate-tectonic-induced migration of the continents, and marine sills/thresholds were lowered and raised. For instance, the lowering of the sill across the Atlantic Ocean between Greenland and Scotland was of immense importance for the flow of the cold deep/intermediate water current between the North and South Atlantic." Still, you know, so much depends on how you interpret the deep-sea oxygen isotope ratios.

54. Ibid., p. 39. As a consequence of this proliferation, "there is a tendency today

to abandon the names of at least the oldest, most-problematic glaciations and use marine isotope-stage numbers." So, welcome to the club!

55. Ibid. Anyone who thinks the ice isn't coming back needs to check this out.

56. Ibid., p. 39.

57. How rapid can be seen in the graph of temperature fluctuations recorded in the core extracted at Vostok Station at the South Pole. See J. Jouzel et al., "Vostok Ice Core: A Continuous Isotope Temperature Record over the Last Climatic Cycle (160,000 Years)" (*Nature, 329*, 1987, pp. 403–408). Here, we are back in the world of Wally Broecker and the Imbries.

58. I can't imagine it hasn't been used before, but I've never seen Afeurasiamerica, though . . . *what else to call it?* The emerged land between North America and Asia is known as Bering Sea Land, Beringland, Beringia. Sundaland includes the Sunda Shelf and the islands associated with it. The Wallacea, or Wallace line, demarks the abrupt drop in the ocean floor that has long separated Asia and Australia.

59. W. S. Laughlin, "Human Migration and Permanent Occupation in the Bering Sea Area," in David M. Hopkins, ed., *The Bering Land Bridge* (Stanford University Press, Stanford, CA, 1967, p. 437).

60. T. B. Kellogg, "Late Quaternary Climate Changes: Evidence from Deep-Sea Cores of Norwegian Greenland Seas" (*Geological Society of America Memoir, 145*, 1976, pp. 77–110).

61. There's a nice pair of maps comparing the climate of Africa in the early Miocene with that of Africa in the present (i.e., in an interglacial) in Peter Andrews "Man and the Primates," in Andrew Sherratt and Grahame Clark, eds., *The Cambridge Encyclopedia of Archeology* (Crown/Cambridge University Press, New York, 1980, p. 59). For more recent shots of more of the world, see COHMAP Members, "Climatic Changes of the Last 18,000 Years: Observations and Model Simulations" (*Science, 241*, 1988, pp. 1043–1052). The maps, on pp. 1046–1047, are in color and quite clarifying. Graedel and Crutzen reproduce a few of these (op. cit., pp. 84–85) in the context of their argument about the superordinate importance of the Milankovitch cycles as the basis for understanding global climate change.

62. See Jürgen Haffer, "Speciation in Amazonian Forest Birds" (*Science, 165*, 1969, pp. 131–136). His subtitle puts the case comprehensively: "Most species originate in forest refuges during dry climatic periods." Terborgh makes Haffer's work more accessible and produces zappy, colorful maps out of Haffer's confusing black-and-white originals (op. cit., pp. 141–147).

63. Andersen and Borns report evidence of four glaciations of Mt. Kilamanjaro during the Cenozoic (op. cit., pp. 99–100).

64. See the brief but comprehensive summary, "Ice Age Environments," by F. Alayne Street, in Sherratt and Clark (op. cit., pp. 52–56).

65. Though I appreciate E. C. Pielou's insistence that "plants did not and do not 'migrate.' What does migrate is only an abstraction, a line on a map representing the margin of a vegetation zone. The plants themselves die" (in *After the Ice Age: The Return of Life to Glaciated North America* [Chicago University Press, Chicago, 1991, p. 83]).

66. For the Miocene extinction event, ~12 million year ago, see Raup (op. cit., p. 176); D. M. Raup and J. J. Sepkowski Jr., "Periodicity of Extinctions in the

Geologic Past" (*Proceedings of the National Academy of Science, 81,* 1984, pp. 801–805); and Stanley, *Extinction* (op. cit., pp. 193–196). The Raup and Sepkowski paper is the original publication of their endlessly cited hypothesis that significant extinction events occur every 26 million years.

67. Not that when he published *Timescale: An Atlas of the Fourth Dimension* (op. cit.) in 1983 Nigel Calder was any kind of expert, but certainly he was, as he modestly put it, "tolerably well-informed," and he sure has us descending from *Dryopithecus* (p. 133) and goes on about *Ramapithecus* as the experts he consulted then encouraged him to. I'm using Calder to stand in for the consensus of informed opinion through the 1970s.

68. *Dryopithecus* no longer has any place in the family tree but sort of floats around as a Miocene sport. "It has no obvious antecedents in Africa or elsewhere and its origins are unknown," Jay Kelly says today ("Evolution of Apes," in Jones, Martin, and Pilbeam [op. cit., p. 225]), though not everyone speaks as definitively as this (see Stanley, e.g. in *Earth and Life through Time* [op. cit., p. 575]).

69. While Kelly explicitly acknowledges the historical importance of *Ramapithecus* in the study of human evolution (op. cit., p. 226), *Ramapithecus* is no longer even a he. "*Ramapithecus*," Elwyn Simons says, "now considered by most to be a female *Sivapithecus*, was long thought to be an ancestor of the hominid *Australopithecus* on the basis of shared characters in teeth and jaws, but its similarities to *Australopithecus* are parallelisms or shared primitive features" (in "The Fossil History of Primates," in Jones, Martin, and Pilbeam [op. cit., p. 207]; though again, see Stanley, *Earth and Life through Time* (op. cit., p. 575). And so it goes in the rapidly evolving world of primate evolution. For a brief, but not entirely dispassionate, review of this history, see Fred H. Smith's "Modern Human Origins" (in Frank Spencer, ed., *History of Physical Anthropology* [Garland, New York, 1997, pp. 661–672]). For a more extended treatment, from sort of the "other side," but which bends over backward to be fair to every current of opinion (or most every current of opinion . . .), see Roger Lewin, *The Origin of Modern Humans* (Scientific American Library, New York, 1993—it includes a couple of boxed pages on "*Ramapithecus*: The Ape That Was Not a Hominid"). What's interesting to me, given the apparent diversity of opinion that generates newspaper headlines and the fights that have famous paleontologists not speaking to each other, is how little the story has changed in recent years. We're still talking about the Pliocene origins of modern man somewhere in Africa. Not that I don't think the details are important, but a lot of this underscores the salience of Margulis and Sagan's observation that "objective scholars, if they were whales or dolphins, would place humans, chimpanzees, and orangutans in the same taxonomic group. There is no physiological basis for the classification of human beings into their own family (Hominidae—the manapes and apemen), apart from that of the great apes (Pongidae—the gibbons, siamangs, gorillas, chimps, and orangutans). Indeed, an extraterrestrial anatomist would not hesitate to put us together with the apes in the same family or even genus" (*Microcosmos* [op. cit., p. 214]).

70. Simons (op. cit., p. 207). The diagram of relationships he provides, as full of question marks as it is, is extremely clarifying (p. 206).

71. The geneology is from Simons (ibid.). Kelly (op. cit., p. 225) observes that *Kenyapithecus* is known from sites that span the Middle Miocene: Maboko, Fort Ternan, Nachola.

72. Quoted by Philip Lieberman in his *Eve Spoke: Human Language and Human Evolution* (Norton, New York, 1998, p. 72). Incidentally, I like Lieberman's book. It's not reasoned as tightly as Morowitz's *Beginnings of Cellular Life* (op. cit.) but it has that same quality of "I've-been-thinking-about-this-for-a-long-time-and-this-is-how-it-looks-to-me-now," at once densely summative and daringly provocative, and really straightforward. What Lieberman has done here is to update his *Uniquely Human: The Evolution of Speech, Thought and Selfless Behavior* (Harvard University Press, Cambridge, MA, 1991) to take into account, and respond to, the whiggery, nay, the reactionary "theorizing" of guys like, specifically, Steven Pinker.

73. While the dates I've assigned within this scheme come from Lewin (op. cit., pp. 12–31), Stanley (*Earth and Life* [op. cit., pp. 574–583]) and Stringer ("Evolution of Early Humans," in Jones, Martin, and Pilbeam (op. cit., p. 250, et passim), the scheme itself is that displayed by L. Luca Cavalli-Sforza et al., as their Figure 2.1.1, to which they attach the note: "Until recently there was considerable uncertainty on the phylogeny of *Australopithecus* and human ancestors (Johanson, 1989). A consensus has now been reached in favor of the phylogenetic structure shown in figure 2.1.1 (Johanson, personal communication)," which is sort of touching, considering the controversy then swirling in which Johanson's was but one voice (if a considerable one). I'm quoting from L. Luca Cavalli-Sforza et al., *The History and Geography of Human Genes* (Princeton University Press, Princeton, NJ, 1994, p. 60). At well over a thousand pages, *The History and Geography of Human Genes* is very much in the class of the Schopf and Klein volume about the Proterozoic biosphere. Still. . . . The Johanson article referred to is D. C. Johanson, "The Current Status of Australopithecus," in G. Giacobini, ed., *Hominidae* (Jaca, Milan, 1989, pp. 77–96). Among other things, Johanson excavated Lucy.

74. For these, see Lewin (op. cit., chapter 1, but especially the authoritative-looking graphic on p. 24). It's hard to produce a graphic for publication embodying the uncertainties beneath it. For seven alternative phylogenies in seven simple diagrams, see Robert Foley, *Another Unique Species: Patterns in Human Evolutionary Ecology* (Longman, Harlow, Essex, UK, 1987, pp. 34–35).

75. Chris Stringer has published an extremely useful graphic that makes this clear. In it nearly a hundred important fossils are located by determined or estimated age and their usual classification. See Stringer (op. cit., p. 250).

76. Lewin (op. cit., pp. 25–35).

77. It's a question of perspective, isn't it? The first 500 million years of the Earth's history must have been pretty crazy too. Still, "Short-time climate and glacier fluctuations occurred through most of the 50 million year [Cenozoic] period," conclude Andersen and Borns, drawing special attention to "the change towards larger ice volume (colder climate) which occurred about 2.5 million years ago and after 0.9 million years ago" (op. cit., p. 21). When describing the changes that transpired 2.5 and 0.9 million years ago, they and others insistently use words like "dramatic" and "drastic"; "changes" become "rapid changes" and "extremely rapid changes"; and "gradual change" gives

way to "steps" and even "prominent steps." Without denying the rapidity and abruptness of these changes, it's important to note that we can resolve this recent record to a degree impossible for earlier glacial periods which very likely were of similar character.

78. Roberts (op. cit., p. 178). It is precisely these rifting-induced developments that have led Yves Coppens to articulate his "East Side Story," essentially that the Rift split the hominids from the rest of the apes: "The population of the common ancestor of humans and apes found itself divided. The western descendants of these common ancestors pursued their adaptation to life in a humid, arboreal milieu; these are the apes. The eastern descendants of these same common ancestors, in contrast, invented a completely new repertoire in order to adapt to their new life in an open environment: these are the humans." See Yves Coppens, "East Side Story" (*Scientific American,* 1994, pp. 62–69).

79. Peter B. DeMenocal and Jan Bloemendal, "Plio-Pleistocene Climatic Variability in Subtropical Africa and the Paleoenvironment of Hominid Evolution: A Combined Data-Model Approach," in Elizabeth S. Vrba et al., eds., *Paleoclimate and Evolution, with Emphasis on Human Origins* (Yale University Press, New Haven, CT, 1995, p. 278). I can't recommend this book too strongly. In a field notorious for its acrimony, the authors of these papers have agreed to disagree in a spirit that is marvelously constructive. The conference, of which this is the result, seems to have been a model of its kind, and the papers have been brilliantly edited to maximize their mutual relevance. Yale is one of my least favorite presses, but the book has been beautifully produced as well.

80. Raup (op. cit., p. 134, but also see his discussion on pp. 134–137). Terborgh's *Diversity and the Tropical Rain Forest* (op. cit.) is a terrific introduction to rain forest ecology, but also check out D. Simberloff, "Are We on the Verge of a Mass Extinction in Tropical Rain Forests?," in D. K. Elliott, ed., *Dynamics of Evolution* (Wiley, New York, pp. 165–180). Thinking about tropical ecosystems from temperate perspectives is impossible. To a degree that's unimaginable in cooler climes, tropical ecosystems are almost ridiculously self-generating.

81. I can't help quoting Terborgh on this point: "We tend to think of trees as permanent fixtures in the landscape, living entities possessed of almost eternal vitality. The huge trees that inspire awe in some tropical forests enhance this impression. For those who live in the forest, the impression can be quite different. At my research station in the Peruvian rain forest, seldom does a week go by that I am not jolted awake by the startling sound, terrifying when close, of roots snapping under tension. A cacophony of cracks and snaps accompanies a prolonged whoosh, culminating in a resounding thump, as a giant trunk slams into the ground, issuing a wave of reverberations that can be felt hundreds of meters away" (op. cit., p. 90).

82. Craig Stanford, *The Hunting Apes: Meat Eating and the Origins of Human Behavior* (Princeton University Press, Princeton, NJ, 1999, p. 16, et passim). Though on reflection I'm not sure Stanford makes much of a point (unless it's the point Glynn Isaac made 20 years ago in "Food Sharing and Human Evolution: Archeological Evidence from the Plio-Pleistocene of East Africa"

[*Journal of Anthropological Research, 34,* 1978, pp. 311–325]), his is a neat enough little book, engaging, and filled with stimulating anecdote. It's annoying, though, that it makes no reference to competing explanations for the size of our brain, which, when you get down to it, is the problem he's concerned with. It saps ones confidence in the strength of his claims. It sure saps mine. In contradistinction, Philip Lieberman has no such hesitations. Though he doesn't take on Stanford per se (the position is hardly unique to Stanford), Lieberman makes a point of *arguing* that hunting is *not* "the key for the evolution of human language and thinking" (*Eve Spoke* [op. cit., p. 30]). I don't think it's a great argument, but at least there's a sense in Lieberman that despite his efforts we really don't have the answer yet. Stanford is pretty sure he does, and you'd never guess from his book anyone thought otherwise: "In this book I argue that the origins of human intelligence are linked to the acquisition of meat, especially through the cognitive capacities necessary for the strategic sharing of meat with fellow group members," and even more emphatically that "the intellect required to be a clever, strategic, and mindful sharer of meat is the essential recipe that led to the expansion of the human brain" (both quotations from p. 5). Note that this is not "Man the Hunter" redux, but "Man the Sharer of the Hunt." Simplemindedness *does* get tiresome. (*Look who's talking!*)

83. J. D. Kingston, B. D. Marino, and A. Hill, "Isotopic Evidence for Neogene Hominid Paleoenvironments in the Kenya Rift Valley" (*Science, 264,* 1994, p. 957). The Messinian salinity crisis refers to the well-documented drop in global sea levels between 5 and 6 million years ago that isolated the Mediterranean and caused it to dry up. The drop is usually attributed to Antarctic glaciation, though it is acknowledged that the cooling was not worldwide. See, inter alia, Stanley, *Earth and Life* (op. cit., p. 561, et passim).

84. Ibid., p. 958. Andrew Hill has argued for some time that there is no evidence in this part of Africa for a pure C4-dominated biomass such as savanna until late in the Pleistocene, absolutely unlike the situation Quade and Cerling have documented in Siwaliks of Pakistan, and which they attributed to a *global* lowering of CO_2 levels. The problem in the game as it is played is that there is so little evidence, that any is taken to mean as much as it can possibly be made to mean, and what this means in practice is that local events are construed as local manifestations of global events (which, of course, they might be). See Hill's really balanced and substantive review in his "Faunal and Environmental Change in the Neogene of East Africa: Evidence from the Tugen Hills Sequence, Baringo District, Kenya," in Vrba et al. (op. cit., pp. 178–193). For the shift from C3 to C4 pathways in the Siwaliks, see J. Quade, T. E. Cerling, and J. R. Bowman, "Development of Asian Monsoon Revealed by Marked Ecological Shift During the Latest Miocene in Northern Pakistan" (*Nature, 342,* 1989, pp. 163–166); and for the extension of these data to changing levels of global CO_2, T. Cerling, Y. Wang, and J. Quade, "Expansion of C_4 Ecosystems as an Indicator of Global Ecological Change in the Late Miocene" (*Nature, 361,* 1993, pp. 344–345).

85. Margulis and Sagan, *Microcosmos* (op. cit., p. 207). And maybe the nuts and fruit were in short supply because monkeys were getting there first. "Some scientists," write Chris Stringer and Robin McKie, "believe the rise and

spread of the monkey, and the corresponding entrenchment of the ape, played critical roles in our own evolution. Faced with creatures that displayed greater flexibility in diet and environmental tolerance, some apes began to adapt to life on the level. Our ape ancestors were forced down from the trees, and once on the ground evolved upright gait and later the large brains and tool technology that are the hallmark of hominid intellect" (*African Exodus* [Jonathan Cape, London, 1996, p. 11]).

86. Terborgh (op. cit., through out, but here especially pp. 73, 103).

87. Ian Tattersall, *The Fossil Trail: How We Know What We Think We Know about Human Evolution* (Oxford University Press, New York, 1995, pp. 155–156).

88. After establishing the absolute salience of our bipedality to our being human, Chris Stringer and Robin McKie ask, "But why did we begin to walk upright?" They conclude that "this is one of the most baffling questions in paleontology today, though there is no shortage of suggested answers" (op. cit., p. 18).

89. This was a *Times* wire story carried in Raleigh's *News and Observer*, April 27, 1999 (p. B8).

90. Which is not to knock Metallica. In fact, I've got the black album on right now.

91. Earlier Maturana had put this as follows: "Living systems are units of interactions; they exist in an environment. From a purely biological point of view they cannot be understood independently of the part of the environment with which they interact, the niche; nor can the niche be defined independently of the living system that occupies it" ("The Neurophysiology of Cognition," in Paul Gavin, ed., *Cognition: A Multiple View* [Spartan, New York, 1970, p. 5]). I like the stark clarity of this formulation.

92. Maturana and Varela (op. cit., pp. 95–96).

93. Maturana and Varela define a unity—a thing, an object, something we can talk about—as follows: "A unity (entity, object) is brought forth by an act of distinction. Conversely, each time we refer to a unity in our descriptions, we are implying the operation of distinction that defines it and makes it possible" (ibid., p. 40). In other words, whenever they're talking about a unity, they're talking about *anything* that can be distinguished from a background or other things. Incidentally, the great clarifying essay on the operation of distinction is G. Spencer Brown's *Laws of Form* (George Allen & Unwin, London, 1969), the theme of which is that "a universe comes into being when a space is severed or taken apart" (p. v). Brown reaches conclusions identical to those of Maturana and Varela from a wholly different set of propositions (mathematicophilosophical).

94. Yes, of course we can intervene in the state of the car, by "fixing" it. But this is no different than "fixing" a person by intervening in his or her state. When we get under the hood and replace things it's like doing surgery. When we put additives in the oil or gasoline it's like prescribing medicine, say an antidepressant. Driving the car differently is like talk therapy. The point is that car and person are similarly *structured systems*.

95. Not that these changes aren't affecting our behavior. They are, though probably in ways way below our threshold of attention. It *has* been a few years, though, since we nicknamed a president "Old Hickory."

96. We've run into Croll before. He was an early predecessor of Milankovitch.

97. Alfred Russell Wallace, "The Measurement of Geologic Time" (*Nature, 1,* 1870, pp. 399–401, 452–455). This two-part note was incredibly prescient. Taking Croll seriously, Wallace noted—130 years ago—the probability that there were 50 or 60 glaciations over the past 3 million years, with all they would entail. He even produced a diagram that looks remarkably like the ones we make based on ice cores.

98. Charles Darwin, *The Descent of Man and Selection in Relation to Sex* (Murray, London, 1871, p. 433).

99. C. K. Brain, "The Evolution of Man in Africa: Was It a Consequence of Cainozoic Cooling?" (*Annals of the Geological Society of South Africa, 84,* 1981, pp. 1–19).

100. Hill (op. cit., p. 188).

101. Beyond the carbon isotope data, Hill marshals a host of (doubtless ambiguous) fossil evidence in his arguments against the turnover pulse hypothesis, at the same time that he problematizes the almost reflex assumption that phenomena reflected in deep-sea isotope data, and so presumably global in character, had local manifestations everywhere: "The African Neogene record essentially monitors very local events, though it may also reflect more widespread or even global ones. But the two categories are difficult to distinguish" (op. cit., p. 189).

102. Tim D. White, "African Omnivores: Global Climatic Change and Plio-Pleistocene Hominids and Suids," in Vrba et al. (op. cit., p. 377). More generally White observes that "the global climatic record is one characterized by a high frequency of variability. The good news is that it is easy to match something on land with some oceanic proxy of global climatic change. The bad news is that most such matches are unlikely to mean very much!"

103. Maturana and Varela (op. cit., p. 51, emphasis added).

104. These are some of the ways organisms and environment interpenetrate, according to Richard Levins and Richard Lewontin in their wonderful *The Dialectical Biologist* (Harvard University Press, Cambridge, MA, 1985, pp. 51–58).

105. Irving Berlin wrote the words and music to this song for Astaire in the 1936 *Follow the Fleet.*

106. This description is stolen from Arlene Croce's 100% right-on description of the dance in her 100% terrific *The Fred Astaire and Ginger Rogers Book* (Galahad, New York, 1972, p. 89). She makes you want to see the movies all over again.

107. John and Mary Gribbin, *Children of the Ice: Climate and Human Origins* (Blackwell, Oxford, UK, 1990, pp. 87–88).

108. Of course they're metaphors: what else? See George Lakoff and Mark Johnson, *Philosophy in the Flesh: The Embodied Mind and Its Challenge to Western Thought* (Basic Books, New York, 1999), especially for this metaphor (pp. 170–234), and particularly for "Causation" ("Causes Are Forces" and "Causation Is Forced Movement" [pp. 184–187]), and "Answering the Causal Concept Puzzle" and the following (pp. 222–234). "Causation and Realism: Does Causation Exist?" is quite helpful. To a great extent, all I'm struggling with in this chapter is the nature and range of this metaphor.

109. I'm far from alone in noting this. "It is one of the challenges of current arche-

ological theory, in all parts of the world, to attempt to bridge macroscale environmental frames with the microenvironmental existence of individual humans going about their everyday activities." So say P. J. J. Sinclair, Thurston Shaw, and Bassey Andah in their "Introduction" to the collection of papers they edited with Alex Okpoko, *The Archeology of Africa: Food, Metals and Towns* (Routledge, London, 1993, p. 15), though I don't see many rising to this challenge.

110. Maurice Sendak, *In the Night Kitchen* (Harper & Row, New York, 1970), not paginated. This is one of the handful of truly great books—I don't care what anybody says—and the fact that it's not on anybody's list of 100 greatest novels (well? what else could it be?) just shows what narrow, stupid nonsense such lists are.

CHAPTER 8

1. Most of the dates again are Lewin's (op. cit., pp. 12–31), Stanley's (*Earth and Life* [op. cit., pp. 574–583]) and Stringer's ("Evolution of Early Humans," in Jones, Martin, and Pilbeam [op. cit., p. 250, et passim]); though see Stanley, *Children of the Ice Age* (op. cit., pp. 160–167), for *rudolfensis*; and Berhane Asfaw et al., "*Australopithecus garhi*: A New Species of Early Hominid from Ethiopia" (*Science, 284,* 1999, pp. 629–635), for *garhi*. Referring to *Homo habilis,* Stanley says, "The latter species, with its smaller brain and apelike limb proportions, does not deserve placement within the human genus" (*Children* [op. cit., p. 160]), an assessment with which Bernard Wood and Mark Collard in their recent reassignment of *habilis* to the australopithecines concur. See their "The Human Genus" (*Science, 284,* 1999, pp. 65–71). That would be the last word, if I didn't have Asfaw et al. opining even more recently that "the position of *A. africanus* relative to the emergence of the genus *Homo* has been particularly difficult to resolve, even in the face of unduly elaborate phylogenetic analysis. One reason for this difficulty is the fundamental disagreement on whether early *Homo* comprises one sexually dimorphic (*H. habilis*) or two (*H. habilis* and *H. rudolfensis*) species" (op. cit., p. 634). It's like a painting. You stand back, the picture's perfectly obvious. You get up close, everything breaks up. Up close, everything's in flux.

2. You know, Cavalli-Sforza makes an interesting point about this, namely, that because bones rapidly disappear in humid forests, we may never be able to know whether the early humans we know from the drier, more savanna-like east side of the Rift Valley lived in the more humid parts of Africa as well (Cavalli-Sforza et. al. [op. cit., p. 93]).

3. We now have *erectus* fossils dated between 1.8 and 1.6 million years ago from the area between the Caspian and Black Seas too (L. Gabuinia and A. Vekua, "A Plio-Pleistocene Hominid from Dmanisi, East Georgia, Caucasus" [*Nature, 373,* 1995, pp. 509–512]). For the Java date, see C. C. Swisher et al., "Age of the Earliest Known Hominids in Java, Indonesian" (*Science, 263,* 1994, pp. 1118–1121).

4. See notes 18 and 19 in this chapter.

5. Joseph Campbell, *The Way of the Animal Powers: Vol. 1. Historical Atlas of World*

294 Notes to Chapter 8

Mythology (Harper & Row, San Francisco, 1983, p. 27). I have refrained from quoting further because of the polluted quality of the text here, which has Campbell referring to such monstrosities as the "Riss–Würm–Wisconsin peak" (far from a glacial peak, this is the Eemian–Sangamon interglacial, currently deep-sea oxygen isotope stage 5e).

6. Images of glacial fluctuations are increasingly common. For an image of the Würm–Wisconsin as a sequence of zingers, see Curt Suplee's, "Unlocking the Climate Puzzle" (*National Geographic*, May, 1998, pp. 58–59). It's wonderful how similar this looks to the diagram Wallace (op. cit.) made of glacial fluctuations 130 years ago.

7. For the most convincing—and indeed it's quite convincing—introduction to the multiregional perspective, see Milford Wolpoff and Rachel Caspari, *Race and Human Evolution* (Simon & Schuster, New York, 1997). Believe me, with the concerted establishment heat directed against Wolpoff it is easy to dismiss him, until you read him. This is a very engaging book, though there's a weirdness about it too—for example, it refers to Wolpoff throughout in the third person. It's as if it were written by Caspari *about* her husband. Its slightly aggrieved tone would be off-putting if it weren't so obvious how much reason Wolpoff has to be aggrieved. Not that . . . he doesn't seem to bring it on himself. But that's something you only get by reading . . . well, almost everybody on the other side.

8. "Unimprovably simple" is Colin Tudge's judgment in his *The Time before History: 5 Million Years of Human Impact* (Scribner's, New York, 1996, p. 220). Tudge is a writer, not a paleontologist, but he has a razor intelligence, and his treatment of this controversy is better than that of the disputants (looser, less dogmatic, more imaginative). Tudge is basically an out-of-Africa man himself, but with frills.

9. At least this is true in the current Wolpoffian model. Earlier versions such as that proposed by Carleton Coon envisioned complete isolation and implied deep genetic divisions between living races. That Coon was a racist cannot be doubted, but this doesn't mean that multiregionalism is. See Wolpoff (op. cit.) for this history (which hints at the salience of his title).

10. For the African transition from *Homo erectus* to *Homo sapiens* in East Africa, see Günter Bräuer, "The Evolution of Modern Humans: A Comparison of the African and Non-African Evidence," in Paul Mellars and Chris Stringer, eds., *The Human Revolution: Behavioral and Biological Perspectives on the Origins of Modern Humans* (Princeton University Press, Princeton, NJ, 1989, pp. 123–154).

11. Or later, since 200,000 years ago is when the split in the mtDNA pool took place, giving us a *terminus post quem* for the separation of the human population. It is this and other genetic evidence that provides the strongest support for the out-of-Africa hypothesis, and that is therefore most problematical from the perspective of the multiregionalists. It is important to note that this date depends on the *assumption* of a constant evolutionary rate of mtDNA and an *estimation* of the rate based on a comparison of the divergence *among* humans with that *between* humans and chimpanzees, with this in turn based on other molecular data; and further that this date is set within an interval of 140,000 to 280,000 years. See Rebecca L. Cann, Mark Stoneking, and Allan C. Wilson,

"Mitochondrial DNA and Human Evolution" (*Nature, 325,* 1987, pp. 31–36) for the original publication of this data, and Cavalli-Sforza et al. (op. cit., pp. 83–93, et passim) for a useful review and contextualization. Lewin treats the "Mitochondrial Eve" story at length (op. cit., pp. 89–113).

12. The best general introduction to the out-of-Africa story is Chris Stringer and Robin McKie's *African Exodus* (Jonathan Cape, London, 1996). I imagine Wolpoff and Caspari (op. cit.) was written as a riposte to Stringer and McKie. Incidentally, both books are chatty, informal, written for the general public without being in the slightest condescending.

13. Tudge (op. cit., pp. 227–230). By introgression Tudge refers to the sharing of genes by remote relatives who have evolved separately to one degree or another, sort of what the multiregionalists have in mind when they talk about weak flows keeping the gene pools in synch. Cann, Stoneking, and Wilson (op. cit.) insist that there is no evidence for this in their mtDNA results, and across a paragraph break tentatively reject introgression and then strongly support a rapid replacement hypothesis.

14. Cavalli-Sforza et al. (op. cit., p. 63).

15. Even this can't be stated unequivocally. Lee Berger believes bipedalism evolved twice, once in East Africa in *A. afarensis,* who eventually died out (perhaps via *A. boisei*); and again in South Africa in *A. africanus,* who evolved into *H. habilis* before migrating north to East Africa. Berger concludes his article in *National Geographic* with these words: "One fact is clear: the paths of human evolution are far more complex than ever imagined" ("Redrawing Our Family Tree?" [*National Geographic,* August, 1998, p. 99]).

16. This is Stanley's big point in *Children of the Ice Age* (op. cit.), especially in "Life among the Lions" (pp. 59–85). Well, it's one of his big points. . . .

17. This is another, justifying his title and its subtitle: *How a Global Catastrophe Allowed Humans to Evolve.*

18. Which involves his third big point, that because human infants mature so slowly, their mothers need their hands free to hold them, and hence cannot be climbing trees all the time as, presumably, *Australopithecus* did to escape predation.

19. See Jean de Heinzelin et al., "Environment and Behavior of 2.5-Million-Year-Old Bouri Hominids" (*Science, 284,* 1999, pp. 625–629).

20. Though it doesn't necessarily make a lot of sense to lay a single global terminology (and the model of development it implies) over what is a process with a gazillion local faces. The idea of "the Neolithic" arose in the European study of European history, and may be ill adapted to other parts of the world. See the interesting comments in the "Introduction" to Shaw, Sinclair, Andah, and Okpoko (op. cit., especially pp. 3–9).

21. The *age* rolls, not the cultural development, which seems generally to have been independently realized, like agriculture.

22. If you just want to waltz through this material, there is no better way to do it than by reading Donald Johanson and Blake Edgar's *From Lucy to Language* (Simon & Schuster, New York, 1996), with its large exquisite photographs (by David Brill) of most of the important skulls, skull fragments, jawbones, and teeth, that is, of the evidence on which the rest of it depends. This is a picture book, but the text, while brief, is clear, straightforward, and as comprehensive

as its length allows. Its scholarly apparatus, while reticent, is formidable and useful. This puts the bones in texts like those of Stringer and McKie, and Wolpoff and Caspari.

23. This is a point explicitly made by de Heinzelin et al. (op. cit., p. 629).

24. The term "Acheulean" derives from the type archeological site of Saint-Acheul in France. Its application to "similar" tool assemblages is as problematical as the use of the term "Neolithic" to describe non-European post-Paleolithic culture (whatever exactly Paleolithic means). This problem gets worse and worse as European type sites are more and more finely differentiated, especially as we move into the Upper Paleolithic and more recent times.

25. Swisher et al. note that the dispersal of *erectus* out of Africa as indicated by the 1.8-million-year-old *erectus* fossil from Java would predate the advent of the Acheulean at 1.4 million years ago, "possibly explaining the absence of these characteristic stone cleavers and hand axes in East Asia (op. cit., p. 1118). They also entertain the possibility that the Java fossils may not be *erectus* or that *erectus* originated outside Africa. I love it.

26. Of course multiregionalists have a different explanation for the variations in these traditions.

27. See J. A. J. Gowlett et al., "Early Archeological Sites, Hominid Remains and Traces of Fire from Chesowanja, Kenya" (*Nature, 294,* 1981, pp. 125–129), but talk about a tentative conclusion!

28. See C. K. Brain and A. Sillen, "Evidence from the Swartkrans Cave for the Earliest Use of Fire" (*Nature, 336,* 1988, pp. 464–466), who are much more assertive about the significance of their find—it's "the earliest direct use of fire in the fossil record"—even though they waffle over whether it was set by australopithecines or *Homos.*

29. See Steven R. James, "Hominid Use of Fire in the Lower and Middle Pleistocene" (*Current Anthropology, 30,* 1989, pp. 1–26); and Lawrence Guy Straus's commentary, "On Early Hominid Use of Fire" (same issue, pp. 488–491).

30. While I first came across a reference to this as a Neandertal practice in Johanson and Edgar (op. cit., p. 97), I first encountered the custom in Fritz Mühlenweg's *Big Tiger and Christian* (Pantheon Books, New York, 1952). One of the *great* books, *Big Tiger and Christian* is a description of the first motorized crossing of the Gobi desert in 1922. They sleep this way out in the desert but, in more civilized settings, on *kangs,* raised sleeping platforms warmed by fires set within them.

31. See the exhaustive description of the Kebara site in Ofer Bar-Yosef et al., "The Excavation at Kebara Cave, Mt. Carmel" (*Current Anthropology, 33,* 1992, pp. 497–550). The conclusion is that these Neandertals were roasting gazelle and deer and maybe parching peas (pp. 523, et passim). The article leaves you with a very heightened appreciation of Neandertal culture in general.

32. But for the skeptical position (and the skeptics of his position) see Robert H. Gargett, "Grave Shortcomings: The Evidence for Neandertal Burial" (*Current Anthropology, 30,* 1989, pp. 157–190).

33. It drives me nuts when otherwise brainy people go on about how much smarter humans are than chimpanzees or monkeys, like smartness was a quality like mass, conserved under every transformation. But smartness is niche-specific. Monkeys are monkey-smart and humans are human-smart and there's

no comparing the two. What would a bacterium do with our big brain? Talk? They've got humans to do that for them.

34. Or, you know, *is it like oxygen*, an unintended by-product lethal to everything around it?

35. At the heart of this argument is the apparent stasis in tool development and the growth in brain size over much of this history, though with respect to the notion that the rise of technology provided the selection pressure for brain growth, Lewin quotes Jerison as saying, "It seems to me to be an inadequate explanation, *not least because tool making can be accomplished with very little brain tissue*" (op. cit., p. 33), a conclusion with which Thomas Wynn, based on his Piagetian analysis of the geometry of Oldawan and Acheulean tools would—if I read him correctly—concur. Wynn is certainly clear that *erectus* and Archaic *sapiens* had full operational intelligence, that is, by 300,000 years ago, with 300 cm³–400 cm³ of growth to look forward to. Of course these early humans had yet to develop speech, and so . . . or at least so goes the argument. Wynn's *The Evolution of Spatial Competence* (*Illinois Studies in Anthropology 17*, University of Illinois Press, 1989) is a pretty amazing little book, from its obvious, but seldom explicitly made, assumption that stone tools are the products of the human mind, to its subtle teasing of the Piagetian stages from the flaking techniques. Jerison is the author of the highly reputable if unavoidably dated *Evolution of the Brain and Intelligence* (Academic Press, New York, 1973).

36. Stanford (op. cit., p. 5).

37. Lieberman, *Eve Spoke* (op. cit., p. 30).

38. After knocking single-factor models, Foley says, "The catch comes in, however, in that when the [multiple-factor] models are built, showing that everything is related to everything, we are left little the wiser" (in *Humans before Humanity* [Blackwell, Oxford, UK, 1995, p. 47]), and of course he's right, more or less. Still, it's the only way to go. The single-factor models—language, art, rock 'n' roll—are self-evidently . . . *wrong*.

39. And of course whole at this point in its evolution in *every* organism, not just, or even especially, us.

40. Lieberman, *Uniquely Human* (op. cit., pp. 109–110).

41. I got the list of flowers from Lewin's fanciful elaboration of what the burial at Shanidar might have been like, if indeed there was a burial (op. cit., pp. 2–3, 177–181).

42. Ibid. (pp. 128–129), but see the carefully reasoned "The Case for Continuity: Observations on the Biocultural Transition in Europe and Western Asia," by G. A. Clark and J. M. Lindly, in Paul Mellars and Chris Stringer (op. cit., pp. 626–676); and on the other side, Richard Klein, "The Archeology of Modern Humans" (*Evolutionary Anthropology, 1*, 1992, pp. 5–14).

43. Just to complicate things and keep them interesting, read this summary comment from Johanson and Edgar: "Barbed bone harpoons and points have been excavated from the Katanda site along Zaire's Semlike River, and these artifacts, which strongly resemble classic Upper Paleolithic bone tools from western Europe, are estimated to be as old as 90,000, which would place them firmly within the preceding Middle Stone Age. And the site of Kapthurin, Kenya, has yielded blades and blade cores reminiscent of Upper Paleolithic technique that may be 240,000 years old" (op. cit., p. 258).

44. For example, Lewin says of the Châtelperronian that it's "the Upper Paleolithic tool culture that appeared at the transition between the Mousterian and
Aurignacian traditions, with characteristics of both," and so on (op. cit., p.
184).

45. Is there any way we can come to see these products of human effort as equivalent to the coccoliths emitted by *Emiliania huxleyi*? That is, to demystify them
as "art" and learn to see them as bioproducts. We need to.

46. "The glacial epoch is not a completed episode of our geological past. We are
in an ice age *now*, and our entire history (at least as the genus *Homo*) has unfolded under its influence," says Stephen Jay Gould in "Abolish the Recent"
(*Natural History*, May, 1991, p. 16).

47. Okay, one thing's for sure: they weren't the same tools, but do allow me the
rhetorical turn which I lift from Campbell (op. cit., p. 30). That the Tasaday
"tools displayed in Manila and shown in photographs were not genuine tools"
seems to an accepted datum in the ongoing debate about the status of the
Tasaday. However, that the Tasaday were a real foraging people I think is also
indisputable. How long it had been since they were in contact with other people when the gatherer–farmer named Dafal ran into them is open to greater
question, as is exactly what we mean when we ask this question. For an interim
report read the 17 (wonderfully contentious) pieces in Thomas N. Headland,
ed., *The Tasaday Controversy: Assessing the Evidence. Special Publication of the
American Anthropological Association Scholarly Series 28* (American Anthropological Association, Washington, DC, 1992). My personal favorite among
these is that by Amelia Rogel-Rara and Emmanuel S. Nabayra in which they
*unmask the Tboli and Manobo farmers who posed as people who had pretended to be
Tasaday*. That is, at least some of the claims that the Tasaday were a hoax were
hoaxes themselves. *I love it*, but feel compelled by my very thrill to end with
Rogel-Rara and Nabayra's conclusion: "We can safely say that the Tasaday exist, that they are making a successful adaptation to the 20th century, and that
they are thriving and increasing in number in the Tasaday Preserve, the last
climax forest of Mindanao, which owes its survival to the Tasaday" (p. 102).

48. John H. Bodley makes this point: "Whether or not a given group is Stone Age
in any technical sense is a question of technology. . . . In fact, only a few places
would fit the definition of Stone Age in its strictly technical sense of autonomous, nonstate cultures operating within a world of similar cultures: Australia
before 1789 and interior New Guinea and Amazonia before 1930" (see his
"The Debate and Indigenous Peoples," in Headland [op cit., p. 198]). So I'd
have to change my date range to c. 50,000 to c. 1930. That's okay. I'll buy it.

49. Apparently this figure appears in the chapter on communications by Jahan
Salehi and Richard Bulliet in Richard Bulliet, ed., *The Columbia History of the
Twentieth Century* (Columbia University Press, New York, 1998), but *I* couldn't
find it there. I came across the figure in Max Beloff's review of the book in
the *Times Literary Supplement* (October 23, 1998, p. 1).

50. Robert Foley, *Another Unique Species: Patterns in Human Evolutionary Ecology*
(Longman, Harlow, Essex, UK, 1987, pp. 126–127). There are tons of problems estimating the sizes of hominid bodies. For some of the (rather grisly)
details, see K. Steudel, "New Estimates of Early Hominid Body Size" (*American
Journal of Physical Anthropology, 52*, 1980, pp. 63–70). It wasn't just brain and

body sizes that were increasing. Everything was growing, including, for example, the length of adolescence. See Christine Tardieu, "Short Adolescence in Early Hominids: Infantile and Adolescent Growth of the Human Femur" (*American Journal of Physical Anthropology, 107*, 1998, pp. 163–178). Tardieu provides a neat summary of many of the interacting factors involved (p. 176).

51. "Metabolic rate and energetics are very dependent on the size of an animal, smaller creatures having higher energy consumption per unit mass. This is probably because so many physiological processes are related to surface: the uptake of gases, diffusion from blood to tissues, food uptake in the intestine, and so on. As the volume of an animal increases, the ratio of surface area to volume falls, and many processes become slower," note Adrian Friday and David Ingram, eds., in *The Cambridge Encyclopedia of Life Sciences* (Cambridge University Press, Cambridge, UK, 1985, p. 88).

52. This is known as "Cope's rule" which, as originally formulated, said that in certain vertebrate lines there is a general trend toward size increase over geological time. It has been subsequently generalized, and of course everything turns out to be far more complicated as soon as you get closer to the data. For a wonderfully readable exposition of the issue, see John Tyler Bonner's *The Evolution of Complexity: By Means of Natural Selection* (Princeton University Press, Princeton, NJ, 1988). The whole book is useful, but look especially at chapter 2, "Evidence for the Evolution of Size Increase (and Decrease) from the Fossil Record," and chapter 3, "The Size of Organisms in Ecological Communities." Of course larger organisms are more likely to become extinct than smaller ones. . . .

53. Robert Foley (op. cit., p. 152).

54. Ibid.; also see p. 160 where he deals with the case of the orangutan.

55. Yes, of course there are exceptions. The panda is a perfectly wonderful example of a large animal that does little but eat and sleep, but as I said, this is broad brush work. The problem with the panda is that it eats the lowest quality foods—bamboo leaves—which it tries to digest with the intestinal tract of a former carnivore. See Tudge's discussion of metabolism and grass (op. cit., pp. 125–129—the panda is treated on p. 126).

56. Foley, *Another Unique Species* (op cit., pp. 206–210).

57. Ibid. (pp. 212–213). What I'm drawing on here is Foley's persuasive argument in his chapter "Hominids in a Seasonal Environment" (pp. 189–221). I should say, though, this is a terrific book.

58. Widely used models hold that territoriality develops only when benefits exceed the costs of defense, which occurs only where resources are simultaneously abundant and predictable. Decreases in both abundance and predictability promote increased dispersion and mobility. See R. Dyson-Hudson and E. Smith, "Human Territoriality: An Ecological Reassessment" (*American Anthropologist, 80*, 1978, pp. 21–41).

59. Laura Ingalls Wilder, *Little House in the Big Woods* (Harper, New York, 1932). The thresher obtrudes into the Edenic life of the Ingallses in the second-to-the-last chapter, "Wonderful Machine." Though in hindsight it's the harbinger of the dependence that is so life-threatening in *The Long Winter* (Harper, New York, 1940), at the time Pa was happy about the machine: "Other folk can stick to old-fashioned ways if they want to, but I'm all for progress. It's a great age

we're living in. As long as I raise wheat, I'm going to have a machine come and thresh it, if there's one anywhere in the neighborhood" (p. 228 in the 1953 edition). Wilder's Little House saga recapitulates man's transition from hunting and gathering to gardening and industrial monoculture.

60. Which is the only form of altruism contemporary sociobiologically driven theorists can even begin to think about. *What's in it for me?* you know. The dominance of Chicago school economic thinking in academia is a scandal.

61. See Isaac's "professional" "Food Sharing and Human Evolution" (op. cit.) and his more influential, more leisurely, and more speculative "The Food-Sharing Behavior of Protohuman Hominids" (*Scientific American, 238,* April, 1978, pp. 90–108). In this paper Isaac advances maternal encumbrance as an important reason for sharing: "Among recent human hunter–gatherers the existence of a division of labor seems clearly related to the females being encumbered with children, a hardship that bars them from hunting and scavenging activities that require speed aloft or long range mobility" (p. 100). This recalls Stanley's (op. cit., *Children of the Ice Age*) argument that australopithecines had to come down from the trees once their hands were filled with babies. It's the fact that all these changes are connected in a kind of *lattice* that dooms the search for single causes. See Stanford (op. cit.) too; he's the most vocal, recent advocate of this whole line of argument.

62. In their work among the contemporary Ache, Hillard Kaplan and Kim Hall were able to reject all hypotheses advanced for food sharing except that food is distributed in such a way that everyone is fed. See their long and difficult "Food Sharing among Ache Foragers: Tests of Explanatory Hypotheses" (*Current Anthropology, 26,* 1985, pp. 223–245). On the other side, see the arguments Kristin Hawkes erects on neoclassical economic principles in her snappish "Why Hunter–Gatherers Work: An Ancient Version of the Problem of Public Goods" (*Current Anthropology, 34,* 1993, pp. 341–361). Her question is, "If one need not give to receive, why give?" (p. 346). Of course its mere propounding drips assumptions about what's . . . *natural.*

63. David Erdal and Andrew Whiten, "On Human Egalitarianism: An Evolutionary Product of Machiavellian Status Escalation?" (*Current Anthropology, 35,* 1994, pp. 175–183, including the replies of Christopher Boehm and Bruce Knauft). I quote from their p. 177 where they propose the term "vigilant sharing" to cover this complex food-sharing behavior.

64. "Reevaluation of the fossil record at Olduvai Gorge," says Richard Potts, "suggests that the concentrations of bones and stone tools do not represent fully formed camp sites but an antecedent to them" see "Home Bases and Early Hominids" (*American Scientist, 72,* 1981, pp. 338–347, quoted on p. 338). Lewis Binford calls them "midday-rest locations." "A midday-rest location," he says, "where bones scavenged from the carcasses of other carnivores prey were carried short distances for processing as marrow bones, results in an association of stone tools and utilized bones, but this is almost certainly not a home base"; see his *Faunal Remains from Klasies River Mouth* (Academic Press, New York, 1984, p. 255). (Incidentally, this is a surprisingly engaging book, contentious, not at all dry, and always alert to what a given inference implies for archeological thinking. There is a good summary treatment of Isaac and the home base problem on pp. 255–266.) Chimpanzees have been observed to

cache stone tools used to smash open palm oil nuts. See classically C. and H. Boesch, "Optimization of Nut-Cracking with Natural Hammers by Wild Chimpanzees" (*Behavior, 83*, 1983, pp. 265–285), but most recently and emphatically Gen Yamakoshi, "Dietary Responses to Fruit Scarcity of Wild Chimpanzees at Bossou, Guinea: Possible Implications for Ecological Importance of Tool Use" (*American Journal of Physical Anthropology, 106*, 1998, pp. 283–295), who documents tool use taking up 32% of total feeding time for a month when the favored fruit pulp was scarce. These chimps depend heavily on tools for their subsistence, and Yamakoshi has no doubt our australopithecine ancestors did too. Furthermore, chimps are more sedentary than the debate over early human sedentism would allow you to guess, as M. C. McGrew points out in "Chimpanzee Material Culture," in R. A. Foley, ed., *The Origins of Human Behavior* (Unwin Hyman, London, 1991, pp. 13–24). Adding this up, it's, like, chimps sort of have home bases and they use tools—lots of them—at need (though they take the tools to the nuts instead of the meat to the tools like hominids), so what's the big deal about imagining early humans doing the same? And maybe it's *hanging around* like chimps (McGrew) → midday rest locations (Binford) → antecedents to camp sites (Potts) → home bases (Isaac) → "permanent" shelters (or at least "starter" or "transition" homes starting at 300 grand), though obviously not all humans have ever had permanent shelters (e.g., the !Kung San, at least until very recently) and some who had gave them up (most pastoralists, including, e.g., Mongols with their yurts). For a contrarian perspective on the value of "permanent shelter," see . . . Thoreau!

65. Foley, *Another Unique Species* (op cit., pp. 184–187).
66. See for instance Binford (op. cit.), but also see K. Scott's "Review of *Faunal Remains from Klasies River Mouth* by L. R. Binford" (*Journal of Archeological Science, 18*, 1986, pp. 89–91) on precisely this point.
67. When I say "set in motion" I mean to imply that each of these—sharing, for example, and the use of shelter—exhibited a history across these 5 million years parallel to that of stone tools, continuous, incremental transformation with all sorts of variants appearing and often enough vanishing as well.
68. Tudge (op. cit., p. 196).
69. There's this lovely literature developing concerned with expanding the idea of domestication to include more than altering the genetic structure of a handful of crops to embrace the full range of ecologically relevant human activity, like the "fire-stick farming" of Australian Aboriginals (though which activity isn't ecologically relevant?). In this view it's all just a kind of niche making and this returns domestication to the larger biotic world, even as, by also thinking about the cognitive dimension of domiculture, it makes it even more radically human. See the encouraging work of A. K. Chase and R. A. Hynes, "Plants, Sites and Domiculture: Aboriginal Influence upon Plant Communities in Cape York Peninsula" (*Archeology in Oceania, 17*, 1982, pp. 38–50), and their subsequent work, and D. E. Yen, "The Domestication of the Environment," in Harris and Hillman (op. cit., pp. 55–75), among others. Sticking a stem in the ground—or in your mouth—is profoundly domestic.
70. Pielou (op. cit.) concludes his discussion of the prehistoric overkill and changing climate hypotheses (pp. 250–266) with this thought: "all that can now be said is that the cause of the great mammal extinctions is still an unsolved puz-

zle" (p. 265). Tudge (ibid.) concludes his discussion (all of chapter 8, "What Difference Do We Make?," pp. 280–314) with this analysis: "Did we kill them? Of course we did. Or at least our immediate ancestors did. . . . The evidence that implicates our ancestors is flawed because it is circumstantial, and yet it seems overwhelming. . . . And now the pace of extinction has increased . . . so now we have the world at our feet—where do we go from here?" (p. 314). Tudge's conviction that humans have been wiping out animals at an increasing rate since the end of the last ice age is what energizes his book (and not at all incidentally connects it to his masterpiece, a cookbook, the earlier *Future Food: Politics, Philosophy and Recipes for the 21st Century* [Harmony Books, New York, 1980]).

71. From an interview with social anthropologist Fox in Alex Shoumatoff, *The Mountain of Names* (Simon & Schuster, New York, 1985, p. 195). Fox is the notorious author of *Kinship and Marriage* (Cambridge University Press, Cambridge, UK, 1983) and perhaps more appositely, with Lionel Triger, of *The Imperial Animal* (Holt, Rinehart &Winston, New York, 1971).

72. Since they also gather for the rest of their hearth group, the women actually work considerably longer, but see Rhys Jones and Betty Meehan, "Plant Foods of the Gidjingali: Ethnographic and Archeological Perspectives from North Australia on Tuber and Seed Exploitation," in D. R. Harris and G. C. Hillman, *Foraging and Farming: The Evolution of Plant Exploitation* (Unwin Hyman, London, 1989, pp. 120–135; the hour and a half figure is given, and contextualized, on p. 126). For an early statement of this theme (dealing with the !Kung San), see Richard Lee, "What Hunters Do for a Living or How to Make Out on Scarce Resources," in Richard B. Lee and Irwin Devore, eds., *Man the Hunter* (Aldine, Chicago, 1968, pp. 30–43).

73. In Richard B. Lee and Irwin Devore, "Problems in the Study of Hunters and Gatherers," in Lee and Devore (op. cit., pp. 3–12).

74. See Tudge (op. cit., pp. 266–267).

75. The same way you're trapped when you find yourself taking a course of action you justify by saying, "I have no choice in the matter."

76. Calder (op. cit., pp. 160–161). Here Calder is fantasizing about the domestication of barley and wheat in the Nile 18,000 years ago as the original publication by Fred Wendorf, Romuald Schild, and others of their finds at Wadi Kubbaniya had encouraged him (e.g., Fred Wendorf et al., "The Use of Barley in the Egyptian Late Paleolithic" [*Science, 205,* 1979, pp. 1341–1347]). The cereal grains were subsequently found to be modern contaminants, and the economy is now characterized as an intensive grass–tuber and fish economy in the Nile some 20,000 years ago (Fred Wendoff et al., "New Radiocarbon Dates and Late Paleolithic Diet at Wadi Kubbaniya, Egypt" [*Antiquity, 62,* 1988, pp. 279–283]). It's still the turn of a trowel. . . .

77. Carl Sauer, *Agricultural Origins and Dispersals* (American Geographical Society, New York, 1952, p. 22). This is a great book, subsequently republished as *Seeds, Spades, Hearths and Herds* (MIT Press, Cambridge, 1969, 1973).

78. Or floodplains. See, "Dump Heaps and the Origin of Agriculture," in Anderson's *Plants, Man and Life* (University of California Press, Berkeley and Los Angeles, 1952, pp. 136–152). This is another of those great books. It's 50 years old, and way out of date, and still worth reading, still . . . *very much* worth reading.

79. This literature is . . . *voluminous.* My tutor here is Bruce D. Smith, *The Emergence of Agriculture* (Scientific American Library, New York, 1995). This is a sort of just-the-facts-ma'am (and *lots* of facts) but very up-to-the-minute tour of the ground. For a little more bite, try the "Introduction" to I. J. Thorpe's *The Origins of Agriculture in Europe* (Routledge, London, 1996, pp. 1–21, but the second chapter's pretty general too). Harris and Hillman (op. cit.) is a terrific anthology. A lot of these guys are made as itchy as I by climatic (and other) stress models driving people to do this or that, and much of the thinking is lively and dialectic. Perhaps not incidentally, Harris and Hillman's is a volume (the 13th) in the One World Archeology series spawned by the notorious World Archeological Congress held in Southhampton, England, in 1986.

80. Jack R. Harlan, "A Wild Wheat Harvest in Turkey" (*Archeology, 20*, 1967, pp. 197–201). He reports on other equally high rates of return on other grains (fescue, corn) in his "Wild-Grass Seed Harvesting in the Sahara and Sub-Sahara of Africa," in Harris and Hillman (op. cit., pp. 80–81) and summarizes the issue generally in *The Living Fields* (op. cit., pp. 8–18, 24–25). Let me say again what an amazing book this last is. I cannot say how heartening it is to read something so free of cant.

81. During the Younger Dryas—between 11,000 and 10,000 years ago—the climate in northwestern Europe went from full interglacial to full glacial and back again. Glaciers advanced in western Scandinavia and a new ice sheet formed over the Highlands of Scotland. Its cause is conjectural (though Wally Broecker has his ideas and you can imagine what they are), but they had nothing to do with supercontinent or Milankovitch cycles. Could this happen again? But of course! Did it have anything to do with Levantine domestication? No reason to think so. There's not much evidence of the cold phase in the region, except that the shores of the Mediterranean were 150 feet lower than they are today, and so, what? Mild desiccation? See Andersen and Borns (op. cit., pp. 82–86).

82. Thorpe's (op. cit., pp. 10–12) review of these climatic stress models, though superficial, is clarifying (perhaps because it is superficial).

83. See Mark Cohen, *The Food Crisis in Prehistory* (Yale University Press, New Haven, CT, 1977), but also see Lewis Binford ("Post-Pleistocene Adaptations," in Sally R. and Lewis R. Binford, eds., *New Perspectives in Archeology* [Aldine, Chicago, 1968, pp. 313–341]); Lewis R. Binford, *In Pursuit of the Past* (Thames & Hudson, London, 1983); and Richard S. MacNeish, *The Origins of Agriculture and Settled Life* (University of Oklahoma Press, Norman, 1992), though Binford's early model depended on sedentism being linked to fishing (à la Sauer [op. cit.]), which seems not to have been the case here.

84. *Against* the seasonal shortage arguments, see Joy McCorriston, "Acorn Eating and Agricultural Origins: California Ethnographies as Analogies for the Ancient Near East" (*Antiquity, 68*, 1994, pp. 97–107), who had earlier advocated a seasonal stress model in Joy McCorriston and Frank Hole, "The Ecology of Seasonal Stress and the Origins of Agriculture in the Near East" (*American Anthropologist, 93*, 1991, pp. 46–69). Apropos the growth–stress arguments, while there is evidence for a growing population, there's none now for any growth-related stress, at least as evidenced in, say, declining health. See, for example, Anna Curtenius Roosevelt, "Population, Health, and the Evolution of Subsistence: Conclusions from the Conference," in Mark Nathan Cohen and George

J. Armelagos, eds., *Paleopathology at the Origins of Agriculture* (Academic Press, New York, 1984, pp. 559–583); and Patricia Smith et al., "Archeological and Skeletal Evidence for Dietary Change during the Late Pleistocene/Early Holocene in the Levant," also in Cohen and Armelagos (op. cit., pp. 101–136), but *how* the factions contend, and the data slip and slide, and the trowels turn!

85. Maturana and Varela (op. cit., p. 51).
86. Ibid., p. 79.
87. Ibid., pp. 181–182.
88. Ibid., p. 193.
89. First came the sporadic Oldawan exploitation of small animals—any that could be caught; then the Oldawan–Acheulean stalking of grazing animals with spears; then the coordinated Mousterian hunting of large mammals; then the Mesolithic–Archaic collecting of small species using nets, weirs, and traps; and finally the Neolithic planting of gardens, the domestication of animals, just the next step in a "patterned directional change" responsive to the periodic episodes of stress that occurred without variation throughout the Pleistocene. See Brian Hayden, "Research and Development in the Stone Age: Technical Transitions among Hunter–Gatherers" (*Current Anthropology, 22*, 1981, pp. 519–548). "Population pressure," he says, "and carrying capacity are sterile, black-and-white ways of looking at problems of cultural evolution" (p. 529). I sort of agree, but what to do with "patterned directional change" responsive to periodic episodes of stress? The problem is that "to explain" is always to show how whatever happened was inevitable (or logical or obedient to physical law or . . .), that is, to make it make sense to us. And you know, that's not a constraint on the behavior of anything.
90. McCorriston and Hole (op. cit.) see the impetus for change arising from the synergistic effects of climate change, anthropogenic environmental change, technical change, and social innovation, which in a way is like begging the question. At heart theirs is climate driven: a hyperarid and highly seasonal Mediterranean climate resulted in the need to store food against the dry times. (However, recall that McCorriston's subsequent work anent California balanophagists has convinced her there probably were no hard times for the Natufians.)
91. Henry proposes a change in Natufian life thanks to their heavy exploitation of cereal and nut resources of the Mediterranean woodlands triggered when the worldwide increase of temperature 12,500 years ago pushed these resources up into the Levant uplands. The incredible resource base induces a change from mobility to sedentism with associated transformations including the sacrifice of population control measures. That's the first transition. The second is a response to rising population pressure on declining resources as Mediterranean woodlands retreat back downslope. Some Natufians embrace a simple foraging life, while others, adjacent to year-round water and upland resources, move toward the incipient cultivation of cereals. Of course there are all kinds of data–fit problems and Henry still requires pushes and pulls to induce change, but I like the clutches between causes and effects (and between the two steps) and the mosaicked quality of the responses. But see *From Foraging to Agriculture: The Levant at the End of the Ice Age* (University of Pennsylvania Press, Philadelphia, 1989). Again, I do not want to deny a role to climate

change. Spurning environmental determinism doesn't mean embracing mysticism.

92. Or as Hodder says, "It is at least clear that the growth of Neolithic symbolism does not occur *after* the domestication of the economy. Indeed, if the symbolic and social parts of the process do not occur before the economic, they are certainly integrally connected with it"; see his *The Domestication of Europe* (Blackwell, Oxford, UK, 1990, p. 32). I'm quite convinced this was an important dimension, and that, indeed, it remains so for us. See my very sketchy (*yes*, even sketchier than this) "Culture Naturale: Some Words about Gardening" (*Landscape Journal, 11*, 1992, pp. 58–66).

93. Brian Hayden, "Nimrods, Piscators, Pluckers, and Planters: The Emergence of Food Production" (*Journal of Anthropological Archeology, 9*, 1990, pp. 31–69) and *Archeology: The Science of Once and Future Things* (Freeman, New York, 1993, especially pp. 192–265).

94. Though I can understand Lewis Binford's revulsion at all this postmodern (or, in archeological terms, *postprocessual*) posturing and "discourse of power talk," I try to imagine Binford reading something like this of Hodder's, "Çatal Hüyük and I, we bring each other into existence" (op. cit., p. 20), and I can understand Binford's insistence that it's "Hodder [who] is involved in a power play, seeking domination for his value-laden ideas" (in Binford's review of Hodder's *Reading the Past*, in Lewis R. Binford, *The New Archeology* [Academic Press, San Diego, 1989, pp. 69–71]). Of course both Hodder and Binford are right, for this is no more than the archeological face of the positivist (processual, New Archeology)/postmodernist (postprocessual) generational conflict that has appeared everywhere in the social (?) sciences (as Binford's New Archeology previously swamped the cultural historical archeology of Robert Braidwood). See Robert Preucel and Ian Hodder, eds., *Contemporary Archeology in Theory* (Blackwell, Oxford, UK, 1996). This is a reader very much . . . in your face!

95. Where this summary of Hodder is Thorpe's (op. cit., p. 14). Harlan says: "One problem I have with all the published models is that they are all conceived by middle-class, university-educated, Industrial Age pragmatists, all looking for some golden bottom line that will explain it all. Input–output studies, optimum foraging strategies and a variety of armchair theories are all products of the modern mind-set" (Harlan, *The Living Fields* [op. cit., p. 25]). YES!!

CHAPTER 9

1. And how should we theorize our descent into automobility? "Toward the end of the second millennium in the common era, a global rise in temperature forced people off horses due to the intrusion of forests into what had been their grazing grounds, at the same time that a crisis in the capitalist mode of production. . . . " Okay. We know *that's* not it. So why can't we start thinking about the change to agriculture like we think about other changes? Is it that we don't want to admit these folks 10,000 years ago were really human? Or, for that matter, that anyone other than us is?

2. Harlan puts it so beautifully. "Suppose," he says, "we were not descended

from tribes of idiots; suppose our ancestors had the same genes we do, the same intelligence and powers of observation; suppose plant-using hunter–gatherers knew all about the life cycles of plants, about flowering, fruiting, seed germination and plant growth; suppose they were economic botanists with an extensive knowledge of plant lore; suppose agriculture began on a basis of knowledge and not ignorance; suppose we were willing to admit that hunting and gathering might be a viable alternative to farming, could we look at the problem more objectively?" (Harlan, *The Living Fields* [op. cit., pp. 7–8]).

3. See Daniel Zohary on this point, in "Domestication of the Southwest Asian Neolithic Crop Assemblage of Cereals, Pulses and Flax: The Evidence from Living Plants," in Harris and Hillman (op. cit., pp. 358–373).

4. As indeed they never did at Ali Kosh where agriculture never did come to dominate. See Frank Hole, Kent Flannery and James Neely, eds., *Prehistory and Human Ecology of the Deh Luran Plain* (University of Michigan Press, Ann Arbor, 1969).

5. Sauer's story is that agriculture had to arise among fishing communities. He understood that whoever domesticated plants was sedentary. He hypothesized that people relying on fish would have had the settled opportunity to experiment with the cultivation of plants, commencing with the cultivation of organic fish poisons (and moving sequentially through asexual plant reproduction to seed technology, all probably in Southeast Asia). Sauer's reasoning is wholly organic and his *Agricultural Origins and Dispersals* (op. cit.) remains worth reading especially since so much spade work has supported so many of his premises (see, e.g., Charles Higham, who documents an almost perfectly Sauerian transition to rice cultivation for the Yangtze, in his "The Transition to Rice Cultivation in Southeast Asia," in T. Douglas Price and Anne Brigitte Gebauer, eds., *Last Hunters-First Farmers: New Perspectives on the Prehistoric Transition to Agriculture* (School of American Research, Santa Fe, NM, 1995, pp. 127–155). So "Sauerian fisherfolk" would be very comfortable, sedentary fishing communities probably experimenting with plant cultivation.

6. This date was established only in 1988 (turn of the trowel). See Yan Wenming, "China's Earliest Rice Agriculture Remains" (*Indo-Pacific Prehistory Association Bulletin*, 10, 1991, pp. 118–126).

7. They were also exploiting a wide range of aquatic resources. Higham characterizes estuarine locations in this part of the world as among the world's most productive ecosystems (op. cit., p. 152) and includes the middle and lower Yangtze in that characterization. Incidentally, he is highly attuned to the impact the Younger Dryas would have had in this part of the world, particularly in these lower lying regions, and while he is convinced rice was domesticated in the Yangtze, he reads the realm from Japan to northern Australia more or less as a single region. Interesting work, provocative, comprehensive. Smith (op. cit.) reviews the debates about the origins of rice (pp. 116–133). For a review inspiring in its ecological and historic breadth (it starts the rice story in Gondwanaland) and its almost breathless brevity, see T. T. Chang's "Domestication and Spread of the Cultivated Rices," in Harris and Hillman, eds. (op. cit., pp. 408–417); and, in the same volume, an even more succinct archeologically oriented review, An Zhimin, "Prehistoric Agriculture in China" (pp. 643–649). Despite Zhimin's confident assertion that "the general outline

of the Chinese Neolithic is now clear," I'm convinced it will be revised by many a turn of the trowel.

8. As Barbara Pickersgill puts it, "The archeobotanical record in the Americas is still so incomplete that additional excavations are liable to change the picture dramatically"; see her "Cytological and Genetical Evidence on the Domestication and Diffusion of Crops within the Americas," in Harris and Hillman, eds. (op. cit., pp. 426–439).

9. These and other neotropical lowlands have long been proposed as important sites for the origins of agriculture in the New World (and by analogy, presumably in the Congo Basin and Southeast Asia)—by Sauer and Donald Lathrap among others. More recently (well, today) Dolores R. Piperno and Deborah M. Pearsall claim neotropical lowland domestication coterminous with that in the Middle East, with large-scale food production in away-from-house-side locations by 7,000 years ago including most of the 100 species domesticated in the New World (including corn, presumably domesticated in the lower reaches of the Rio Balsas). See their *The Origins of Agriculture in the Lowland Neotropics* (Academic Press, New York, 1998) for a *spirited* and convincing defense of these (minority) propositions.

10. Smith (op. cit., pp. 159–160).

11. As it is with respect to the calabash or bottle gourd. For both, see the review by Charles B. Heiser, "Domestication of Cucuritaceae: Cucurbita and Lagenaria," in Harris and Hillman, eds. (op. cit., pp. 470–480). Heiser argues that *Cucurbita pep* was independently domesticated in Mexico and the eastern United States. Smith agrees (op. cit., pp. 192–196, but see also pp. 163–170), and at greater length, in "Seed Plant Domestication in Eastern North America," in Price and Gebauer, eds. (op. cit., pp. 193–213). This extends and develops the line of reasoning Edgar Anderson opened up in "Dump Heaps and the Origin of Agriculture," in Anderson (op. cit.), where he most strongly advocated for the importance of the "open" habitat (especially floodplains, and hence Smith's "floodplain weed theory of plant domestication") and the significance of weeds. Smith's work is an effort at filling in Anderson's history of weeds: "The history of weeds," wrote Anderson, "is the history of man, but we do not have the facts that will let us sit down and write much of it" (p. 150). Which is why Smith has been crawling around floodplains all over the eastern United States documenting the floodplain habitat of *Cucurbita*, to get us back to the beginning of this note!

12. Note that I haven't included corn on this list though its independent domestication in the Andes has vocal proponents (who happen to be vigorous anti-diffusionists). See Duccio Bonavia and Alexander Grobman, "Andean Maize: It's Origins and Domestication," in Harris and Hillman, eds. (op. cit., pp. 456–470). This is also as good a source as any for slipping into the ever-hot debate over the origins of corn. By the way, there are some wild early dates for potatoes, like 10,000 years at Tres Ventanas. See Donald Ungent et al., "Archeological Potato Tuber Remains from the Casma Valley of Peru" (*Economic Botany, 38*, 1982, pp. 417–432). These are old-fashioned radiocarbon dates, however, which remain to be redated using the accelerator mass spectrometer (on which see John A. J. Gowlett, "The Archeology of Radiocarbon Accelerator Dating" [*Journal of World Prehistory, 2*, 1987, pp. 127–170]).

13. Note that quinua is *Chenopodium quinoa*, and that another variety was domesti-

cated in Middle America. *Chenopodium* was independently domesticated in at least all three of these locations.

14. But the rice as much as a couple of thousand years later than the millet and the sorghum, though there is a great deal of uncertainty about exactly when the African rice was domesticated (though certainly before the introduction of the Asiatic variety), all this according to Bassey W. Andah, "Identifying Early Farming Traditions of West Africa," in Shaw, Sinclair, Andah, and Okpoko, eds. (op cit., pp. 240–254, especially p. 252 for rice).

15. See John Barthelme, "Early Evidence for Animal Domestication in Eastern Africa," in J. Desmond Clark and Steven A. Brandt, eds., *From Hunters to Farmers: The Causes and Consequences of Food Production in Africa* (University of California Press, Berkeley and Los Angeles, 1984, pp. 200–205). His seems to be a minority opinion, however, as everyone else seems to think the evidence is all up in the air. See Juliet Clutton-Brock, "The Spread of Domestic Animals in Africa," and Roger Blench, "Ethnographic and Linguistic Evidence for the Prehistory of African Ruminant Livestock, Horses and Ponies," both in Shaw, Sinclair, Andah, and Okpoko, eds. (op. cit., pp. 61–70 and 71–103, respectively—though Clutton-Brock does says it's possible cattle were locally domesticated from the wild aurochs endemic in North Africa). Even though Africa is comparatively impoverished, archeologically speaking (especially south of the Mediterranean littoral), the literature is still off-puttingly enormous.

16. "Early humans," Johanson and Edgar point out, "began to use stone tools as extensions of their bodies to modify or manipulate other objects or elements of their environment, and this became a significant part of our ecological adaptation, to the extent that paleolithic technology seems to have evolved in parallel with the expanding brain and the enhanced social behavior of Homo" (op. cit., p. 250), or in other words, used stone tools to shape up their niche.

17. Note, *areas*, not centers, not only because, as Harlan says, "not a single modern study has agreed with Vavilovian theory" (Harlan, *The Living Fields* [op. cit., p. 54]), which is where the idea of *centers* of domestication is rooted (incidentally, Harlan has the utmost respect for Vavilov and would have studied with him had it not been for Lysenko); but because the areas where crops were domesticated (like sub-Saharan Africa [almost *all* of sub-Saharan Africa]) look like mosaics, not points. Harlan calls these "noncenters," and says of the African one: "The geographic pattern is noncentric in that no one region stands out. The pattern is rather a mosaic of cultures, crops and farming practices" (ibid., 143). *My* point is that this is true of the world.

18. C. Grigson, J. A. J. Gowlett and J. Zarins, "The Camel in Arabia—A Direct Radiocarbon Date Calibrated to about 7000 B.C." (*Journal of Archeological Science, 16*, 1989, pp. 355–362).

19. And I'm not counting the agricultural systems brought into being by ecosystemic manipulation, by foraging and fire, prodding and pruning, such as those of the Australian Aboriginals, though, as Yen says (op. cit., p. 71), "Within the span of human cultural development, and assuming it has a future, the 10,000 years of subsistence agriculture was a temporal episode in the domestication of the world environment, and domestication, in the normally used genetic sense, a moment."

20. I get this comparison from T. Douglas Price, Anne Brigitte Gebauer, and Lawrence H. Keeley, "The Spread of Farming into Europe North of the Alps," in Price and Gebauer, eds. (op. cit., pp. 95–126). While I accept their contention that the spread of farming is the primary means through which farming has become the basis of human subsistence, and therefore the importance of keeping the difference between colonization and adoption in mind, I think they overstate the rarity of acts of domestication (i.e., the origins of agriculture).

21. If it doesn't take long to make things *forever*, it doesn't take long to make them *never* either.

22. As I make achingly clear in my "The Spell of the Land," in George Thompson, ed., *Landscape in America* (University of Texas Press, Austin, 1995, pp. 3–13).

23. Summing up his observations about the origins of agriculture, Harlan says: "First we will not and cannot find a time or a place where agriculture originated. We will not and cannot because it did not happen that way. Agriculture is not the result of a happening, an idea, an invention, discovery or instruction by a god or goddess. It emerged as a result of long periods of intimate coevolution between plants and man. Animals are not essential: plants supply over 90% of the food consumed by humans. The coevolution took place over millennia and over vast regions measured in terms of thousands of kilometers. There were many independent tentatives in many locations that fused over time to produce effective food production systems. Origins are diffuse in both time and space" (*The Living Fields* [op. cit., pp. 239–240]). This should be, but probably has no chance of being, the last word on the topic.

24. Fundamentally the difference between *here* and *there* establishes a gradient along which stuff can be moved. The whole process can be modeled as in any chemical system, though biologically it amounts to a simple tropism. The effect is to increase the size of the niche and, by bridging biomes, to enrich the resource base. This is the simplest possible formulation of the logic underwriting trade, though typically trade implies some sort of reciprocal movement (which may be through third, fourth, or further hands). In "5000 Years of World System History: The Cumulation of Accumulation," Barry K. Gills and Andre Gunder Frank put it like this: "However, [the alluvial plains of Egypt, Mesopotamia, and the Indus] were deficient in many natural resources, such as timber, stone, and certain metals. Therefore, they had an ecologically founded economic imperative to acquire certain natural resources from outside their own ecological niches in order to 'complete' their own production cycles" (in Christopher Chase-Dunn and Thomas D. Hall, eds., *Core/Periphery Relations in Precapitalist Worlds* [Westview Press, Boulder, CO, 1991, p. 72]). Despite the mystical handwaving ("complete"), this comes to much the same thing. *I'm* obviously situating the origins of trade, and so the contemporary world system, in the Paleolithic (and see below); and whereas Gills and Frank acknowledge (p. 71) that the origin of the world system could be pushed back to the Neolithic, they prefer to begin—in their metaphor—"further downstream" with the confluence of Egypt, Mesopotamia, the Indus Valley, and Syria and the Levant, which they date to 2,700 to 2,400 B.C. Here they're amending David Wilkinson's account, which places the origin with the confluence of Mesopotamia and Egypt around 1,500 B.C. (see his "Cores, Periph-

eries, and Civilizations" in Chase-Dunn and Hall, eds. [op. cit., pp. 113–166]). This note also looks ahead to an argument I have yet to make, so hang on to it.

25. Having corrected for dehulling and wastage and moisture content and annual fluctuations, Harlan says, "More than three-quarters of our food comes from cereals. We are eaters of grass seeds like canaries" (p. 241). Here are some figures: wheat accounts for 23.4% of our consumption, corn 21.5%, rice 16.5%, barley 8.0%, soybeans 4.4%, with everything else tending rapidly toward zero (potato: 2.7%). You can quibble. It won't change the picture. Here's another terrifying tidbit: "About 83% of our food resources come from regions of long dry seasons, either Mediterranean or tropical savannas" (p. 243). Did I mention what a regular ag-and-life science guy Harlan is? Crop scientist at a Midwestern university and all? All from Harlan (op. cit., "What Does the World Eat?," pp. 240–250).

26. Johanson and Edgar (op. cit., pp. 89–90).

27. I say "moved" because the simple fact that we find some obsidian from there here doesn't mean that it was traded here. It certainly doesn't mean it was directly traded here. Folks from here could have gone there to get it and brought it back. This is called "direct access." Or it could have reached here through down-the-line exchange. This, of course, is trade, but not long-distance trade. For an almost obsessive analysis of one such instance, see the articles in the chapter on the obsidian trade—or whatever it was—in Colin Renfrew and Malcom Wagstaff, eds., *An Island Polity: The Archeology of Exploitation in Melos* (Cambridge University Press, Cambridge, UK, 1982, pp. 182–221).

28. Katherine Ozment, "Journey to the Copper Age" (*National Geographic*, April, 1999, pp. 70–79).

29. See Anthony Harding's discussion of trade in "Bronze Age Chiefdoms and the End of Stone Age Europe," in Göran Burenhult, ed., *The People of the Stone Age* (Harper, San Francisco, 1993, pp. 106–122).

30. Ozment (op. cit., pp. 74–75).

31. Joanna Hall Brierley, *Spices: The Story of Indonesia's Spice Trade* (Oxford University Press, Oxford, UK, 1994, p. 2). This is one of those "romance of . . . " books but a neat enough little sketch (for more detail, see G. Buccellati and M. Kelley-Buccellati, "Terqa: The First Eight Seasons" [*Annales Archéologiques Arabe-Syriennes, 33*, 1983, pp. 47–67]). For the general context, see Robert M. Adams's well-known "Anthropological Perspectives on Ancient Trade" (*Current Anthropology, 15*, 1974, pp. 239–258). Having observed how little we know (which is still the truth), he says, "On the basis of what is already known, however, there appears to be little doubt that long-distance trade was a formidable socioeconomic force. This was so in spite of its being confined largely to commodities of very high value in relation to weight and bulk because of high transport costs, and in spite of its directly involving only a small part of the population" (p. 247).

32. Note that I'm treating trade, raid, and tribute as equivalent only in terms of their gross effects *in reshaping the niche*. That trade implies a kind of reciprocity different from that implied by raid and tribute I understand, but what here and in the preceding paragraphs I've hoped to establish is the gradual transformation of the land from one with human niches here and there, to one in which these fuse into an increasingly larger and better integrated niche. I take

for granted the continuously shifting economic modalities and political structures. As Gills and Frank would have it, at least of the Egypto–Mesopotamian–
Indic realm from which they trace the origin of the world system, "Urban civilization and the state required the maintenance of a complex division of labor,
a political apparatus, and a much larger trade or economic nexus than that
under the direct control of the state. Thus, the ecological origins of the world
system point to the inherent instability of the urban civilizations and the states
from which it emerged. This instability was both ecological–economic and
strategic. Moreover, the two were intertwined from the beginning. Economic
and strategic instability and insecurity led to efforts to provide for the perpetual acquisition of all necessary natural resources, even if the required long distance trade routes were outside the direct political control of the state. This
was only possible through manipulated trade and through the assertion of direct political controls over the areas of supply," and so on, to their conclusion
that this created a dynamic of perpetual rivalry (op. cit., pp. 72–73). See in the
same volume the extremely fruitful thought of Jane Schneider on the less than
intuitive *similarities* between at least trade and tribute in her critique of
Wallerstein's treatment of the historic role of luxury trade (in "Was There a
Precapitalist World System?," in Chase-Dunn and Hall, eds., *Core/Periphery Relations* [op. cit., pp. 45–67]).

33. For anyone who imagines the world was really, really different in the Bronze
Age—like a wasteland with these little pinpricks of light in the darkness—
Shereen Ratnagar's *Encounters: The Westerly Trade of the Harappa Civilization*
(Oxford University Press, Oxford, UK, 1981) is a salutary corrective. Harappa,
he notes, "is but one of several bronze age cultures scattered through Asia be-
tween the Euphrates and Gujarat. We will be considering here the lower
Mesopotamian plains (Sumer and Akkad), the plains of Khuzistan with their
adjoining mountain region (Elam), the Arabian/Persian Gulf (the Barbar and
Umm an Nar cultures), the southern Zagros and the mountain zone south
of Zerman (the Yahya culture), Seistan and the lower Helman valley (with
Shahri-Sokhta as the most prosperous settlement), and Persian and Pakistani
Baluchistan (the Bampur and Kulli cultures); this brings us to the western
fringes of the Harappan zone, spread eastward up to the Indo-Gangetic divide
and present-day Gujarat" (pp. xiii–xiv). See also the neat map in Philip Kohl's
"The Balance of Trade in Southwestern Asia in the Mid-Third Millennium,
B.C." (*Current Anthropology, 19,* 1978, p. 467), showing connections among
these places. What impresses me is the way the semiarid Near East during the
Bronze Age ends up sounding like Jersey between Philadelphia and New
York—I mean, it was *that* densely rich with cultural life. The whole book is a
revelation this way, but see p. 80 for the trade in foodstuffs with these Gulf
ports.

    That, of course, is all from a spatial perspective, but there's a similar kind
of continuity from a temporal perspective too. Wilkinson, for example, argues
that the Harappan entity is continuous with the Gangetic (or Vedic), and that
what has been mistaken for the collapse of one civilization and the rise of another was actually the dispersal and shift of a core. He advances this as an empirical hypothesis, but it's got all the support in the world. See Wilkinson's
long footnote 7 in his "Cores, Peripheries, and Civilizations" (op. cit., p. 162).

34. And there is *no* question that cars and jets have changed the way we live (but

then what hasn't?). On the other hand, in the long view they're to transportation as sneakers are to shoes. All the heavy lifting was done when leather was
first strapped to the sole.

35. Maybe. . . . Of course there's no doubt at all that the Indo-Pacific peoples, the
Melanesians, the . . . Malaysians, Indonesians, and Polynesians, sailed to and
inhabited all those islands in the Pacific: the Bismarcks, Solomons (perhaps as
early as 28,000 years ago), the Fijis, Cooks, Marquesas (say 2,000 years ago),
Easter Island and the Hawaiians (by 1,500 years ago), so there's no doubt
about their ability—and willingness—to sail east. Colin McEvedy tells this story
beautifully in his swell *The Penguin Historical Atlas of the Pacific* (Penguin
Books, New York, 1998 [damn! the guy can put an atlas together!]), but also
check out Peter Bellwood, *The Polynesians: Prehistory of an Island People*
(Thames & Hudson, New York, 1987). On the other hand, whether they made
it to South America and if they did, when, is more suppositional. Though
without a shred of physical evidence, Stephen C. Jett thinks they did and that
they spread the blowgun, among other things, among the inhabitants of the
lowland tropics. In "Further Information on the Geography of the Blowgun
and Its Implications for Early Transoceanic Contacts" (*Annals of the Association
of American Geographers, 81*, 1991, pp. 89–102), Jett—who is by no means
alone—supposes that the sailing-savvy Indonesians were forced to migrate by
the post-Pleistocene rises in sea level and the consequent inundations of low-
lying Sundaland. While the climate forcing makes me squeamish, and the
diffusionism makes me anxious, I kind of like his argument that the Indonesian blowgum trait complex is not simple, and is substantially reproduced in
the lowland tropics of the Americas. It's like a case of plagiarism: how many
parallels can you stomach? I think James Blaut's sarcastic dismissal of Jett's argument ("Native Americans probably were not inventive enough to think up
these things on their own") unreasoning (and essentially polemical in motivation). But see Blaut's *The Colonizer's Model of the World* (Guilford Press, New
York, 1993, p. 11). I like Blaut's justly angry—and largely right-on—book, but
Blaut's thoroughly justified rejection of diffusionism as a general rule can't
obscure the fact that much human culture has, in fact, diffused (I live in it).
On the other hand, as my friend Arthur Krim has pointed out, the dates are
all wrong for Jett's argument, noting, that "if the blowgun is taken as a trait of
tropical hunting and gathering culture as Jett suggests, then the antiquity of
the regional economy appears to predate any trans-Pacific Polynesian contact
by several millennia" ("On 'Geography of the Blowgun and Early Transoceanic Contacts' by Jett [*Annals of the Association of American Geographers, 82*,
1992, pp. 315–319], including Jett's reply). Paul Tolstoy makes a similar but
more sophisticated argument about trans-Pacific contacts and the origin of
Mesoamerican papermaking (over 300 variable features in the papermaking
process and a cladogram analogue!). See his "Paper Route: Were the Manufacture and Use of Bark Paper Introduced into Mesoamerica from Asia?" (*Natural
History*, June, 1991, pp. 6–14). There's an endless amount of this stuff. But my!
it's an ugly, contentious literature—who got where and did what first (even
without the likes of Thor Hyerdahl—*Kontiki!*—and his analogues and their adherents and detractors)—very reminiscent of the contention between the out-
of-Africaners and the multiregionalists.

36. Kerri Westenberg, "From Fins to Feet" (*National Geographic*, May, 1999, pp. 114–127, p. 120).

37. And achieving precisely the reduction of risks by smoothing over irregularities in harvest hypothesized to have driven the trade in the first place. In the Euphrates Valley, for instance, Adams observes that "under Hammurabi's successors, private entrepreneurs even undertook to insure the kingdom of Babylon against fluctuations in the harvests from which the state's income was derived by taxation, a highly remunerative form of usury requiring enormous reserves of capital not similarly tied to the vagaries of the harvest" but in fact derived from trade (op. cit., p. 247). Yes, dude: all the way back then!

38. Although my inclination to see human culture as a global phenomenon is rooted in the global perspective I've brought to this history of the land, I've been heartened to find a quiet but growing inclination even on the part of others less grandly oriented than I to see it the same way. As Gills and Frank have said, "New historical evidence suggests that economic connections through trade and migration, as well as through pillage and conquest, have been much more prevalent and much wider in scope than was previously recognized. They have also gone much farther back through world history than is generally admitted. . . . Historical evidence to date indicates that economic contacts in the Middle East ranged over a very wide area even several thousand years before the first urban states appeared." They go on to document a merger of the Middle Eastern, Indian, and Chinese nexuses into a single system better than 4,000 years ago. I couldn't agree more, especially with their construction of these connections as *mergers* rather than as assimilations. See also Frank's solo "Bronze Age World System Cycles" (*Current Anthropology, 34,* 1993, pp. 383–430).

39. Also known as Gilolo. I know 5,000 years ago sounds like *way* back, but it's actually only 3,000 B.C., which, okay, is back far enough. The classical description of this trade is due only to Pliny (1st century A.D.), but there's archeological evidence for all of the parts, some of it, from Mesopotamia, documentary. Really it only sounds like a long time ago because we still can't get used to the idea that these weren't aliens or craggy-browed Neandertals, but people "just like" you and me. Okay, their "consciousness" was undoubtedly different, but . . . how different? Of course, "There's no 'reading' of the genetic code that is itself not part of the organism's development in its environment," as Tim Ingold most appositely reminds us (in "The Evolution of Society," in A. C. Fabian, ed., *Evolution: Society, Science and the Universe* (Cambridge University Press, Cambridge, UK, 1998, p. 97), and yet . . . and yet. . . .

40. For the regional trade in lapis lazuli, see Ratnagar (op. cit., pp. 130–138, et passim). Exactly what goods were going where at what times is not yet really clear, but that there were complicated movements of etched carnelian beads, lapis, pearls, lead, silver, shells, mother-of-pearl, and dice is certain. Yes, dice. "It may have been that the game of dice attained popularity over a world united to a greater or lesser extent by commercial interaction," Ratnagar hazards (p. 146), presumably through the mechanism of the Arab middlemen from Dilmun (Tilmun).

41. Note that there are known Chalcolithic sites in abundance along both the Eastern and Western Ghats, though south of the Vindhya there's not a lot of

evidence of contact with Harappa or Mohenjo-Daro. A useful way to acquire a synoptic view of all these connections is by letting your eyes lose focus while staring at the plates in some of the modern historical atlases. Look at *The Harper Atlas of World History* (HarperCollins, New York, 1992), for instance, with its cartography by Jacque Bertin; or *The Times Atlas of World History* (Hammond, Maplewood, NJ, 1993), with so many of its maps oriented to local perspectives.

42. We're talking about serious trade. For example, Christopher Edens documents "massive amounts of barley entering the Gulf from Mesopotamia" (p. 127), referring to the barley as a "subsistence staple," this in a discussion of category shifts of tradestuffs from luxury to necessity; in the case of barley, at the end of the 3rd millennium B.C. (in "Dynamics of Trade in the Ancient Mesopotamian 'World System' " [*American Anthropologist, 94*, 1992, pp. 118–139]).

43. The phrase is T. W. Beale's from his "Early Trade in Highland Iran: A View from the Source Area" (*World Archeology, 5*, 1973, pp. 133–148).

44. This is Colin Renfrew's "down-the-line-exchange," from his *The Emergence of Civilization: The Cyclades and the Aegean in the Third Millennium B.C.* (Methuen, London, 1972, pp. 465–466).

45. It's because of my absolute conviction that this is how trade self-organizes (not to get into all the empirical evidence) that I have so much trouble with even world system theorists when, like Wilkinson, they insist on conceptualizing the system as originating *somewhere* and *growing on* to encompass the world. See, for instance, Wilkinson's series of nine authoritative-looking maps displaying the growth of what he calls "the Central civilization," that is, Egypt and Mesopotamia. Amusingly (pathetically, stupidly, infuriatingly), this "Central civilization" fails to include China until the present time (Wilkinson [op. cit., pp. 144–148]). For a radically different, and reassuringly more, Sino-centric perspective, see Frank's scathing *ReOrient: Global Economy in the Asian Age* (University of California Press, Berkeley and Los Angeles, 1998), where the Asian Age—and this is more like it—is more or less the last 2,000 or so years. This is a cool book. You will appreciate that I especially liked his writing: "We are badly equipped to confront our global reality when we are misguided into thinking that our world is only just now undergoing a belated process of 'globalization.' Our very language and its categories reflect and in turn misguide our thinking when they lead us to suppose that the parts came first and then only combined to make a whole" (p. 38). Throughout the book he bemoans our and his own ability to encompass, or even have the words for, a more global reality.

46. The full quote is "The world is full of power and energy and a person can go far by just skimming off a tiny bit of it"; it's from Neal Stephenson's *Snow Crash* (Bantam Books, New York, 1993, p. 33).

47. Is this hard to get? Preparing to launch our nanotechnology into the universe to a Euro-hiphop techno-rap-house beat, we're reading the Bible, written between two and three millennia ago in this very part of the world, with ever greater frequency. That is, *we* perfectly embody this madly experimenting deeply conservative way of being.

48. See Brierley (op. cit., pp. 11–16) for an evocative description. But for serious

doubts about the routes of all this trade, check out the revisionist efforts of Patricia Crone in *Meccan Trade and the Rise of Islam* (Princeton University Press, Princeton, NJ, 1987). She doubts both the inland route and that the cinnamon came from Indonesia, arguing an East African source. For a succinct but thorough review of what we know about Punt, see K. A. Kitchen's "The Land of Punt," in Shaw, Sinclair, Andah, and Okpoko, eds. (op. cit., pp. 587–608); and on Aksum (or Axum), see Stuart Munro-Hay, "State Development and Urbanism in Northern Ethiopia" (ibid., pp. 609–621). This is a part of the world with respect to which practically any flipping of the page at all—to say nothing of a turning of the trowel—is sure to enhance and alter our understanding.

49. See the chapter "The Unification of Afroeurasia: Circa 500 B.C.E.–1400 C.E.," in Christopher Chase-Dunn and Thomas D. Hall's *Rise and Demise: Comparing World-Systems* (Westview Press, Boulder, CO, 1997, pp. 149–186). For a really quick idea of what they're about here, just check out the maps (pp. 153, 155, 165, 169, 173, and 179), especially that on p. 153 showing nine Afroeurasian circuits of trade. This happens to be Abu-Lughod's map of the eight circuits of the *thirteenth-century* world system (which suggests something of the conservatism of these routes) amended to include the trans-Saharan salt–gold circuit (see Janet Abu-Lughod's *Before European Hegemony: The World System* A.D. *1250–1350* [Oxford University Press, New York, 1989, p. 34]). The map fails to include the Americas—where there were trade circuits similar to these (to cite just *one* example, the salt trading spheres of the Classic Maya period, i.e., c. 300–1,000 C.E., mapped out in Anthony Andrews's *Maya Salt Production and Trade* (Arizona University Press, Tucson, 1983, p. 117)—as well as the east–west Sahel trade (and for that matter, any trade with the East African coast, i.e., with Sofala, with Malindi), *but* here's (1) my China Sea circuit, (2) the Bay of Bengal circuit, (3) the circuit around the Arabian Sea, (4) that around the Persian Gulf, (5) that along the Red Sea and the Nile, (6) the circuit across the Sahara, (7) that around the eastern Mediterranean, (8) through Inner Asia, and (9) that along the Burgundian corridor between northern Italy and the Lowlands, this last "belatedly joined [to], or at least cemented its previously looser ties with, an already existing world economy and system" (Frank, *ReOrient* [op. cit., p. xxiii]). *Rise and Demise* is maybe the best introduction to world's system thinking I've read, though Abu-Lughod's is, as they say, a wonderful read.

50. Though exactly what we mean by "control" is an interesting question.

51. In *Leading Sectors and World Power: The Co-Evolution of Global Politics and Economics* (University of South Carolina Press, Columbia, 1996), George Modelski and William R. Thompson take the Silk Roads—which for them include the southern maritime linkages I'm calling "the cinnamon routes" as well as the Inner Asian land routes—as the basic framework of the world economy, for the period from 100 B.C. through 1500 A.D. (This is the fourth of their world economy periods: Bronze, Fertile Crescent, Iron, Silk Roads, Market Economy, and World Market). At its most enduring, the Silk Road system for them reduces to a structure linking the four major regions of the classical world system: China, Europe, India, and the Near East, that is, say, c. 100 B.C., Lo-yang ↔ Rome ↔ Alexandria ↔ Muziris ↔ Lo-yang. See especially their discussion on

pp. 126–133 and following. I'm arguing that this structure, and these routes, are incipient at the opening of their *first* world economy period (which they set 5,500 years ago).

52. For a classical idea of the density of this network by the time of the Roman Empire, see J. Innes Miller's *The Spice Trade of the Roman Empire, 29 B.C. to A.D. 641* (Oxford University Press, Oxford, UK, 1969).

53. Prior to which the trade was mostly east–west along the Sahel.

54. Now, *this* is a specialist literature, but see J. Devisse's general and brilliant review, "Trade and Trade Routes in West Africa," in M. Elfasi, ed., *General History of Africa III: Africa from the Seventh to the Eleventh Century* (UNESCO, Heinemann, University of California Press, Berkeley and Los Angeles, pp. 367–435). As should come as no surprise, especially in African studies, "Nearly all recent work has cast great doubt on findings that were taken for granted two decades ago" (p. 367), and indeed the accepted image of the Ghanese exchanging gold and salt measure for measure has been trashed, the south's "dire need" for salt turning out to have been an artifact of monopolist trade practices and the sad model making of "northern" historians. Devisse treats the origin of 11th-century European gold (pp. 399–400). Also check out the more schematic treatment of S. A. M. Adshead in *Salt and Civilization* (St. Martin's Press, New York, 1992, pp. 15–23), and the more nuanced, regional history of Paul Lovejoy in *Salt of the Desert Sun: A History of Salt Production and Trade in the Central Sudan* (Cambridge University Press, Cambridge, UK, 1986).

55. I mean, it's not like they didn't have gold of their own to *make* things with. See, among others, Giulio Morteani and Jeremy P. Northover, *Prehistoric Gold in Europe: Mines, Metallurgy and Manufacture, NATO ASI Series E: Applied Sciences, Vol. 280* (Kluwer, Dordrecht, The Netherlands, 1995).

56. Devisse (op. cit., p. 399).

57. What amazes me, given *my* secondary education, is that today in the 10th grade standard literature texts actually include Du Fu and Li Bai. Not a lot, and little Chinese literature beyond the Dong, but *holy moly* it's different. What else is different is having to get used to saying Du Fu, Li Bai, and Dong instead of Tu Fu, Li Po, and T'ang. Change, ceaseless change.

58. "This is what it comes to then, / A sick old man in a smelly daycoach, / Riding nowhere through the night, / Without a lousy dime to his name," as Kenneth Patchen put it in "Garrity the Gambling Man Grown Old."

59. By saying *"en route"* I mean to imply that writing arose out of the need to connect the brains of traders at opposite ends of trade routes, that is, out of the need to transmit bills of lading; and so, not coincidentally, writing arose only once trade really got going, say between 5 and 6 thousand years ago (yes, precisely along with the copper and the cloves): in Mesopotamia about 5,500 years ago; maybe a few hundred years later or maybe simultaneously in Egypt (if Günter Dreyer's new radiocarbon dates for the tombs at Abydos are accepted (though much skepticism has been politely expressed)); a smidge later (okay, maybe a thousand years later) in the Indus Valley (though Jonathan Mark Kenoyer and Richard H. Meadow are interpreting as protowriting marks on Indic potsherds dated to 5,300 years ago [but again . . . skepticism has been expressed]); maybe a thousand years later still in China (but watch for those

trowels); and a thousand years after that among the Olmecs, Zapotecs, and Mayas. I say . . . *trailing trade,* that is, *trailing domestication* (what if we thought about writing as the *linguistic face of domestication?*). A big break in figuring out writing came with the work of Denise Schmandt-Besserat in the 1970s (notably "An Archaic Recording System and the Origin of Writing" [*Syro-Mesopotamian Studies, 1,* 1977, pp. 1–32]) which is most thoroughly presented in her beautiful book, *Before Writing: From Counting to Cuneiform* (University of Texas Press, Austin, 1992). I think the style and maybe even the general thrust of her argument is relevant to the question of the origins of the Chinese and Mesoamerican scripts, but it's also clear that their emergence took a different path. When I say the *"linguistic face of domestication"* I guess what I'm talking about is the domestication of humans, which is a beating-around-the-bush way of saying writing is about controlling what people do, how people live, that is, it's about power. No, surprise, then, that "states" begin to emerge about the same time. See the extension of this line of thinking in my book about maps as forms of social control, *The Power of Maps* (Guilford Press, New York, 1992). I know it may seem outrageous to leave so "big" an issue in a note about language, but you know, it doesn't seem like such a big deal in the company of the Big Bang, the self-organization of the protocell, symbiogenesis and the evolution of eukaryotic cells, even the evolution of bipedalism and use of stone tools. . . . As Schmandt-Besserat (op cit., p. 195) quotes Claude Lévi-Strauss in an epigraph in her book: "Writing appeared in the history of humanity some three or four thousand years before the beginning of our era, at a time when humanity had already made its most essential and fundamental discoveries . . . agriculture, the domestication of animals, pottery-making, weaving—a whole range of processes which were to allow human beings to stop living from day to day as they had done in paleolithic times, when they depended on hunting or the gathering of fruit, and to accumulate. . . . We must never lose sight of the fact that certain essential forms of progress, perhaps the most essential ever achieved by humanity, were accomplished without the help of writing."

60. By which I mean to encapsulate the evolution of printing, telegraphy, telephony, radio, television, the Web. . . .
61. Harlan, *The Living Fields* (op. cit., p. 240).

CHAPTER 10

1. If you don't believe this check it out in Thomas Ryan's *American Hit Radio: A History of Popular Singles from 1955 to the Present* (Prima Publishing, Rocklin, CA, 1996). This is another of those improbably great books written by an obsessive–compulsive of broad discernment, impeccable taste, unending knowledge, and great good humor. You try to look up some little thing, and 3 hours later you're still reading.
2. For an extended meditation on *Tristes Tropiques,* and in particular on this sense of paradise violated (paradise lost), see Jacques Derrida's *Of Grammatology* (Johns Hopkins University Press, Baltimore, 1976), especially "The Violence of the Letter" (pp. 101–140).

3. And that was in 1968. Here's a more recent picture: "To give you an idea of how old it was, the 45s were listed under the following five categories: Rhythm & Blues, New Releases, Country Music, Old Favorites, and Polka," William Bunch writes in his *Jukebox America: Down Back Streets and Blue Highways in Search of the Country's Greatest Jukebox* (St. Martin's Press, New York, 1994, p. 133). When he gets a chance to fill out a title strip, he has no idea into which categories to put "Sloop John B" and "Wouldn't It Be Nice." "Old Favorites," I'd guess.

4. Neither does Bunch (ibid.), but he does develop, in a romantic way, a regional *jukebox* geography, looking hard but unsystematically at jukeboxes in New Jersey (Sinatra), Illinois (Kenny Rogers), Chicago (Chicago blues), Virginia (Patsy Cline), Louisiana (Cajun), Mississippi (Delta blues), Baltimore (Elvis), Seattle (grunge), and Detroit (where he watches interracial couples dance to Al Green on what he decides is America's greatest jukebox).

5. Of course it was in the air. George Carney writes that he was slipping lectures about music into the introductory cultural geography course he was teaching at Oklahoma State in the late 1960s (in the "Preface" to his *The Sounds of People and Places: A Geography of American Folk and Popular Music* (3rd ed., Rowan & Littlefield, London, 1994, p. xviii). A year after I proposed my study, Peter Nash published "Music Regions and Regional Music" (*Deccan Geographer, 6,* July–December, 1968, pp. 1–24). But it would be 1973 before Carney started teaching a course in the subject (and another 20 years before Nash at the University of Waterloo established a second), and *by then* a couple of students had written (unpublished) theses on the topic at Penn State, among them Jeffrey Gordon's *Rock-and-Roll Music: A Diffusion Study* (1970), and Ben Marsh's *Sing Me Back Home: A Grammar of Places in Country Music Song* (1971); Larry Ford had published, "Geographic Factors in the Origin, Evolution, and Diffusion of Rock and Roll Music" (*Journal of Geography, 70,* November, 1971, pp. 455–464); and Wilbur Zelinsky had "called for" the study of folk music in his influential *The Cultural Geography of the United States* (Prentice-Hall, Englewood Cliffs, NJ, 1973, p. 107).

6. An equivalent phenomenon today would be rappers pressing just enough CDs to sell out of the trunk of a car to make a reputation in the "hood." This is even sung about. See, inter alia, the lyrics to Lauryn Hill's "Superstar": "Why you beef with freaks as my album sales peak? / All I wanted was to sell like 500 / And be a ghetto superstar since my first album Blunted. . . ., " and so on (*The Miseducation of Lauryn Hill* [Ruffhouse CD CK 69035, 1998]). As we'll see, in both cases, rap and *la música criolla*, such localism can constitute a basis for claims of authenticity.

7. Especially in that in the hands of these professionals the music remained folkloristic, but was no longer folk music. "The *aguinaldo* is a lyric composition with strongly rhythmic accompaniment in which something is offered or asked for. The songs are usually of a lively sensual nature based on a happy mode or atmosphere and set in the jovial spirit of Christmas Eve, New Years and Three Kings Day," says James McCoy in his unpublished dissertation, *The Bomba and Aguinaldo of Puerto Rico as They Have Evolved from Indigenous, African and European Cultures* (Florida State University, 1968, p. 47). More accessibly (and generally), see the entry in Ronald Fernandez, Serafín Méndez

Méndez, and Gail Cueto's *Puerto Rico Past and Present: An Encyclopedia* (Greenwood Press, Westport, CT, 1998). Their source is the extended treatment in Francisco López Cruz's *La Música Folklorica de Puerto Rico* (Troutman Press, Sharon, CT, 1967), pp. 183–199, et passim), which is in turn indebted to his own *El Aguinaldo y el Villancico* (Instituto de Cultura Puertorriqueña, San Juan, Puerto Rico, 1956), all of which is taken account of by McCoy, whom see especially for the form's—via the Andalusian *villancico*—Moorish ancestry (from the Arab *Zéjel*, or even the Persian *dubait* [pp. 70–82]), as well as the African origin of its rhythmic pulsation (p. 82, et passim, and 129). Listen to *aguinaldos*, "cantada y tocada por gente del pueblo, y grabada sobre el terreno en distintas regiones del país," on *Música del Pueblo Puertorriqueño*, a long-playing record put out in 1965 by Troutman Press with great liner notes by López Cruz (the notes dated 1966). Even earlier *aguinaldos*—from the mid-'40s—can be heard on *Folk Music of the Americas, Album XVIII: Folk Music of Puerto Rico* (Music Division, Library of Congress) with useful notes by Richard Waterman. The more commercial music of the professional *aguinalderos* is available at any Puerto Rican music outlet. *Aguinaldos* are more broadly situated in the history of Puerto Rican music in María Luisa Muñoz's *La Música en Puerto Rico: Panorama Histórico-Cultural* (Troutman Press, Sharon, CT, 1966); and in the culture in Cesareo Rosa-Nieves's *Voz Folklorica de Puerto Rico* (Troutman Press, Sharon, CT, 1967).

8. Or earlier. McCoy (op cit.), traces them to Andalusian *décimas* of the 16th century (p. 31). Technically, *décimas* have lines of eight syllables (as heard in the seis, a closely related song type), the *decimillas* of the *aguinaldos* lines of six. There's also a rhyme scheme.
9. I wrote about these experiences in "To Catch the Wind" (*Outlook, 46*, Winter, 1982; epitomized in *Newsletter*, International Association for the Child's Right To Play, *8*(3), October, 1982); and "Doing Nothing" (*Outlook, 57*, Autumn, 1985), and the related, "Doing Nothing (Extracts)" (*Children's Environments Quarterly, 2*(2), Summer, 1985).
10. It's precisely the same with our Christmas carols, continuously reconstructing that golden past of rosy-cheeked children in the choir or the snow—that snow forever drifted in the corners of the tiny panes of those windows overlooking either an early 19th-century English city street, filled with bustling shoppers (top hats, long scarves) or a pond dense with skaters.
11. That was the pinch, that was the punch in the mix for El Jíbaro del Yumac, El Jíbaro de Lares, El Indio de Bayamon, El Gallito de Manatí, El Pico de Oro de Bayamon, Los Alegres Bayamoneces, Los Alegres de Hato Tejas, El Trio Caborrogeño, and El Trio Vegabajeño. The parallel here with rap is the rappers passing themselves off as thugs. Here the issue is credibility as well as authenticity.
12. Again, the use of "sign" here is not to take a position in the "cipher–clue" debate Tim Ingold rehearses (Ingold, *The Perception of the Environment* [op. cit., esp. pp. 16–18]). If anything it is to muddy Ingold's distinction, particularly where he maps that between *cipher* and *clue* over that between *communicating* and *showing* (and *information* and *knowledge*). I acknowledge the distinction as important, but *showing* can be *communicating*, and after all, songs are unavoidably simultaneously things-in-the-world *and* communications.

13. "Native" refers to those resident in paradise prior to the intrusion. Native and natural are pretty interchangeable in this realm of the great mystification. But see Tim Ingold's useful parsing of this problem in *The Perception of the Environment* (op cit., pp. 132–151).

14. The quotation is from the unsigned liner notes to *Juan Rios Ovalle/Arturo P. Pasarell, Antología de la Danza Puertorriqueña, Volumen VI*, Instituto de Cultura Puertorriqueña, San Juan, no date (but mid-1960s), but the expression is rampant. Thirty years later, Fernandez, Méndez, and Cueto (op. cit.) open their article on the *danza* with "a music form that represents the identity, values and pride of Puerto Rican society." But this is just a mantra.

15. For an introduction to the *danza*, see Peter Manuel, "Puerto Rican Music and Cultural Identity: Creative Appropriation of Cuban Sources from Danza to Salsa" (*Ethnomusicology, 38*[2], 1994).

16. The quotation here is from López Cruz, liner notes to *Música del Pueblo Puertorriqueño* (op. cit., p. 9). Fernandez, Méndez, and Cueto (op. cit.) cite his use of " 'the backbone' of Puerto Rican native music" in López Cruz (op. cit., 1967, p. 3). This too is a mantra, perhaps a countermantra.

17. Maybe literally. Ricardo Alegría estimates there were 30,000 Taínos living on the island they called Boriquen when Columbus encountered them, and that 60 (that is, *sixty*) were alive in 1550. Estimates vary widely (Alegría's is extreme), but none do the Spanish credit. See Alegría's *Ballcourts and Ceremonial Plazas in the West Indies* (Yale University Press, New Haven, CT, 1983); his children's book, *History of the Indians of Puerto Rico* (Colección de Estudios Puertorriqueños, San Juan, Puerto Rico, 1983); and Irving Rouse's *The Taínos: Rise and Decline of the People Who Greeted Columbus* (Yale University Press, New Haven, CT, 1992). For a sketch of the Taíno worldview, see Antonio M. Stevens-Arroyo, *Cave of the Jagua: The Mythological World of the Taínos* (University of New Mexico Press, Albuquerque, 1988).

18. These were all types of songs played on the Noche del Folklore of the Fiesta de la Música Puertorriqueña, October 21 and 22, 1966, at the Teatro Tapia, the first such celebration of Puerto Rican music held by the Instituto de Cultura Puertorriqueña. James Blaut, one of my faculty advisors, had a copy of this program which, heading into the field work, was my only guide.

19. This was sheer naiveté on my part. The only song with *jazz* after the title was Ricardo Ray's "Queguancoen" (with "On the Scene" on the flipside); the only song labeled *fox* was—so much for imperialist corruption—"De Mi Tierra" on the back of "Mania," a *mazurka*, by Santos Rolon y Su Contijo in the jukebox in the Cafetin el Corozon on the way to Vega Alta from Morovis, and I'll bet Señor Rolon knew the owner of the *cafetin*; and the only songs described as *rock* were—again!—"Que Linda Es Mi Tierra" and "Vacio," flip sides of a single by La Lloroncita, a young *jíbarita* who had hit it big the previous year with "Papito en Vietnam," a *seis*. Even with "I'm a Believer" and "Gloria" I barely had enough songs or jukeboxes with *fox*, *jazz*, and *rock* on them to notice, much less find a pattern in.

20. Yes, this is the very band that, this evening—September 5, 1998—as I type this in Raleigh, North Carolina, is playing the 23rd annual Salsa Festival at Madison Square Garden in New York. This is a band with a worldwide reputation that's been around almost four decades. And here's what's amazing about the

size and richness of the world: most of you reading this have never heard of El Gran Combo before.

21. English titles were everywhere, at the same rate, a low-level hum like the background radiation left over from the Big Bang, reflecting the diaspora to the Northeast mainland that wrenched the island equally everywhere. The titles came from El Gran Combo ("Shake It Baby"), Pepe Rodriquez ("Oh That's Nice, Part 1" ["Part 2" was on the flipside]), Joe Cuba ("Push, Push, Push"), and Tito Puente ("Have a Ball Baby"). They didn't come from the Monkees and the Shadows of Night. Had I studied jukeboxes in San Juan or outside the military bases doubtless I would have found something different, but in the central part of the island the songs they were listening to with English titles were almost all written and sung by Puerto Ricans. This is, of course, a vexed subject. The bottom line, though, is that by 1968 most Puerto Ricans spoke some English and very many spoke an awful lot. Whether this was terrible or terrific I don't propose to engage—it *was*—but my sympathies lie with thinkers like Juan Flores, John Attinasi, and Pedro Pedraza who back in 1981 noted that "when the dimensions of class and nationality serve as the basis for analysis, what had seemed the pitiable loss and absorption of the inherited cultural tradition appears as its extension and enrichment through a process of internationalization" ("La Carreta Made a U-Turn" [*Daedalus, 110*(2), Spring 1981, p. 213]); and more recently by Raquel Rivera, who writes that "the need of purists to preserve national culture is not shared by rappers. Culture is not 'lost' even if it changes with time; it merely evolves," and "cultural purity, as defined by the thesis of cultural imperialism, has virtually no place in the contemporary context. Therefore, cultural cannibalism, or adopting different cultures to enrich one's own, is seen in rap as a common process" ("Rapping Two Versions of the Same Requiem," in Frances Negrón-Muntaner and Ramón Grosfoguel, eds., *Puerto Rican Jam: Essays on Culture and Politics* [University of Minnesota Press, Minneapolis, 1997, pp. 246–47]). That the idea of Puerto Rican *purity* is and always has been ironic (not to say oxymoronic) seems to occur to few however.

22. Raphael Cortijo founded this band in the 1940s. It pioneered the fusion of a sort of Cuban *son*-inflected *jazz band* sound with the rhythms of the *bomba* and *plena* that would ultimately provide a platform for *salsa*. Following Cortijo's incarceration on drug possession charges, the band's pianist, Rafael Ithier, spun off El Gran Combo.

23. Or maybe *sí*. Certainly nobody was talking about *salsa* in the middle of the island in 1968, or maybe they just weren't talking about it to me, but I think seeing *salsa* in 1968 is an instance of retrospective reconstruction. On the other hand, Cesar Miguel Rondon—who should know—clearly sees *salsa* as such by the mid-1960s, at least in New York. His *El Libro de la Salsa: Cronica de la Música del Caribe Urbano* (Editorial Arte, Caracas, Venezuela, 1980) is another one of those great obsessive surveys that if you're foolish enough to start you just can't stop reading. And intelligent insightful history too.

24. Percentages ranged from > 1% (a lonely *mazurka* on a 104-selection box in the Colmado Bar Universitario in Barranquitas) to < 61% (32 of the 52 songs on the box at the Paisaje del Via Jante at Km. 11.1 south of Cayey on Rt. 15—and then you come out of the Sierra de Cayey down to Guayama and the percent-

age drops to zero), with an average of 13% on the jukeboxes with any *criolla*. The average percentage *criolla* on all 93 jukeboxes was 9%.

25. Okay, the exceptions. There are always exceptions. The jukebox at La Playa Mar Chiquita. It bore the same resemblance to other jukeboxes that its beach bore to other beaches. None. It had all kinds of *criolla*—over 20%—including the only *bomba* on any of the jukeboxes. Against this there were the jukeboxes in the mountains—on the plaza in the center of town—that had no *criolla* at all, in Orocovis, Morovis, Corozal, and Naranjito, smack dab in the heart of *criolla* country. But there're always exceptions. It's not like the land climbs evenly either. There's a valley here and a peak over there, but as you climb—going up, then down, then up again—the probability continuously rises that you will find yourself at a higher elevation. The rule is probabilistic, it's like Schrödinger's wave function, ψ the electron isn't *here* . . . or *there*, it's lost in a cloud of probabilities, but . . . the probability of running into it rises as you move toward the nucleus. As you rose up into the mountains the probability rose that you would find an *aguinaldo* on a jukebox.

26. This is the gospel according to James L. Pindell and Stephen F. Barrett, "Geological Evolution of the Caribbean Region: a Plate Tectonic Perspective," in G. Dengo and J. E. Case, eds., *The Caribbean Region: The Geology of North America, Volume H* (Geological Society of America, Boulder, CO, 1990). In an introductory historical overview to the volume, Dengo and G. Draper note that "although most modern workers embrace the precepts of plate tectonics, the tectonic evolution of the Caribbean is still among the least understood of any of the world's regions, and is hotly debated in the following chapters." Indeed, in the chapter following Pindell and Barrett's, Anthony E. L. Morris, Irfan Taner, and Howard A. and Arthur A. Meyerhoff (the former Meyerhoff had been dead for 8 years at the time of publication) advance an argument based on "mantle surge tectonics" ("Tectonic Evolution of the Caribbean Region: Alternative Hypothesis"). There is essentially no overlap between the explanations. In their editorial foreword, Dengo and Case list three other models of tectonic evolution that have not, in their view, received adequate attention: those based on the "eugeosyncline–miogeosyncline (tectogene)" concept, those based on phase changes (basaltification), and those based on the concept of "suspect or tectonostratigraphic terranes." I have followed Pindell and Barrett as the most mainstream exposition (see also, inter alia, J. F. Stephan, R. Blanchet, and B. Mercier de Lepinay's "Northern and Southern Caribbean Festoons Interpreted as Pseudo-Subductions Induced by the East–West Shortening of the Peri-Caribbean Continental Frame," in F. -C. Wezel, ed., *The Origin of Arcs: Developments in Geotectonics 21* [Elsevier, Amsterdam, 1986], for a global perspective that transcends the peculiarities of the Caribbean situation); but before chuckling at (and chucking out) Meyerhoff et al.'s "surge channels," recall how laughable continental drift once sounded, and be warned that the elder Meyerhoff's field experience was unrivaled. The literature is immense (which isn't these days?), and almost as varied and dynamic as the terrain. (Or is that *terrane?*)

27. Friedrich Beinroth, "An Outline of the Geology of Puerto Rico" (*Bulletin, 213*, Agriculture Experiment Station, Mayagüez Campus, University of Puerto Rico, February, 1969, p. 6).

28. Margulis and Sagan (op. cit., p. 173). Yes, as though *this* weren't uncontested. In this morning's paper, in fact, there's a story about a fossil find of a flowering plant dating 146 million years.

29. So broadly put, this is relatively uncontested, but in *The Sugar Industry: An Historical Geography from Its Origins to 1914* (Cambridge University Press, Cambridge, UK, 1989), J. H. Galloway writes: "There has been disagreement over the number of species of sugarcane as well as over the time and place of domestication. In recent years authorities have found some area of common ground, but discussion continues and present conclusions may well be revised" (p. 11). Galloway is a good guide to the big picture and to the—it goes without saying—voluminous literature. Sydney Mintz (*Sweetness and Power: The Place of Sugar in Modern History* [Viking Press, New York, 1985]) and Philip Curtin (*The Rise and Fall of the Plantation Complex* [2nd ed., Cambridge University Press, Cambridge, UK, 1998]) are perhaps more *interesting*—indeed, these are both terrific books—but unless otherwise noted, Galloway's the source for what follows.

30. See Edwin Schafer's mellifluent *The Golden Peaches of Samarkand: A Study of T'ang Exotics* (University of California Press, Berkeley and Los Angeles, 1963, p. 152).

31. Andrew Watson, *Agricultural Innovation in the Early Islamic World* (Cambridge University Press, Cambridge, UK, 1983). Watson is completely convincing about the easy movement of ideas and practices through the Islamic world, and the institutional support Arab rule gave innovation in so many fused spheres of culture, amounting in the Middle East to a revolution in agriculture (including the introduction of tropical and subtropical plants, advances and expansion in irrigation, and changes in land tenure, among others).

32. See Charles Verlinden, *The Beginnings of Modern Colonization* (Cornell University Press, Ithaca, NY, 1970, pp. 26–32).

33. And so forth and so on. I'm obviously simplifying. The second chapter in Galloway (op. cit.) is . . . replete with detail.

34. How much did it dominate it? This from Dudley Smith and William Requa's *Puerto Rico Sugar Facts* (Association of Sugar Producers of Puerto Rico, Washington, DC, 1939): "Sugar production is frequently called the life blood of Puerto Rico. It provides about one-half of the employment and two-thirds to three-fourths of the income of local people. Truly this industry is the backbone of the insular economy" (p. v). This was written from an industry perspective as part of a campaign to get import quotas reduced. The terrifying consequences of such dependency are sketched in Belinda Coote's *The Hunger Crop: Poverty and the Sugar Industry* (Oxfam, Oxford, UK, 1987).

35. There is disagreement about the exact figure—big surprise!—but that given in S. Drescher and S. L. Engerman, eds., *Historical Guide to World Slavery* (Oxford University Press, New York, 1998) for the period between 1450 and 1900 is 11.7 million (p. 171). In this they follow the revisions Paul Lovejoy (*Transformations in Slavery: A History of Slavery in Africa* [Cambridge University Press, Cambridge, UK, 1983]) made to the estimates Philip Curtin made in *The Atlantic Slave Trade: A Census* (University of Wisconsin Press, Madison, 1969), still the necessary foundation, though Curtin is himself indebted to the pioneering work of Noel Deere, *The History of Sugar*, 2 vols. (Chapman & Hall,

London, 1949–1950). There is something delicious about this chain of citations since Curtin opens his book with a critique of a similar chain of citations which it was his self-appointed task to interrupt. Curtin is an exemplary historian and his census is, believe it or not, a very good read. Great graphics too.

36. Drescher and Engerman (op. cit.) are relatively circumspect when they say that "the heaviest mortality occurred in regions of America with strong concentrations on the production of sugar on large plantations" (p. 173). The high mortality rate and the relative paucity of females to males meant that slaves failed to reproduce themselves over much of the region. In *The Rise and Fall of the Plantation Complex* Curtin puts it like this: "Deaths exceeded births on the West Indian plantations from the sixteenth century on, and the slave trade supplied the deficit. The migration of the slaves was not, therefore, a one-time event. The plantations needed a continuous supply of new labor, if only to remain the same size. Growth required still more" (op. cit., 1998, p. 17). The issue is complicated and locally various. See the careful analysis in Francisco Scarano's *Sugar and Slavery in Puerto Rico: The Plantation Economy of Ponce, 1800–1850* (University of Wisconsin Press, Madison,1984), especially pp. 120–143.

37. Fernandez, Méndez, and Cueto (op. cit., p. xv). Notice how the conflation of "people" to those living "en la isla" by being contrasted to the "Spanish import" are rendered *native*. Note as well how this turns "en la isla" into paradise. Subsequent arrivals are powerfully capacitated to establish the nativity of earlier arrivals.

38. María Luisa Muñoz (op. cit., p. 46), who in dismissing the society dances as no more than an echo of European ones is quoting Manuel Alonso, *El Jíbaro* (Edición del Colegio Hostos, Rio Piedras, 1949).

39. Curtin (1969, op. cit., pp. 32–34 and 42–44); Scarano (op. cit., pp. 120–143 and 163–164).

40. Scarano (op. cit., p. 162).

41. Named after the two drums—one high, one with a lower pitch—originally made from barrels that with the maracas comprise the traditional instrumentation of the *bomba*. Except for the self-consciously anachronistic performances at the Santiago Apostol festival in Loíza Aldea, the *bomba* is no longer living music, long since having given itself up to the music that would in time become *salsa*. In 1968, McCoy (op. cit.) predicted the *bomba*'s complete eradication within 25 years (p. 124). In addition to the sources cited for the *aguinaldo*—all of which with the exception of López Cruz (1956) also deal with the *bomba*—see Gregorio Toro, *Bomba and Plena: The African Roots of Puerto Rican Traditional Music* (Connecticut Commission for the Arts, Hartford, 1994), and Halbert Barton's unpublished dissertation, *The Drum-Dance Challenge: An Anthropological Study of Gender, Race and Class Marginalization of Bomba in Puerto Rico* (Cornell University Press, Ithaca, NY, 1995). At stake, among other issues (in Barton's dissertation, *many* other issues), is how *African* as opposed to *Antillean*—especially French Caribbean—are the *bomba*'s roots, related in turn to the Antillean component of Puerto Rican slavery. That this *is* an issue is one of the issues in Barton's dissertation, which among other things, probes popular ideologies that rank various "musical nationalisms"—*danza*, *salsa*, *plena*, *bomba*—along dimensions of race, class, and gender. The *bomba*, by the

way, is not to be confused with the *bombas* that pop up in *criollo* Christmas celebrations, often in the context of *seises*, though increasingly no more than interpolated call-and-response patter in any Christmas songfest, with amusingly pointed, often sexual connotation: "Blah blah blah," one says. "Bomba!" say the others. See the texts of some of these transcribed by McCoy from Christmas records (op. cit., p. 144, et passim).

42. "Junto al seis—música de la montaña—y a la bomba—música de la costa—figura la plena, nuestra música del arrabal," writes the unsigned author of the notes to *Plenas* (no date), a long-playing record the Instituto de Cultura Puertorriqueña devoted to the *plena*: "Along with the *seis*—music of the mountains—and the *bomba*—music of the coast—we have the *plena*, music of the . . . " *what?* Suburbs doesn't cut it. Outskirts? What it was was the cities of the coast outside of San Juan—"the slums of the coastal towns" as Waterman so abruptly put it in his notes to the Library of Congress anthology—but what our unsigned author's trying to catch in his formula is the sense, so strong among all commentators, of the way the *plena* seems to draw inspiration equally from criollo and African sources. As he adds, "Written in 2/2 time, combining Spanish melody with rhythms of a clearly African origin, it is ours, Antillean, 100% native Puerto Rican" (that this is a nostalgic perspective of the later 20th century goes without saying). Origins, I might add, completely obscure and highly controversial: never heard before 1915 *or* datable to before the American occupation of 1898; the name "plena" referring to the frenzy associated with the full moon (*luna plena*) *or* derived from the English phrase "play, Anna." Un huh! What's clear is that lyrics are popular, topical, and that during the 1980s *plenas* regained a genuine popularity that continues to the present. See all the foregoing sources.

43. The piece by Raquel Z. Rivera (op. cit.) is not only all about Puerto Rican rap, but reframes, in concert with the volume's other pieces, the discussion of Puerto Rican life, jettisoning both nationalist and colonialist discourses for one that, refreshingly, focuses on race and class. And in her case, music. Rap, of course, radically forces such a move, since it's not about being Puerto Rican, but about being simultaneously marginalized (from the projects, that is, the *caserío*) and cool. As Puerto Rock and Tony Boston, both members of Latin Empire, rap, "I'm into hip-hop as well as *plena*. . . . "

44. Subject of yet another unfathomably rich—certainly extensive—literature. To sample the range try Adrian Bailey and Mark Ellis, "Going Home: The Migration of Puerto Rican-Born Women from the United States to Puerto Rico" (*Professional Geographer*, 45[2], 1993), for a "hard-eyed" quantitative analysis; and Alberto Sandoval Sánchez, "Puerto Rican Identity Up in the Air: Air Migration, Its Cultural Representations, and Me 'Cruzando el Charco' " (in Negrón-Muntaner and Grosfoguel [op. cit.] for something a little more au current. While you're in that volume look at Agustín Lao's less directly relevant but unbelievably sharp "Islands at the Crossroads: Puerto Ricanness Traveling between the Translocal Nation and the Global City," which opens on another air shuttle, and includes a brilliant analysis of a Banco de Puerto Rico Christmas television advertisement (pp. 170–171). Way less cool is my own contribution, "The Shape of a World with Many Centers: The World View of Some Puerto Rican Adolescents, 1969–1970," (*Children's Environments Quarterly*,

5[2], Summer, 1988, pp. 30–38). Carlos Antonio Torre, Hugo Rodriguez Vecchini, and William Burgos edited *The Commuter Nation* (University of Puerto Rico, Rio Piedras, 1994), which is where to start.

45. Or maybe he's just pretending to be a gangster to play up the Mafia echoes in *Cosa Nuestra*. Still, I think this is an inescapable reading, especially given Colon's self-conscious confrontation of *salsa*'s relationship to the *jíbaro* Christmas heritage in the "experimentales" *Asaltos Navideños* of 1971 and 1973. (Doesn't "El Malo" from his first album—*El Malo*—it was 1967, he was 15!—sound like something from the heyday of rap: "El malo de aquí soy yo / Porque tengo corazón"? I guess it must be genetic.)

46. Actually Martin was a founding member of Menudo, the 1980s, and counting, "Latino boy band" (as it is usually described), but he didn't go huge until he woke up the Grammy Awards ceremony, February 24, 1999, with his production of "La Copa de la Vida," scoring the only spontaneous standing ovation of the evening, and rocket-propelling his "Livin' La Vida Loca" to the number one position on *Billboard's Hot 100* singles chart by mid-May. Not that at the time he hadn't previously sold 13 million copies of his records, which goes to show the size of the market, since he was nevertheless "unknown" to "American" audiences. Chayanne, another "unknown" Puerto Rican pop star who's sold millions of records, costarred in the 1998 Vanessa Williams vehicle *Dance with Me*. The relationship of this Puerto Rican Latino pop to *la música criolla* is moot, but it's got to be *unbelievably* complicated.

47. Though generally I'm convinced attention alone isn't enough. In this case this dispossession was accompanied—or caused—by an influx of West African slaves, as well as other sharp dislocations throughout the island, throughout the Caribbean, the Spanish-speaking world, and West Africa, indeed elsewhere as well; and in the island, in any case, it was a time not so much forgotten as ignored, buried beneath a nostalgia for a subsequent "golden age" which alone was what the *jíbaro*, as well as the official organs of culture, wished to remember. Certainly my predisposition to experience life as fresh and whole played into this, but what it played into was a naturalization that a whole machinery labored to maintain, of postcards, advertising logos, souvenirs, travel brochures, and of course the songs themselves. The naturalization of the marginalization simultaneously masked and celebrated by these songs required this broad conspiracy about the history of the island, a history committed to . . . *not going back too far*. It required that the mists of time close around a . . . relatively *recent* past since the one celebrated as eternal wasn't all that old. Since history is about processes we're in the middle of that are too big to get our eyes around, since it's about unveiling the dynamics at work in the present (*not* about fossilizing things that happened in the past), history had to be . . . dispensed with. This wasn't a story people liked.

48. Hunter S. Thompson, "The Price of Rum: Fear and Loathing to Go to Puerto Rico" (*New Yorker*, September 7, 1998, p. 72). This is vintage Thompson but others have heard this too. Here, for instance, is Wenzell Brown writing in the mid-1940s: "The words are so slurred I cannot catch them all—only phrases. . . . One man strums the guitar as he sings in a high-pitched wail. His skin is gray, tinged with blue" (*Dynamite on Our Doorstep* [Greenberg, New York, 1945, p. 280]).

49. Of course "whine" *may* no more than refer to the clenched throat vocalization Tariq ibn Ziyad brought with him in the 8th century when he led the Moors into southern Spain (or which the Persian musician, Ziryab, brought with him in the early 9th century when he joined the court of 'Abd ar-Rahman II). Here's McCoy (op. cit.) on Puerto Rico's Spanish heritage: "Since the *conquistadores* and early settlers came from maritime districts of Southern Spain, they brought southern Spanish speech characteristics with their Moorish elements rather than the Castilian, and this southern Spanish origin accounts for the many sudden reminiscences of the Orient which are found unexpectedly in Spanish American music, such as the tight constricted way the mountain people have of singing" (p. 32). What Thompson heard may have been no more than an unfamiliar ... orientalism (more specifically, an Iraquism). On the other hand, McCoy does feel that the mode of "melancholy happiness" characteristic of the *aguinaldo* is purely Puerto Rican (p. 49).

50. That *la música jíbara* is rooted in the "*mescla de tres razas*"—the mix of Spanish, African, and Taíno that is supposed to be Puerto Rican—is a claim increasingly and increasingly self-consciously made. Check out these lyrics from a recent *seis milonguero* by Arturo Santiago: "Unequaled tradition / mixture of the white of Spain / of the black who planted cane / to the sound of whip and sable / of the unconquerable Taíno Indian / of bronze and wheat-colored skin / three races in an island land / white, black and aborigines / these were your origins / Puerto Rican culture." In another song Santiago writes of "the mixture of three races / with joined hands / Spaniards, Taínos / and the African" (Edwin Colón Zayas y Su Taller Campesino, *¡Bien Jíbaro!*, Rounder CD 5056, 1994, translations by Daniel Sheehy). Given that the Taínos have been extinct for 350 years and that the Africans arrived in chains, one wonders exactly what Santiago means by "with joined hands."

51. Maurice Sendak (op. cit.). I say again, you need to know this book!

Index

"n" following a page number indicates a note

Atomic Age, tool use during, 156
Atoms
  Big Bang and, viii
  formation of, 270n13
  as history, 53–54
  interconnection and, 110, 122–125,
    163–165
  primeval atom model, 248n34
  process of forming into molecules, 56
Aurignacian technology, 162, 169,
  298n44. *see also* Tool use
Australia, 115. *see also* Continents
Australopithecines
  butchery and, 168
  description, 287n69
  evolution of, 131–132, 144–146
  phylogeny of, 288n73
  tool use and, 154–159
Australopithecus aethiopicus, 132, 151
Australopithecus afarensis, 132, 151, 160,
  293n1
Australopithecus africanus, 132, 151
Australopithecus anamensis, 132
Australopithecus boisei, 132
Australopithecus garhi, 132, 151, 168
Australopithecus robustus, 132

Babylon, taxation in, 313n37
Bacteria
  coupling of, 61–62, 95–96
  emergence of, 59–60
  immortality of, 284–285n46
  importance of, 276n47
  and oxygen in the atmosphere, 101
  sun and, 62–63
Baltica, 113–115, 125. *see also*
  Continents
Baryon Genesis epoch. *see* Inflationary
  epoch
Basalt, relationship to granite, 266–
  267n48
Belief, vs. "buying into," 39
Bible, and how we conceive history, 14,
  21, 24
Big Bang. *see also* Universe
  beginning with, vii
  beliefs and, 242n4
  description, 42–44, 91–92
  interconnection and, 44–47, 205–206
  origins of the term, 241n1

prior to, 38
  radiation and, 249n39
  story of, 38, 40
Bipedalism
  climate and, 141–143
  evolution of, 295n15
  onset of, 133–135
  turnover pulse hypothesis of, 140–141
Bitung, 3–4
Boisei, australopithecus, 132
Bomba, 324–325n41
Brain, evolution of
  language and, 182
  onset of speech and, 159–163
  size of humans and, 166
  tool use and, 297n35
Breath of the land, 149–150, 186, 187–
  188, 209
Bronze Age, 156, 311n33
Butchery, evolution of, 168–168

Cambrian period, 93, 117, 281n15, 281–
  282n17
Canada, 113–115. *see also* Continents
Carbon cycle, interconnection and, 122–
  125
Carboniferous period, 117
Caribbean Plate, 2, 5, 17–18, 229,
  322n26. *see also* Plate tectonics
Cenozoic period, 117, 127–128, 281n16
Cereals, agriculture and, 310n25
Chalcolithic period, tool use during,
  155–156
Chenopodium, 307–308n13. *see also*
  Agriculture
Chlorophyll, in bacteria, 62–63
Chronocentricity, invisibility of, 20
Climate
  atmospheric circulation and, 73–77,
    261n21
  bipedalism and, 141–143, 145–146,
    167
  bi-stability of, 268–269n4
  change and, 304n90
  description, 258n1, 259n5
  history of, 77–80
  during the Ice Age, 130
  mantle plume model of, 280n10
  mountains and, 86–87
  in northwestern Europe, 303n81

About the Author

Denis Wood is a writer living in Raleigh, North Carolina. He received a BA in English from Case Western Reserve University and MA and PhD degrees in geography from Clark University. From 1974 to 1996, he was Professor of Design at North Carolina State University. In 1992 he put together the award-winning Power of Maps exhibition for Cooper-Hewitt, National Design Museum, for which he wrote the best-selling *The Power of Maps*; the exhibit was remounted at the Smithsonian Institution in 1994. His other books include *Home Rules* (with Robert Beck) and *Seeing Through Maps* (with Ward Kaiser).